MECHANICAL ENGINEERING
A Decade of Progress 1960-1970

MECHANICAL ENGINEERING

A Decade of Progress
1960-1970

Edited by E. G. SEMLER, BSc, CEng, FIMechE

The Institution of Mechanical Engineers

LONDON

First published 1971

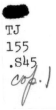
First published as a series of articles in *The Chartered Mechanical Engineer* between 1960 and 1970. Reprinted by photolitho in this form by Compton Printing Ltd, Aylesbury, Buckinghamshire, on cartridge paper and bound by G. and J. Kitcat Ltd, London, England

Made and printed in Great Britain

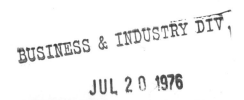

CONTENTS

PREFACE

by Professor J. L. M. Morrison, CBE, DSc

President of The Institution of Mechanical Engineers, 1970-71

As the sum total of human knowledge increases, so inevitably does the need for specialisation. No one man can be expert over more than a tiny fraction of the whole field. Even in what used to be thought of as one discipline, engineering science, specialisation is inescapable, and though educationalists may today struggle to preserve generality in their teaching of basic principles and to some extent succeed, the further the student advances the greater is the temptation, and indeed the necessity, to concentrate on particular aspects of learning.

Nevertheless, the intelligent man resents the thought that his knowledge should be so restricted, and insists on retaining his width of interest in the world around him. In areas remote from his specialism, general reading, lectures and discussion, radio and television may be sufficient to keep him tolerably well informed, but in less remote areas, where his depth of understanding must be far greater, such sources are, for the most part, too superficial. In these regions he—and the less-specialised managerial types—needs accounts of progress which fall somewhere between the research paper, of which he has neither the time nor the need to understand the minutiae, and the popular account written for the layman, which would indeed simply waste his time and add little to his knowledge. It was to fill this gap in the available literature that the present series of articles was conceived: articles written in non-specialist language for mechanical engineers in general and definitely not for the experts in the disciplines concerned, and published over a period of years in *The Chartered Mechanical Engineer* under the general heading 'Review of Progress'.

There is an insistent demand for review articles. These often take the form of a review written by specialists for specialists and are, of course, immensely valuable. I know of no other attempt such as this, however, at a systematic coverage of mechanical engineering as a whole for the benefit of the non-expert with a view to helping him to cross the interdisciplinary barriers. It was, in fact, as a result of the widespread demand from readers of *The Chartered Mechanical Engineer* that the decision was made to collect material in this volume, which could reach a readership very much wider than the membership of our Institution. Unfortunately, exigencies of space have prevented the inclusion of all the articles in the original series but the editor has concentrated on those subjects of most general interest to mechanical engineers.

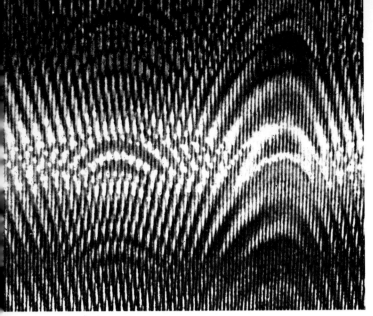

Fig. 1. Not unlike embroidery, this shows the surface of a gun barrel which vibrated during machining. The theory is explained in the Jrnl. of Mech. Engg. Science, March 1961, p. 7

Most mechanical engineers are at one time or another faced with problems in vibration. Few may be aware of the research and analysis which has been devoted to the solution of such problems. The field is so diverse that many different treatments are required. New approaches have become possible due to improved experimental techniques and the facilities offered by modern digital computers. This article summarizes some of the recent work done in this field.

Mechanical Vibration

by R. N. Arnold, DSc, PhD, DEng, MS, CEng, FIMechE

Mechanical vibration occurs in so many forms that few engineers can hope to evade it. Literature on the subject is expanding rapidly and as machine speeds increase vibration problems multiply. It may lead to failure by fatigue, inaccuracy in the functioning of machines or be a source of intense irritation. Fig. 1, for example, shows the effect that vibration can have on the finish of a machined surface.

Though expressed in similar mathematical form, the problems which arise normally require their own unique solution.

For example, there is little in common between the vibration of ball-races, control systems, suspension bridges, ship hulls, chimneys, propellers, surge tanks and machine tools, on which research has been conducted in recent years. The subject, in fact, is so diverse that it has attracted specialists from many fields. Moreover, it is often difficult to appreciate the physical significance of vibration phenomena and this has necessitated a mathematical approach. Many problems, in fact, have required so much computation that, without the advent of the digital computer, they might have remained unsolved.

To attempt a general survey of progress in vibration research during recent years would be a major undertaking* and the result might well become a catalogue of references. Fortunately some aspects have already been discussed in a Thomas Hawksley Lecture[1] and in what follows examples of progress in other directions will be given. Though it is hoped that a general pattern may emerge this will naturally be influenced by the author's personal preferences and experience.

*Upward of 5000 abstracts of publications on the subject have appeared in *Applied Mechanics Reviews* alone during a decade.

Professor R. N. Arnold was Regius Professor of Engineering at Edinburgh University since 1946. Following early research experience in Britain and U.S.A. (1932-36) he was Assistant Lecturer at the Royal Technical College, Glasgow (1936-40); member of Research Staff, Metropolitan - Vickers Electrical Company Limited (1940-44); and Professor of Engineering at University College, Swansea (1944-46). He is a graduate of the Universities of Glasgow, Sheffield and Illinois, and a Member of this Institution and the Institution of Civil Engineers. His publications, many of which have appeared in the Proceedings of this Institution, have been concerned mainly with problems in dynamics and mechanical vibration. Among his awards are: Thomas Lowe Gray Prize (1940); Thomas Hawksley Gold Medal (1946); T. Bernard Hall Prize (1957). He died in 1963.

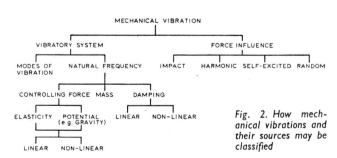

Fig. 2. How mechanical vibrations and their sources may be classified

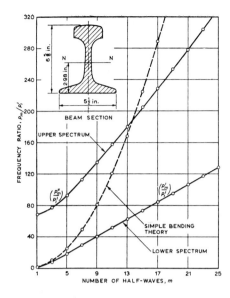

Fig. 3. Two natural frequencies exist for every nodal arrangement of a freely supported beam; full lines show these in terms of lower fundamental frequency; dotted line applies when bending only is assumed as the form in which the beam is deflected

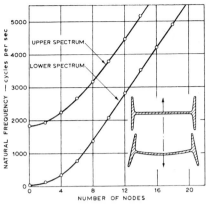

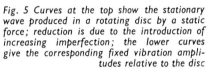

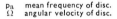

Fig. 4. Natural frequencies of 3 in. × 1½ in. R.S.J. × 60 in. long, on flexible end supports; the double spectrum is due to a combination of lengthwise bending and distortion of the web

Fig. 5 Curves at the top show the stationary wave produced in a rotating disc by a static force; reduction is due to the introduction of increasing imperfection; the lower curves give the corresponding fixed vibration amplitudes relative to the disc

p_a mean frequency of disc.
Ω angular velocity of disc.

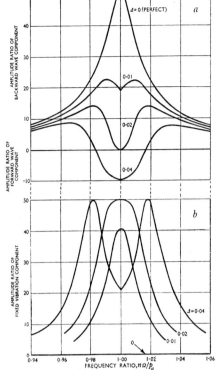

Vibrating systems

Mechanical vibration depends essentially on a system capable of vibrating and a force to initiate or sustain the motion, as indicated in Fig. 2. Suppose these are examined separately. All mechanical devices possess elasticity and mass and, however rigid they may appear, are capable of vibration. The time-honoured example is that of a mass suspended on a spring and confined to move vertically. If the spring has no mass and the mass no elasticity—an imaginative ideal—the system possesses one degree of freedom and one natural frequency of vibration. In any real system, such as a shaft, mass and elasticity are inseparable and an infinite number of natural frequencies are possible. Moreover, each frequency is associated with a particular pattern of movement called a normal mode of vibration.

Laborious computation is normally required in obtaining frequencies of all but the simplest of elastic bodies. Approximate solutions exist for a variety of symmetrical bodies, including thin cylinders, shafts and beams but for more practical components such as compressor blades or conical shells, only isolated results are available. Again, it is difficult to adapt this work to aircraft wings, ship hulls and the like, where cellular construction is combined with distributed mass.

For small deflection it is usual to assume that the load-deflection curve is linear within the limits of the motion. This may cease to be true for large vibrations in which case the frequency becomes dependent on amplitude and we enter the field of non-linear vibration. This subject has engaged the ablest of minds but few general solutions have emerged and then only for relatively simple systems. Nevertheless, work in this subject has led to practical results, particularly in the isolation of equipment from shock.

All normal free vibration is subjected to some form of damping and subsides in time. The damping may result from internal hysteresis or may be introduced by design as in the shock-absorbers of an automobile. It is rarely simple and is usually a function of both velocity and displacement. Its effect is to lower the natural frequency of the system but if it exceeds a certain critical value, free vibration becomes impossible. When a critically-damped system is deflected and released, it moves asymptotically towards its equilibrium position.

The tendency of a system to vibrate is not necessarily dependent on elasticity. The simple pendulum, for example, is controlled by gravity and there are a large variety of complicated systems which belong to a similar category. A simple gyroscope in double gimbals will vibrate, if disturbed, and control systems are also susceptible.

When vibration exists every part of the system takes part in the motion. Components can only be treated in isolation if they are attached to infinitely rigid supports. Thus it is wrong to consider the wings of an aircraft as cantilevers with fixed ends for both the wings and the fuselage must vibrate together. However, the vibration characteristics of a number of components assembled together can be predicted from a knowledge of their individual characteristics.

Force influences

The natural frequencies and modes of vibration of systems are essentially dormant qualities. They tell us what might happen if the system were disturbed from equilibrium. Consider the types of disturbance which may arise, as shown in Fig. 2. The simplest to visualize is impact, which deforms the system rapidly at some isolated point. The resulting motion, while it lasts, may be extremely complex, for large numbers of modes may be excited and each will subside independently at its own frequency as in the ringing of a church bell. The damped motion of an automobile after negotiating a bump, on the other hand, shows the effect of impact on a simple system.

A common source of forced vibration is out-of-balance in rotating machinery. This exerts a harmonic force and the system on which it acts vibrates at the forcing frequency. If this coincides with a natural frequency a state of resonance occurs, the amplitude becomes large and structural damage may result.

In the general case a periodic force may contain a large number of harmonic components as, for example, the torque transmitted by a multi-cylinder diesel engine. This depends on engine speed, firing order and the pressure changes within each cylinder, but since it is periodic it may be expressed as a Fourier series. Each component torque in the series is a potential source of resonance of the torsional modes of the engine.

There is a class of vibration which itself generates the forces by which it is sustained and systems showing this characteristic are said to be dynamically unstable. The vibration can be initiated by a small disturbance and grows with time, its amplitude being limited only by the inherent damping of the system. Flutter in aircraft and chatter in machine tools are examples of self-excited vibration but control systems may also exhibit such behaviour.

Vibrations cannot always be expressed in terms of periodic functions. Road forces acting on an automobile, for example, are obviously random in character but if the road is of uniform roughness the result is not entirely indeterminate. Such a condition is called a stationary random process and can be defined by statistical parameters from which it is possible to predict the vehicle response. This is a relatively new approach to vibration phenomena but such results are now being used in the study of damage by fatigue.

Analysis

A note may be added on theoretical progress which has been greatly assisted by electronic devices. The analogue computer is based on the mathematical equivalence of mechanical and electrical systems. The ease with which electrical characteristics may be displayed on an oscilloscope and changes made by the turning of a knob has made this instrument invaluable, particularly in the study of non-linear vibrations and systems with many degrees of freedom. A more universal but much more expensive device is the digital computer. This is essentially an arithmetical machine which produces arithmetical answers. Its speed of operation, however, is so rapid that it can perform prodigious calculations in short intervals of time and can be programmed to compare results and make decisions for its subsequent calculations. Its ability to explore results over a wide range of conditions has revolutionized theoretical analysis.

Many mathematical approaches have been devised for the solution of vibration problems but few lend themselves to verbal explanation. The method of 'receptances'[2] is perhaps an exception. This approach is particularly useful for the calculation of natural frequencies of complicated linear systems. Suppose we apply a force $P_1 \cos \omega t$ at any point 1 of an undamped system capable only of movements in the direction of the force. Let the resulting amplitude at point 1 be $x_1 = a_{11}P_1$. The quantity a_{11} is called the direct receptance of the system at point 1. Expressed mathematically

it will contain the constants of the system and the variable ω. Now resonance will occur when ω is such that $a_{11} = \infty$ and this will define the natural frequencies of the system. The essential problem is, therefore, the determination of a_{11}.

Now suppose we divide the system into two parts at point 1, replace the forces which were acting and retain the same amplitudes at frequency ω. Each part will then be oblivious of what has transpired. Denoting the new motions by $x_a = \beta_{11}P_a$ and $x_b = \gamma_{11}P_b$, where β_{11} and γ_{11} are the receptances of each part at point 1, then $P_1 = P_a + P_b$, $x_1 = x_a = x_b$ and we obtain $\dfrac{1}{a_{11}} = \dfrac{1}{\beta_{11}} + \dfrac{1}{\gamma_{11}}$. This means that the overall receptance (and thus the natural frequencies) of the system can be found directly from the receptances of its parts.

The above refers to a particularly simple problem. Extenion to more complicated problems, including the presence of damping, is naturally possible.

Vibration of elastic bodies

Considerable theoretical and experimental research has been done in recent years on the vibration of elastic bodies[3] such as beams, cylinders, shells and plates.[4] Although engineering components are often more complex in shape than these simple forms, the results are nevertheless of great value. From the theoretical standpoint, the equations of motion have become more and more precise. In early work on beam vibration it was considered sufficient to include only the effects of bending. For a freely-supported uniform beam this revealed that the natural frequencies varied inversely as the square of the half wave-length. Thus, by calling the fundamental frequency unity, the frequencies ascended in the order, 1, 4, 9, 16. . . .

A closer approximation was made by including both the effect of shear and rotatory inertia[5] but the full significance of this was not at the time apparent. That the presence of shear and rotatory inertia introduced a new spectrum of frequencies was indicated in the discussion of a paper on impact on beams.[6] This has led to investigation of what is now called the Timoshenko beam.[7, 8] If we consider a mode of vibration in a freely-supported beam, then two almost identical sinusoidal deflections with, say, n nodes are possible, by positive bending coupled with either positive shearing or negative shearing. If released from either deflection free vibration would result though it might be difficult to distinguish between the two results. However, in the latter case the initial strain energy in the beam is much greater than in the former and the natural frequency is therefore greater. This means that two natural frequencies exist for every nodal arrangement. The results for a railroad section forming a freely-supported beam are reproduced in Fig. 3. The frequencies for odd numbers of half waves are plotted in terms of the lower fundamental frequency p_1 and the full lines show the manner in which they increase. The dotted line indicates the results obtained when bending only is considered.

Quite apart from the existence of the upper frequency spectrum, which is not predicted by the simple theory, we see the wide divergence in the frequencies of the higher modes. Only in slender beams, where shear plays a minor part, is this effect imperceptible. Similar doubling of the frequency spectrum occurs in other cases. For example in Fig. 4,[8] are shown two spectra of experimental frequencies

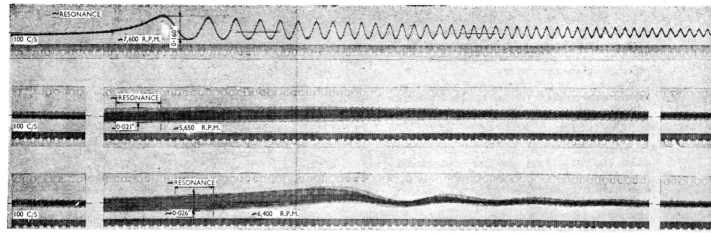

Fig. 6. Top trace shows development and collapse of a stationary wave in an almost perfect disc, due to static force from an air jet; the other traces show (to a larger scale) the effects of removing metal from diametrically opposite sides of the same disc

obtained from a 3 in. × 1½ in. R.S.J. beam, vibrating flexurally in its plane of greatest flexibility. In this case the double spectrum is due to a combination of the usual lengthwise bending of the beam and distortion of the cross-section in its plane, resulting in cross-bending of the web and torsion of the flanges as indicated.

The lower spectrum is mainly associated with the lengthwise bending and the upper with the cross-bending.

It would be difficult to say how many frequency spectra exist for a given elastic body. For example in thin cylinders[9] there are at least three possible arrangements of radial, longitudinal and tangential motion of an element which can support vibration in a given nodal pattern.

But progress has not been confined to studying the frequencies of bodies of uniform section or shape. In the case of beams the effects of taper,[10] pre-twist,[11] initial curvature[12] and torsional flexural coupling[13] have been investigated while work has been done on tapered, skewed and stiffened plates[14] as well as on grillages.[15] Shells have also received considerable attention, particular interest being shown in thick-walled cylinders,[16] stiffened cylinders[17] and shells of conical[18] and shallow spherical shape.[19]

A further factor which affects vibration is environment such as a centrifugal field due to rotation, immersion in fluids[20] or sudden changes in temperature.[21] Space permits consideration of only one example, that of rotating discs. Many turbine disasters have been caused by disc failure following the formation of what is termed a stationary wave. This travels around the disc in the reverse direction to that of rotation and at the critical speed becomes stationary in space. Under such conditions any static force, due to unbalance in steam flow, is capable of increasing the amplitude of the wave.

Research has shown that all discs, however accurately machined, possess two natural frequencies for each nodal configuration.

For example, with n nodal diameters, the frequencies are associated with two separate configurations fixed in position in the disc, whose nodes are dispersed between each other. In a perfect disc this could not occur for the nodal positions, due to symmetry, would be arbitrary. The amount of imperfection present determines how far the two frequencies

are apart, while the interaction of the two modes of vibration creates the standing wave which is virtually a forced vibration.

As imperfection is increased the frequencies separate and the interaction of the modes is reduced. This has important consequences, for each vibration component which ceases to combine appears as a vibration fixed to the disc. By means of controlled imperfection it is therefore possible to subdue the standing wave and obtain in its place two vibrations rotating with the disc which are almost unaffected by static force, and are heavily damped by aerodynamic action.

These effects are illustrated in the theoretical curves of Fig. 5.[22] The upper curves show the reduction in the stationary wave (backward relative to disc) with increasing imperfection Δ while the lower curves show the corresponding fixed vibration amplitude. The high aerodynamic damping of the fixed vibrations is not included in Fig. 5 but results obtained from experimental discs are reproduced in Fig. 6. The trace at the top shows the development and subsequent collapse of a stationary wave in an almost perfect disc subjected to a static force from an air-jet. The remaining traces (photographed to a larger scale) show the fixed vibration which develops in each mode after the same disc has been made imperfect by the removal of metal at diametrically opposite positions. It will be seen that the double amplitude of the top trace is 0·160 in. compared with 0·021 in. in the centre and 0·026 in. in the bottom trace.

Ground vibration

There is one type of elastic body which is associated with the science of seismology—the semi-infinite solid, sometimes called the elastic half-space.[23, 24] We may think of this as simulating the ground on which we live. Unlike finite elastic bodies it possesses no natural frequencies of its own. A disturbance on the surface only results in waves being propagated radially; these are comparable to the water waves which emanate from a stone dropped in a pond. When a rigid circular mass is attached to the surface of the solid the mass immediately assumes four natural frequencies of vibration corresponding to (i) vertical translation, (ii) horizontal translation, (iii) rotation about the vertical, (iv) rotation about the horizontal.[25]

The vibration amplitude of such a system, however, must always be finite even if no damping exists. As vibration proceeds waves are continually propagated outwards and their energy becomes distributed throughout the half-space. Since there are no boundaries this energy can never be reflected and can never return.

Much work has been devoted to this subject and in Fig. 7 experimental results for vertical motion of a circular cylinder of radius r_o under forced vibration are shown.[25] These amplitude–frequency curves all relate to cylinders of equal mass but differing radii, defined by the quantity b, which is the ratio of cylinder mass to the mass of volume r_o^3 of elastic material. As b decreases (the cylinder tends to become a disc) the resonance peak becomes less marked and, when $b = 0$, entirely disappears.

The above work provides a fundamental background for a study of the vibration of machine foundations[26] and buildings subjected to earthquakes.[27] The properties of ground, however, are very complex compared with those of an ideal elastic material.[28] The ground is not homogeneous, has non-linear elasticity and exhibits hysteresis. Moreover, its properties may change radically due to a night of frost or a shower of rain. Some results of ground vibration are shown in Fig. 8.[28] The resonant frequencies for vertical motion with different weights on two sizes of base are given on the left. The two results are for identical tests, with a 10-day interval, at precisely the same location. On the right is shown the way in which the resonant frequency varies with amplitude due to non-linearity. It is not surprising that this aspect of the subject is only partially understood, yet it is a problem which constantly confronts the engineer.

Some years ago the author was asked to investigate a mysterious vibration which affected an office block in a large city. It was discovered that the vibration, which had a frequency of seven cycles per second, only existed when the ram of a large hydraulic accumulator in the yard of a nearby steelworks was rising to its upper position. Investigation showed that the ram was operated by a triple-acting high-pressure reciprocating pump and that the second order component of the pulse was virtually coincident with the natural frequency of the ram and water-reservoir system. Vibration of the ram resulted in a vertical harmonic force of considerable magnitude being transmitted to the ground and the waves emanating from this source had sufficient energy to produce forced vibration in the office block.

Further features of the above are the effects of blasting[29] and traffic vibration on buildings. These subjects involve so many factors that they are still improperly understood. The former has caused much concern to householders, but there is little correlation between the vibration measured at a given house and the internal damage caused to plasterwork. This is probably due to lack of information on the mechanism by which plaster cracks under dynamic action.

The velocities of waves in a solid depend on the properties of the material and considerable attention has been devoted to their study in the case of concrete. Information from pulse-techniques is now used for studying the strength and constitution of concrete roads and runways.[30]

Random vibration

The interest of the mechanical engineer in random vibration is of recent origin.[31, 32, 33] It is important to realize that the quantities used do not relate to a unique vibration but rather to an infinite number of possible vibrations which,

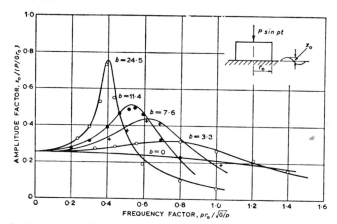

Fig. 7. Amplitude-frequency curves for several cylinders of equal mass but different radii, supported on a semi-infinite elastic solid; the resonance peak disappears when the mass is distributed over a large area

though not identical, have something in common. In general, we think of a stationary random vibration in which the statistical characteristics remain invariable. This is, in fact, much easier to visualize than to define. Suppose we have a stationary random curve, plotted to a base of time and that its mean value is zero. If we select any part of this curve representing some time T and consider it to be repeated indefinitely then the result could obviously be expressed by the Fourier series

$$f(t) = \sum_{1}^{\infty} a_n \sin (\omega_n t + a_n)$$

where $\omega_n = 2\pi n/T$, and we could find and plot the values of a_n to a base ω_n. The diagram would be composed of discrete lines of height a_n at positions $\omega_1, \omega_2, \omega_3, \ldots$. This would give some indication of the character of the vibration but greater accuracy would obviously be achieved by increasing the time T of the chosen sample. As T increased, so the number of lines per unit frequency would increase until eventually when $T \to \infty$, a solid spectrum would result. If now we assume the values of the phase angles a_n to be arbitrary this allows the spectrum to define a vast series of curves, all different, of which the part we have examined over time T is representative. The vibration is partially defined by this frequency spectrum though it is normally plotted as $\sum a_n^2/\Delta\omega$ to a base ω (where the summation is over a finite interval $\Delta\omega$) and called the power spectral density.

Of all curves drawn from this characteristic no two, in general, will be identical. Some, in fact, may contain an isolated peak due to a large number of harmonics reaching a maximum at the same instant. This would naturally be a rare occurrence but, if our curve represented stress, it would be important to know what were the chances of a high stress occurring within the life of a component. This is where amplitude distribution becomes important, a characteristic which can be defined approximately by mathematical statistics. The Gaussian distribution is often assumed, which is of the form

$$p(x) = \frac{1}{\sigma \sqrt{2\pi}} e^{-\frac{1}{2}\left(\frac{x}{\sigma}\right)^2}$$

where σ is the root-mean-square of $f(t)$ and can be calculated

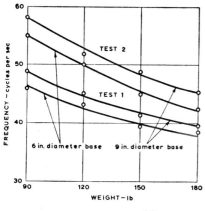

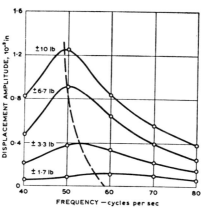

Fig. 8. Forced vibration frequencies of mass resting on soil; on the left are two sets of test results obtained under identical conditions except for a ten-day interval between them; the way resonant frequency varies with amplitude due to non-linearity is shown on the right for a 90 lb weight on a 9 in. dia. base

The Gaussian distribution is found to be a good approximation to many physical random effects. If a linear system is subjected to a random influence of Gaussian distribution, the response will also be Gaussian.

Jet engine noise

A large variety of engineering phenomena involve random vibration. We need only mention cavitation in water turbines, forces on road vehicles,[34] earthquake effects on structures,[27, 35] wave action on ships and sound from rocket motors and jet engines.[36] The latter, in fact, has created serious problems in aircraft. The noise intensity near the efflux of a jet engine is so high that the pressure variations acting on the panels of the fuselage can cause failure by fatigue.

From the design standpoint it is important to know the spectral density of pressure propagated by a jet. If then the response of a panel to harmonic pressure variation is known over a wide frequency range its response under the random pressure can be calculated. The amplitude response then leads to a value for σ which defines the shape of the Gaussian distribution. Thus the proportion of time at which a given stress level is likely to occur can be studied in relation to the known fatigue properties of the panel material. Criteria for the damage resulting from interaction of fatigue stresses of various levels[37] and under random loading[38] have been studied in recent years.

The noise emitted by a jet engine gives a spectral density which is almost constant over a considerable frequency range.[36] This is called 'white' noise and is characteristic of many random vibrations, including electrical phenomena.

If a system is subjected to a force of this type the spectral density of its response will be identical in shape to the square of the amplitude response to forced vibration. The actual response will be a random vibration whose spectral density will have peaks close to the natural frequencies of the system. If damping is small, the motion of the system will mainly be composed of the system frequencies and the amplitude envelope will display a random variation. The experimental records of input and response of such a system to random vibration is shown in Fig. 10.

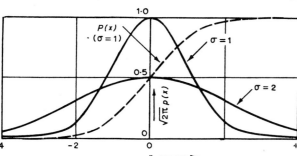

Fig. 9. Gaussian distribution (solid curves) is usually assumed for amplitudes of random forces; the probability of $f(t) < x$ is shown dotted

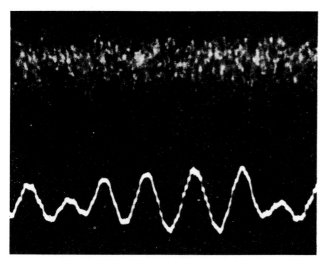

Fig. 10. Motion of the centre point of a beam (lower trace) due to random central force (upper trace)

from the power spectral density. Here $p(x)$ is called the probability density function and has the shape shown in Fig. 9. A function which has a simpler physical interpretation is $P(x) = \int_{-\infty}^{x} p(x)dx$ which is also plotted in Fig. 9. This is called the probability distribution function and indicates the probability of $f(t)$ being less than x. Thus the chance of a high positive or negative value occurring is much more remote than that of a low one.

Gyroscopes

The gyroscope, because of its unique behaviour, has become of immense importance[39] particularly in the field of space navigation. Its high momentum tends to maintain its axis pointing in a fixed direction. But due to friction and other causes, small extraneous torques are inevitable, and these cause the axis to wander from its original position. Elaborate sensing and control mechanisms are, therefore, incorporated in the gyroscopic units used in inertial navigation. One cause of trouble is nutation in which the axis of the gyroscope describes a cone (Fig. 11) at a frequency depending on the rotor speed n and the moments of inertia of the system. This motion is composed of simultaneous vibrations about axes OX and Oy and at position A the precessional velocity $\dot{\theta}_1$ about OX is a maximum and is exerting a maximum torque about Oy tending to increase θ_2. At position B the velocity $\dot{\theta}_2$ about Oy results in a torque tending to reduce θ_1 and so the motion proceeds in the direction indicated. Now small steady-state vibrations about a mean position would be acceptable so long as axis Oz' remained undisturbed. But practical gyroscopes require suspensions and the double-gimbal system of Fig. 11 is the usual way of providing the necessary freedom. In such cases, due to nutation, axis Oz' may change its orientation. This depends on the alignment of the inner and outer gimbals.[10-42] Consider, for example, the case in which axis Oz' is inclined by α_o to OZ. Since no energy is supplied, there can be no mean torque about axes OX or Oy. Now if speed n remains constant, the torque about Oy is proportional to $\dot{\theta}_1 \cos \theta_2$ and its mean value over a cycle must be zero. This will only be true if $\dot{\theta}_1$ has a mean value other than zero since the mean value of θ_2 is α_o. The resulting motion therefore comprises nutation superposed on a constant angular velocity about OX. In the case shown, the average numerical value of $\dot{\theta}_1$ must be greater from B to D than from D to B so that the mean velocity about OX will be in the negative direction of θ_1. This effect can be serious in navigational instruments.

The nutational frequency of a gyroscope is affected by the elasticity of the rotor shaft and gimbals. A symmetrical system comprising a free gyroscope with a flexible shaft and rigid gimbals has, in fact, three such frequencies of vibration.[43] Consider the rotor at rest: due to shaft elasticity, it can vibrate by twisting relative to the inner gimbal either about Ox or Oy (Fig. 11). As the gimbal inertias about these axes are unequal two natural frequencies will be present but nutational modes, which depend on gyroscopic coupling, will be absent. Rotation not only affects the above frequencies but introduces nutation. The frequency of each mode is affected by rotor speed and, in Fig. 12, the manner in which the three frequencies vary with speed n is indicated. This work has revealed that, since the frequency p_1 (corresponding to the mode of nutation with a rigid shaft) is no longer linear, it is possible under certain conditions for it to coincide with the rotor speed. In such a case, a small unbalance might produce resonance.

Whirling of shafts

This is a problem which, though partially under control, is still receiving considerable attention.[44] When the rotational speed approaches one of the shaft's natural transverse frequencies, there is a large deformation, the centre of any shaft section describing a closed curve about the centre-line of the bearings. A simpler system is that of a disc mounted on a light shaft and supported in rigid bearings. If the damping is zero and the centre of mass G of the disc has an eccentricity e, then at speeds below the critical, G lies radially outward from the deformed centre-line of the shaft. In passing through the critical speed G changes its position to one which is radially inward and at very high speeds coincides with the centre-line of the bearings. In the latter case, the whirling is small and its deformation is equal to the eccentricity e.

Most practical systems differ substantially from the above. For example, the mass of a rotor is usually distributed along its length; gyroscopic effects may be present; the form of damping may influence the motion; the bearings may have flexibility of a complicated nature; there may be asymmetry

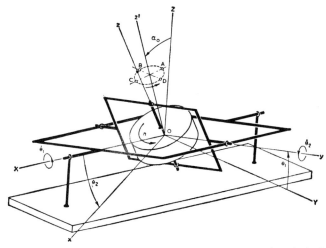

Fig. 11. The double gimbal system shown here is the usual method of providing the necessary degrees of freedom for gyroscopic instruments: due to nutation, axis Oz may drift, depending on the alignment of the gimbals

in shaft or bearings. Recent research has done much to make possible the inclusion of these factors in analysis.

The single disc theory may be extended to the case of a continuous heavy shaft by expressing displacement in terms of the natural modes of vibration. The difference between the mass axis and geometrical axis, then takes the place of the single eccentricity e. The equations of motion derived from this approach, though expressed in a number of normal coordinates, are mathematically identical to those obtained for the single disc.[45-47]

In most practical cases whirling is the result of resonance rather than instability; the amplitude increases to a maximum as the critical speed is reached and thereafter subsides. It is then safe to run the shaft at speeds beyond the critical. Sometimes, however, true instability occurs in which the amplitude is only limited by non-linear effects. This can be shown to occur in an unsymmetrical shaft. It may also exist in a symmetrical shaft with ideal bearings at a critical speed determined by the ratio of internal damping (fixed in the shaft) to external damping (fixed in space).[45]

If, as is likely, the bearings have different flexibilities in the horizontal and vertical directions, two sets of critical speeds will exist, corresponding to the natural transverse frequencies in each plane.[48] The result bears some resemblance to the interaction of vibration modes in rotating discs. For two corresponding modes, one in each plane, there is a speed range between their frequencies at which whirling occurs in the opposite direction to shaft rotation. If the frequencies are close to each other, the whirling may be large with consequent danger of fatigue, since the stresses under this condition vary harmonically. Other phenomena arise due to the bearing characteristics which are not always simple to define.

Lubricated bearings give rise to hydro-dynamic forces which can lead to whirling at half the speed of shaft rotation[49] while ball-bearings are found to produce a number of effects[50] which will await adequate explanation.

It may also be mentioned that since the forms of whirl for a shaft with distributed masses differ for each critical speed, normal balancing procedure may not be effective. The possibility of balancing for each critical speed has been examined[46] and it has, in some cases, been successfully carried out.

Panel flutter

This is a particularly difficult problem on which there is still considerable controversy.[51] It has the characteristics of self-excited vibration and occurs in a panel, one face of which is subjected to a high velocity air-stream. Though much theoretical work has been devoted to the subject, the amount of experimental work is small and correlation of results inconclusive.

One objective of aircraft and rocket designers is to keep the thickness of structural panels to a minimum. It is therefore important to know the limitations imposed by flutter. Curiously enough, early work[52] indicated that all panels, regardless of thickness, were unstable for air speeds with Mach numbers M between 1 and $\sqrt{2}$. While later theory has modified this result, it is generally agreed that within this speed range critical conditions exist.

Later theoretical work[53] has examined the phenomenon in relation to the shape of the middle surface of the panel,

distinguishing between initial lack of flatness and deformation imposed by movement of the supports or temperature effects. It is shown that for $M > \sqrt{2}$, panel flutter is possible under certain conditions. Thus if flutter exists at a given deformation and air speed, it may be eliminated by (i) thickening the panel, (ii) reducing that part of the deformation produced by external load. In this range of supersonic flow the normal pressure acting on the panel is proportional to the slope of the surface. If the panel is deformed towards the air-stream in a half sine wave it will be subjected to increased pressure on the leading half and decreased pressure on the trailing half.

Obviously at some critical speed the plate will become statically unstable (the oil-can effect) and it is not difficult to appreciate that this will be greatly influenced by the initial stress in the panel. Flutter is, in fact, associated with this instability, though it is difficult to understand its physical significance.

Experimental evidence appears to indicate two distinct types of panel flutter.[54] Tests conducted in the region of $M = 2\cdot18$ on a panel with clamped ends (capable of lateral adjustment) showed that, when it was slightly buckled an irregular low-amplitude flutter occurred. As buckling was increased, this gave place to a more regular high-amplitude flutter which persisted at higher deformations. Fig. 13[53] attempts to compare the above results with theory. Each line represents a given initial, unstressed deformation and is a boundary above which flutter exists. This deformation is defined by λ which is $\sqrt{3}$ times the ratio of arch height to panel thickness.

The ordinate is the ratio of dynamic air pressure to panel stiffness and is only valid for $M > \sqrt{2}$ while the abscissa λ_1 denotes the total deformation under stress. Unfortunately the initial warping of the panel was not measured in the experiments so that it is difficult to make a direct comparison. There is a tendency, however, for the points to follow the curves, particularly those belonging to the high-amplitude flutter.

Conclusion

The above discussion has emphasized the diverse nature of mechanical vibration though it must not be imagined that all vibration phenomena are necessarily complicated. Many problems met with in practice are due to simple forced or self-excited vibration which, once identified, can readily be eliminated. In such cases an elementary knowledge of the subject may be of value. Where complication exists, vibration measurement is normally required and this is best left to the expert.

Much theoretical work remains to be done on the vibration of elastic bodies but this will require to be adapted to meet the needs of specific engineering problems. Moreover, new phenomena will appear from time to time which may necessitate prolonged experimentation. Progress, however, will depend to a large extent on the digital computer for time alone will be a determining factor in understanding and solving the problems of the future. More and more engineers will require to be trained in the technique of programming these machines.

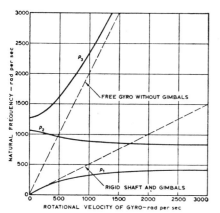

Fig. 12. How natural frequencies of a gyroscope with flexible shaft vary in relation to speed; solid curves correspond to the three modes of vibration

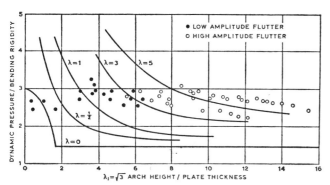

Fig. 13. Comparison of experimental evidence (which suggests two distinct types of panel flutter) with theory. Each line represents an initial deformation and is a boundary above which flutter occurs

Acknowledgement

My sincere thanks are due to Dr J. D. Robson, Professor L. Maunder and Dr A. D. S. Barr for assistance so freely given in the preparation of this review.

REFERENCES

1. DEN HARTOG, J. P., 'Vibration: A Survey of Industrial Applications' 1958, *Proc. Instn mech. Engrs, Lond* , vol. 172, p. 8.
2. BISHOP, R. E. D. and JOHNSON, D. C., *The Mechanics of Vibration.* 1960, Cambridge University Press.
3. MIKLOWITZ, J., 'Recent Developments in Elastic Wave Propagation'. 1960, *Appl. Mech. Reviews*, vol. 13, No. 12, p. 965.
4. MINDLIN, R. D., 'Waves and Vibrations in Isotropic Elastic Plates'. 1960 Structural Mechanics, *Proc. 1st Symp. on Nav. Struct. Mech.*, Pergamon Press, p. 199.
5. TIMOSHENKO, S., 'On the Correlation for Shear of the Differential Equation for Transverse Vibrations of Prismatic Bars'. 1921 *Phil. Mag.* (Ser. 6), vol. 41, p. 744.
6. CHRISTOPHERSON, D. G., 'Effect of Shear in Transverse Impact on Beams'. 1951 *Proc. Instn mech. Engrs, Lond.*, vol. 165, p. 184.
7. ANDERSON, R. A., 'Flexural Vibrations in Uniform Beams according to the Timoshenko Theory'. 1953 *J. Appl. Mech.*, vol. 20 (*Trans. A.S.M.E.*, vol. 75), p. 504.
8. BARR, A. D. S., 'Some Notes on the Resonance of Timoshenko Beams and the Effects of Lateral Inertia in Flexural Vibration'. 1957 *Proc. 9th Int. Cong. Appl. Mech.*, Brussels, vol. 7, p. 448.
9. ARNOLD, R. N. and WARBURTON, G. B., 'Flexural Vibrations of the Walls of Thin Cylindrical Shells having Freely Supported Ends'. 1949 *Proc. roy. Soc.* Ser. A, vol. 197, p. 238.
10. CRANCH, E. T. and ADLER, A. A., 'Bending Vibrations of Variable Section Beams'. 1956 *J. Appl. Mech.*, vol. 23 (*Trans. A.S.M.E.*, vol. 78), p. 103.
11. ANLIKER, M., 'Biegeschwingungen verwundener, einseitig eingespannter und am andern Ende gelenkig gelagerter Stäbe'. 1956 *Zeit. fur ang. Math. und Physik*, vol. 7, No. 3, p. 248.
12. MORLEY, L. S. D., 'Elastic Waves in a Naturally Curved Rod'. 1960 *R.A.E. Rep. No.: Structures 255.*
13. MENDELSON, A. and GENDLER, S., 'Analytical Determination of Coupled Bending-torsion Vibrations of Cantilever Beams by means of Station Functions'. 1950 *N.A.C.A. Tech. Note TN 2185.*
14. THORKILDSEN, R. L. and HOPPMANN II, 'Effect of Rotating Inertia on the Frequencies of Vibration of Stiffened Plates'. 1959 *J. Appl. Mech.*, vol. 26 (*Trans. A.S.M.E.*, vol. 81), p. 298.
15. ELLINGTON, J. P. and McCALLION, H., 'The Free Vibration of Grillages'. 1959 *J. Appl. Mech.*, vol. 26 (*Trans. A.S.M.E.*, vol. 81), p. 603.
16. LIN, T. C. and MORGAN, G. W., 'A Study of Axisymmetric Vibrations of Cylindrical Shells as affected by Rotatory Inertia and Transverse Shear'. 1956 *J. Appl. Mech.*, vol. 23 (*Trans. A.S.M.E.*, vol. 78), p. 255.
17. MILLER, P. R., 'Free Vibrations of a Stiffened Cylindrical Shell'. 1960 *Aero. Res. Committee, R. and M. 3154 ARC 19338.*
18. HERRMANN, G. and MIRSKY, I., 'On Vibration of Conical Shells'. 1958 *J. Aero. Sci.*, vol. 25, p. 451.
19. JOHNSON, M. W. and REISSNER, E., 'On Transverse Vibrations of Shallow Spherical Shells'. 1958 *Q. Appl. Math.*, vol. 15, no. 4, p. 367.
20. BLEICH, H. H., 'Dynamic Interaction between Structures and Fluids'. 1960 Structural Mechanics, *Proc. 1st Symp. on Nav. Struct. Mech.*, Pergamon Press, p. 263.
21. BOLEY, B. A., 'Thermal Stresses'. 1960 Structural Mechanics, *Proc. 1st Symp. on Nav. Struct. Mech.*, Pergamon Press, p. 378.
22. TOBIAS, S. A. and ARNOLD, R. N., 'The Influence of Dynamical Imperfection on the Vibration of Rotating Disks'. 1957 *Proc. Instn mech. Engrs, Lond.*, vol. 171, p. 669.
23. EWING, W. M., JARDETZKY, W. S. and PRESS, F., *Elastic Waves in Layered Media.* 1957, McGraw Hill, Chap. 2.
24. BYCROFT, G. N., 'Forced Vibrations of a Rigid Circular Plate on a Semi-infinite Elastic Space and on an Elastic Stratum'. 1956 *Phil. Trans. roy. Soc., Lond.*, (A), No. 948, vol. 248, p. 327.
25. ARNOLD, R. N., BYCROFT, G. N. and WARBURTON, G. B., 'Forced Vibrations of a Body on an Infinite Elastic Solid'. 1955 *J. Appl. Mech.*, vol. 22 (*Trans. A.S.M.E.*, vol. 77), p. 391.
26. BYCROFT, G. N., 'Machine Foundation Vibration'. 1959 *Proc. Instn mech. Engrs, Lond.*, vol. 173, p. 469.
27. ERINGEN, A. C., 'Response of Tall Buildings to Random Earthquakes'. 1958 *Proc. 3rd U.S. Nat. Cong. Appl. Mech.*, Amer. Soc. mech. Engrs, p. 141.
28. ROBSON, J. D., 'Effects of Non-linearity on the Resonant Frequency of a Body on Soil'. 1957 *Proc. 9th Int. Cong. Appl. Mech.*, Brussels, vol. 7, p. 344.
29. EDWARDS, A. T. and NORTHWOOD, T. D., 'Experimental Studies of the Effects of Blasting on Structures'. 1960 *Engineer*, London, vol. 210, p. 539.
30. JONES, R. and GATFIELD, E. N., 'Testing Concrete by an Ultrasonic Pulse Technique'. 1955 *Road Research Technical Paper No. 34*, H.M. Stationery Office.
31. CRANDALL, S. H., 'Random Vibration'. 1959 *Appl. Mech. Reviews*, vol. 12, no. 11, p. 739.
32. 1958 *Random Vibration*, edited by S. H. Crandall, Technology Press, Cambridge, Mass.
33. ERINGEN, A. C., 'Response of Beams and Plates to Random Loads'. 1957 *J. Appl. Mech.*, vol. 24 (*Trans. A.S.M.E.*, vol. 79), p. 46.
34. KANESIGE, I., 'Measurement of Power Spectra of Vehicle Vibration and Vehicle Road Roughness'. 1960 *Proc. 10th Japan Nat. Cong. for Appl. Mech.*, p. 371.
35. MORAN, D. F. and CHENEY, J. A., 'Earthquake Response of Elevated Tanks and Vessels'. 1960 *Trans. Amer. Soc. civ. Engrs.*, vol. 125, p. 503.
36. Jet Efflux—A Symposium. 1957 *J. roy. aero. Soc.*, vol. 61, p. 103.
37. MINER, M. A., 'Cumulative Damage by Fatigue'. 1945 *J. Appl. Mech.*, vol. 12 (*Trans. A.S.M.E.*, vol. 67), p. A-159.
38. HEAD, A. K. and HOOKE, F. H., 'Random Noise Fatigue Testing'. 1956 *Proc. Int. Conf. on Fatigue of Metals, Instn mech. Engrs*, London, p. 301.
39. ARNOLD, R. N. and MAUNDER, L., *Gyrodynamics and its Engineering Applications.* 1961, Academic Press, New York.
40. MAGNUS, K., 'Beiträge zur Dynamik des Kräftefreien, kardanisch gelagerten Kreisels'. 1955 *Zeit. fur angewandte Math. und Mech.*, vol. 35, p. 23.
41. PLYMALE, B. T. and GOODSTEIN, R., 'Nutation of a Free Gyro subjected to an Impulse'. 1955 *J. Appl. Mech.*, vol. 22 (*Trans. A.S.M.E.*, vol. 77), p. 365.
42. GOODSTEIN, R., 'A Perturbation Solution of the Equations of Motion of a Gyroscope'. 1959 *J. Appl. Mech.*, vol. 26 (*Trans. A.S.M.E.*, vol. 81), p. 349.
43. MAUNDER, L., 'Natural Frequencies of a Free Gyroscope supported in Gimbals on an Elastic Shaft'. 1960 *10th Int. Cong. of Appl. Mech.*, Stresa, Paper II-52-1.
44. DIMENTBERG, F. M., *Flexural Vibrations of Rotating Shafts.* 1961, Butterworths, London.
45. BISHOP, R. E. D., 'The Vibration of Rotating Shafts'. 1959 *J. mech. engng Sci.*, vol. 1, p. 50.
46. BISHOP, R. E. D. and GLADWELL, G. M. L., 'The Vibration and Balancing of an Unbalanced Flexible Rotor'. 1959 *J. mech. engng Sci.*, vol. 1, p. 66.
47. GLADWELL, G. M. L. and BISHOP, R. E. D., 'The Receptances of Uniform and Non-uniform Rotating Shafts'. 1959 *J. mech. engng Sci.*, vol. 1, p. 78.
48. DOWNHAM, E., 'Theory of Shaft Whirling', Parts 1 to 5. 1957 *Engineer*, London, vol. 204, p. 518 et seq.
49. HAGG, A. C., 'Influence of Oil Film Journal Bearings on the Stability of Rotating Machines'. 1946 *J. Appl. Mech.*, vol. 13 (*Trans. A.S.M.E.*, vol. 68), p. A-211.
50. YAMAMOTO, T., 'On Critical Speeds of a Shaft supported by a Ball-Bearing'. 1959 *J. Appl. Mech.*, vol. 26 (*Trans. A.S.M.E.*, vol. 81), p. 199.
51. GOODMAN, L. E. and RATTAYYA, J. V., 'Review of Panel Flutter and Effects of Aerodynamic Noise: Part 1, Panel Flutter'. 1960 *Appl. Mech. Reviews*, vol. 13, No. 1, p. 2.
52. MILES, J. W., 'Dynamic Chordwise Stability at Supersonic Speeds'. 1950 *North American Aviation, Inc., Rep. no. AL-1140*, Oct. 18.
53. FUNG, Y. C., 'On Two-Dimensional Panel Flutter'. 1958 *J. aero. Sci.*, vol. 25, no. 3, p. 145.
54. EISLEY, J. G., 'The Flutter of a Two-Dimensional Buckled Plate with Clamped Edges in a Supersonic Flow'. 1956 Ph.D. Thesis, Guggenheim Aero. Lab., California Inst. of Tech., Published as AF OSR TN 56-296.

Stress Analysis

by J. R. Dixon, BSc, CEng, FRAeS

*Stress analysis must continue
to play a role of increasing importance
in economic and efficient
design. The limiting factors
in many structures are
the stress concentrations arising
at discontinuities. These
may depend on the shape of
the component but also on the
material itself and its homogeneity.
Many theoretical and experimental
techniques have been developed
to tackle the complex
problems connected with stress
distribution.*

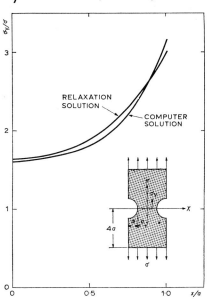

Fig. 1. Using a computer for direct numerical determination of stress distribution as compared with a relaxation solution

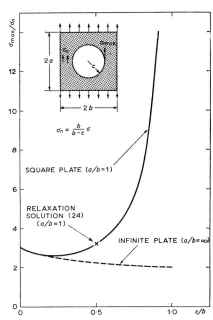

Fig. 2. The importance of plate size: a warning against overenthusiastic application of data beyond the range for which they were intended

The subject of stress analysis was formally initiated by Galileo in 1638 and his theory of the bending of beams developed by the French workers in the 18th century still forms the basis of many present-day elastic calculations.[1] On the experimental side, by the mid-19th century, mechanical extensometers were in use in Britain; Maxwell had already shown how stresses could be measured photoelastically, but his work was not appreciated by the engineers of his time. Probably the earliest reference to what is now known as brittle lacquer coatings can be found in the reports by Clarke[2] in 1850 on the Britannia and Conway tubular bridges; Clarke noticed the flaking away of paint from the metal in regions of high strain.

Methods of theoretical and experimental stress analysis have developed considerably since those early days. By the complexity of much of modern design, especially in aircraft and space engineering, and the frequent need to design for minimum weight and maximum fatigue strength, a tremendous impetus has been given to stress analysis.

The term 'stress analysis' is often used loosely to include almost anything concerned with the mechanics of solids. The present survey is confined largely to the treatment of stress concentrations and their importance in mechanical engineering.

Stress concentrations are local accumulations of stress that arise in the presence of discontinuities of various kinds. The principal sources of such disturbances are: geometric discontinuities such as holes, cavities, notches, fillets, grooves and cracks; material discontinuities, such as inclusions or the reinforcing fibres of composites; load discontinuities, such as concentrated surface loads.

In practice, most stress concentrations are the result of design, but there are the invidious types which result from poor execution of design, for example, inferior machining, tool marks and scratches. The failure of a part normally occurs in a region of stress concentration, particularly under cyclic loading.

During the past 30 years several societies, devoted to stress analysis, have come into existence; perhaps the best known is the Society for Experimental Stress Analysis in the USA. In Britain the Stress Analysis Group of the Institute of Physics was formed in 1946. In 1960 the Joint British Committee for Stress Analysis was formed and in 1964 a British Society for Strain Measurement followed. Two new British journals devoted almost entirely to stress analysis commenced publication in 1965: *Strain*, a BSSM publication, and the *Journal of Strain Analysis*, published by the Institution and sponsored by JBCSA.

Mathematical techniques

Although mathematical solutions are generally restricted to relatively simple shapes, one should not underestimate their importance. They enable a problem to be generalised and the effect of varying parameters can be found merely by

11

substituting their different values into formulae which make it easier to optimise a design.

The representation of the solution of the plane problem of the theory of elasticity with the aid of real or complex potentials has resulted in a well-developed theory and is still the best method of solution for this class of problems. The work of Stephenson,[3] the translation of Muskhelishvili's book[4] and, more recently, the publication of Milne-Thomson's books[5, 6] have done much to stimulate the current interest in the use of complex-variable functions in elasticity problems. The power and usefulness of this method is demonstrated well by Savin[7] who shows the complex-variable theory applied to static problems of holes of various shapes in plates subjected to inplane or bending loads. The application of complex-variable methods can be extended to the dynamic equations of the plane theory of elasticity and leads to as lucid a presentation as for the static equations.[8, 9]

The class of holes for which exact solutions have been found is however rather limited and a numerical procedure is required for the investigation of the stress distribution round a hole of irregular shape. This requires first the solution of the associated problem of mathematically mapping such a hole into a circle. An iterative procedure using a computer, together with a method for the subsequent automatic computation of the stress distribution, has been developed by Sobey.[10] He gives the results of a worked example of the stress distribution round a square hole with rounded corners in a plate under tension. Further work on numerical methods is given by Dollimore.[11] Solutions of problems in plane elasticity may also be obtained by means of the calculus of variations.[12]

Three-dimensional problems generally present considerably greater analytical difficulty than two-dimensional ones, and progress has been correspondingly slower. Exact methods usually depend on the use of stress functions to solve the appropriate equations. For elastic solutions of static axisymmetric problems involving cavities and notches, the best source of information is still Neuber.[13] Further discussion on this class of problem is given by Sternberg.[14] Three-dimensional stress concentrations in thick plates have so far been mainly restricted to a circular cylindrical hole in an infinite plate of finite thickness, subject to an applied bending moment or in-plane applied stresses at infinity.[15] During recent years increasing effort has been devoted to the theory of shells.[16, 17] Flugge's book[18] on shells was written with the design engineer and stress analyst in mind.

The mathematical solution of plane elasticity problems requires the satisfying of the stress-equilibrium and strain-compatibility equations, together with the boundary conditions. Using finite difference equations across the nodal points of a mesh covering the region in question, approximate solutions to engineering problems can be obtained directly by an iterative procedure known as Southwell's relaxation method.[19] The calculations are carried out on a desk calculating machine and the amount of work depends to some extent on the intuition of the operator. When high precision is required, the amount of work can become prohibitive. However, with the advent of the high-speed electronic computer, the dread of long hours at a desk calculator is mitigated. A method using an electronic computer for direct numerical solutions of plane elasticity problems is given by Griffin and Varga.[20] As an example they computed the stress distribution on a tensile specimen with semi-circular edge notches for 66 interior mesh points, and compared the results with Southwell's relaxation solution, using 116 interior mesh

points of the same problem; the distributions of the stress across the net section are shown in Fig. 1.

There is a need in modern stress analysis to tackle idealised problems which simulate real engineering problems. Engineering problems generally have boundary conditions which are difficult to satisfy in an exact theoretical analysis and may be equally difficult to simulate experimentally. This is where numerical methods come into their own. A good example of this is given by Schlack and Little[21] who considered the plane problem of a central circular hole in a square plate under a uniform tensile stress applied to two opposite edges of the plate. They expressed the appropriate stress function as an infinite series and obtained solutions numerically using a computer. Results are shown in Fig. 2 and compared with corresponding results by Howland[22] and Koiter[23] for a plate of finite width, but of infinite length in the direction of the load. For ratios of hole-diameter (2c) to plate width (2b) > 0·2, the difference in maximum stress at the hole boundary for 2a (plate length)/2b values of one and infinity, respectively, becomes appreciable. Fig. 2 shows the importance of the aspect ratio a/b in this type of problem and should serve as a warning against an over-enthusiastic application of data sheets, many of which relate only to large values of a/b beyond the range for which they were intended. The solution of this problem by Spencer,[24] obtained by relaxation methods for the particular case of c/b = 0·5, a/b = 1, is also included in Fig. 2.

Plastic strains

So far mainly elastic solutions have been discussed, that is, solutions which will strictly be applicable to engineering hardware only when the material is obeying Hooke's law. Although, in some engineering components or structures, a small amount of plastic deformation may be tolerated, it is often necessary to restrict the plastic strains because of strength considerations or in order to ensure that the shape of the structure is not seriously altered. Hence the need to know stress and strain distributions with plastic flow.

The problem is of considerable analytical difficulty. Analytical solutions are possible only in a very limited number of cases and to obtain solutions at all it is usually necessary to use numerical methods. The difficulties of these elastic-plastic problems are due partly to the non-linearity of the equations of plasticity and partly to the fact that a deformed body contains, in general, both elastic and plastic regions, separated by interfaces which have to be determined as part of the solution.

Relaxation methods for the numerical solution of problems concerned with an elastic-perfectly plastic (non-work-hardening) material were introduced almost 20 years ago by

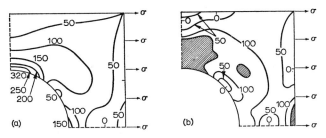

Fig. 3. The elastic solution of plane strain in a square plate with a central circular hole in uniform tension is shown on the left. At higher stresses the shaded regions on the right undergo plastic deformation

Allen and Southwell.[19] More recently Spencer[24] has shown how the computation can be simplified and has extended the method to bring within its scope multiply-connected elastic-plastic bodies (eg, finite plates with holes). For example, Spencer solved the plane strain problem of a square plate containing a central circular hole and loaded by a uniform tensile stress applied to two opposite edges of the plate. Fig. 3a gives the map of contours of $\tau(100/\sigma)$ for $\sigma < 0.31k$, that is, the elastic solution; τ is the maximum shear stress, σ the applied stress and k the yield stress in simple shear. Fig. 3b gives the distribution $\tau(100/\sigma)$ for $\sigma = 0.80k$; the shaded regions of the solid have now undergone some plastic deformation.

Analogue models

Electrical resistance and conducting paper analogue methods have been developed for solving elastic problems. In the former method[25, 26] the differential equations governing the extension or flexure of flat plates or torsion of shafts are described in terms of a stress function for which electrical resistance analogues are available. A network of electrical resistances which represents a region of material automatically solves the finite difference form of the differential equation. In torsional problems an extension of the analogue permits study of the growth of local plastic regions.

The conducting paper analogue[27] is based on the fact that the distribution of a steady-state potential in a thin conducting sheet of constant thickness and uniform resistivity is governed by Laplace's equation. The voltage distribution is analogous to the distribution of the sum of the principal stresses in two-dimensional elastic plate problems or to the distribution of the warping function in the torsion of prismatic bars. In practice, a sheet of conducting paper is cut to the outline of the actual component and external boundary voltages applied through wire staples; voltages within the sheet are measured.

A generalisation of structural analysis, known as the finite element method, makes possible the analysis of two and three-dimensional continua by the same procedures used in the analysis of ordinary framed structures. Advantages of the method are ease of treating variation of properties throughout the material, complex loadings and difficult boundary conditions. The finite element method is based on the idealisation of the actual body as an assemblage of discrete structural elements, interconnected at joints or nodal points. The elements are formed by imagining the original continuum cut into a number of appropriately shaped pieces, retaining the properties of the original material.

In the analysis, these assumed structural elements are dealt with as entirely equivalent to the beams and girders of an ordinary framed structure.

The finite element method has been shown to provide an effective means for the analysis of stress distributions in two and three-dimensional continua.[28, 29]

Strain gauges

Techniques of experimental stress analysis, or more accurately strain analysis, range from the direct measurement of strain on the surface of a structure to the use of scale models.

The old type of mechanical extensometer is often inadequate because of its comparatively long gauge length, its method of attachment to the specimen, and its weight. A partial improvement is the bonded electrical-resistance strain gauge, in very wide use today, which has been much developed since its introduction in 1938.[30-32] Its popularity lies in its availability in small sizes, its low cost, ease of application and direct readout.

The most commonly used form of resistance strain gauge is made of fine wire wound so that most of its length lies parallel to one direction; or of a grid etched or stamped from thin foil. Various types are available for measuring uniaxial or biaxial strains and principal strain directions. The length of the wire grid changes in proportion to the surface strain of the structure to which it is bonded and this results in a change of electrical resistance, usually measured on a Wheatstone Bridge.

Extremes of temperature, however, present problems of material and bonding agents.[33, 34]

The most significant recent advance has been the development of semi-conductor strain gauges.[35, 36] This advantage is their high sensitivity, about 70 times that of their metal counterparts; but they have an inherently non-linear strain/resistance characteristic and special circuits may be required.

Another development is the self-adhesive strain gauge.[37] The sensitive element is a thin film of gold, vacuum deposited on to a thin sheet of Melinex. The transmission of strain depends on the adhesion of the smooth surface to the structure. The gauge has a linear response up to 5×10^{-3} in/in and a sensitivity of the same order as conventional resistance gauges. It is ready for use immediately after being placed on the structure and can be peeled off and used again elsewhere.

To overcome temperature problems Hickson[38] developed a miniaturised pneumatic extensometer for use at high temperatures, which is shown diagrammatically in Fig. 4. The variable orifice or valve is closed or opened by a linkage operated by the extension or compression of a gauge length

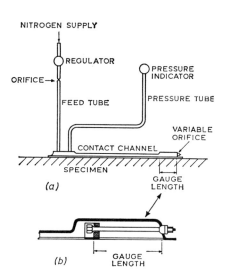

Fig. 4. This miniature pneumatic extensometer, based on the principle of an air gauge, overcomes temperature problems: the orifice size depends on strain along gauge length

Fig. 5. Each fringe in this photoelastic pattern corresponds to a line of constant maximum shear stress. Compare with moiré patterns, also for circular hole, shown in Fig. 8

on the structure. There are no temperature restrictions apart from the strength of the material itself.

Coatings and photoelasticity

Brittle coatings are among the most useful of the modern experimental techniques.[39, 40] They work on the principle that, under strain, cracks will first appear normal to the direction of the maximum tensile principal stress. The strain level at which cracks just appear in the coating is determined by reference to a calibration bar. Although quantitative information can be obtained, the technique is more often than not used qualitatively on large structures to locate regions of stress concentration and the directions of principal stresses, prior to a more detailed study.

There is much room for improvement of coating materials as regards accuracy and ease of application. Resin-based coatings, for application at room temperature or only slightly above, are in general use, for example, *Stresscoat*. Unfortunately they are sensitive to humidity and temperature and this can present some difficulties; also there can be toxic and fire risks. In an attempt to avoid the latter and to find a sensitive lacquer which could be applied to a structure without heat, Hickson[37] developed a lithium tetraborate water-based brittle coating which is also quick-drying at normal room temperatures.

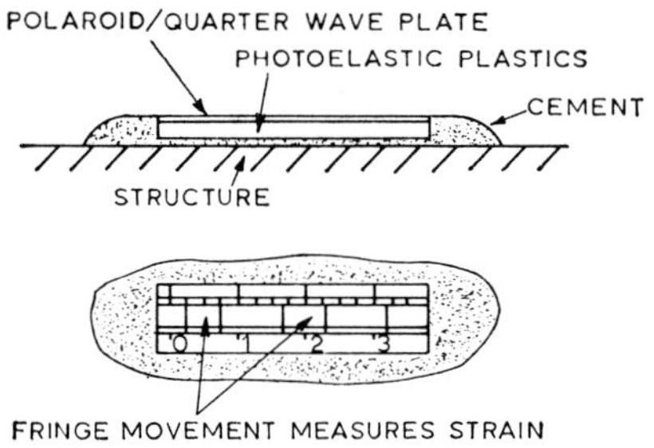

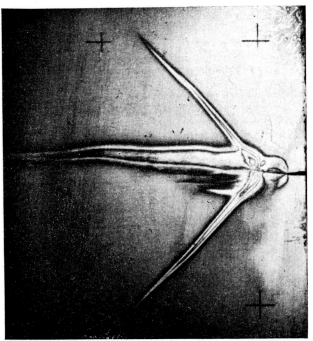

POLAROID/QUARTER WAVE PLATE
PHOTOELASTIC PLASTICS
CEMENT
STRUCTURE

FRINGE MOVEMENT MEASURES STRAIN

Fig. 7. The photoelastic strain gauge, a logical application of photoelastic surface coating, dispenses with additional instrumentation but its accuracy is limited

Fig. 6. This fringe pattern at one end of a crack in uniaxially loaded MS sheet was obtained by the photoelastic surface coating technique

Of all the techniques of experimental stress analysis, photoelasticity has certainly been most in the limelight during the past 40 years, when it has been developed from what was regarded as an academic toy to a versatile engineering tool. The availability of more suitable materials has led to the use of larger models, resulting in higher accuracy. Recently frequent excursions have been made outside the conventional use of the technique for static elastic stress analysis into the field of dynamic, thermal and non-elastic problems.

In conventional photoelastic analysis, stressed structures are simulated by models made from suitable transparent plastics. When viewed in polarised light, the loaded model displays interference fringes from which the stress distribution can be obtained. A typical fringe photograph of a circular hole in a uniaxially loaded sheet is shown in Fig. 5. For elastic behaviour, stresses in the prototype can be calculated from those in the model by simple dimensional analysis. Essentially, the method analyses stresses inside two-dimensional bodies but stresses in a three-dimensional structure can be found by heating the loaded model to a critical temperature and then slowly cooling it. The model can then be sliced (with no relief of stress) into a series of two-dimensional sections for analysis and a three-dimensional picture can be built-up. For an 'engineering' introduction to the subject there are several textbooks, for example, Jessop and Harris[41] and Frocht.[42] For those who wish to delve deeper into the theory, a second edition of the classic work of Coker and Filon,[43] revised by Jessop, was published in 1957.

Improvements have come from the development of new materials.[44] Large castings, comparatively free from residual stresses, can now be made in the laboratory from readily available resins. Models simulating composite structures made from dissimilar materials, for example, solid propellant rocket motors[45] and reciprocating engines, present special problems of satisfying similarity conditions by choice of suitable photoelastic materials, especially for three-dimensional work.[46, 47] Much scope still remains for development of photoelastic materials.

The most notable extension of the photoelastic method during the past ten years has been in the field of photoelastic coatings: a layer of suitable birefringent material is bonded to the surface of the structure. When load is applied the resulting strains on the surface are followed by this layer, polarised light is reflected from the surface of the structure, passing twice through the coating, and fringe patterns are obtained as in the normal photoelastic method. The patterns are interpreted in terms of strain to provide continuous strain measurement over a comparatively large area. One of the major advantages of the technique is that measurements are not limited to elastic strains as with the normal photoelastic method, but will measure total surface strain, that is, the sum of the elastic and plastic parts. A photograph of the fringe pattern around a crack in a uniaxially loaded mild steel sheet is shown in Fig. 6.

The photoelastic coating technique is now well-developed, especially in France and the USA, where Zandman[48] has been largely responsible for its popularity under the trade name *Photostress*. In the UK, industry in general has been slower but the method has been used for some time in research laboratories.[49-51] It is fairly straightforward when applied to the measurement of static strains on flat surfaces; but it is by no means limited to these.[52] It can be used at slightly elevated temperatures,[53] but the practical aspects become more difficult and the interpretation of the results requires

Fig. 8. Produced by optical interference from a ruled grating, this moiré fringe pattern represents lines of constant displacement, parallel to direction of loading. Compare with Fig. 5

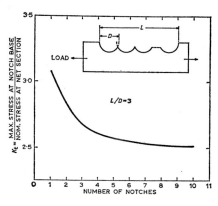

Fig. 9. The combined effect of several stress raisers can be smaller than that of a single notch so that a number of additional grooves can serve to relieve stress concentration

caution. Corrections are necessary if the coating is too thick.[54, 55]

A logical application of the photoelastic surface coating technique is the photoelastic strain gauge.[56] A sketch of a uniaxial type is given in Fig. 7. Such gauges provide a simple means of measuring strain with no additional instrumentation provided a high degree of accuracy is not required. However, current gauge lengths are not much less than an inch and, of course, the gauge must be accessible to visual or photographic inspection. The future may see further developments of photoelastic strain gauges, possibly for high temperature work. Heat-resistant glass, for example, is suitable photoelastically and could be used at high temperatures.

Three further extensions of the photoelastic method are dynamic photoelasticity, photothermoelasticity and photoplasticity. The first is concerned with the study of transient stress systems.[57, 58] Fringe patterns can be recorded by high-speed photography but it is essential to determine the dynamic mechanical and optical properties of the photoelastic materials, particularly their dependence on strain rate.

Photothermoelasticity is concerned with thermal stress fields.[59] Conventional two-dimensional techniques are employed but the loading is thermal, rather than mechanical. Three-dimensional thermal problems are appreciably more difficult but some progress has been made using embedded polariscopes (in a sense, analogous to embedded strain gauges).

The solution of non-linear stress problems using photoelastic models is called photoplasticity.[60, 61] Model materials are used that exhibit non-linear stress/strain characteristics corresponding to those of the structures to be investigated. Some materials, such as magnesium oxide, exhibit a photoelastic effect which can be used to study deformation mechanisms in the material itself.[62]

Future progress in these three fields will surely be rewarding, for they deal with classes of stress problems that are very difficult to solve analytically, even in this era of the computer.

Grid methods

Distortions of regular rulings (lines or circles) either on the stressed component or on a screen observed by way of reflections or transmission from the component, are used as indicators of strain or displacement. Moiré techniques involve localised differential and interference effects. A review of these methods is given by Duncan.[63]

Although the use of moiré fringes as a method of strain measurement is not new, there has been renewed enthusiasm for it during the past five years.[64] A fine parallel grid (usually between 200 and 2000 lines/inch) is formed on the surface of the specimen. When the specimen is loaded, the grid deforms. The super-imposition of the deformed grid on a master or undeformed grid produces an optical phenomenon known as the moiré effect, an interference fringe pattern which can be observed and photographed. A moiré fringe is a line of constant displacement, and from its distribution the strains can be derived. A typical moiré fringe photograph of a hole in a uniaxially loaded sheet is shown in Fig. 8.

An advantage of this method is that it is purely geometrical, measuring strain directly and not via intermediate physical properties such as electrical resistance. It has no reinforcing effect on the structure nor any 'thickness' effects. It can be applied to a wide range of strain measurements, elastic and plastic,[65] thermal[66] and dynamic.[67]

An extension of the normal grid method, developed for static strain measurement by Hickson,[38] is known as the replica technique. A network of fine scratches is made on the surface by drawing rouge paper across it. Replicas of the scratch pattern are made by means of an alloy casting before and during loading. Matching of the replicas in a special microscope comparator permits the measurement of relative displacements of lines to a precision of half a micron. The replica technique has all the advantages of a grid method with the additional advantage of recording and measuring local strain distributions in full detail over a wide range of strain.

Other, perhaps not so well known, methods of stress analysis have undergone recent development. The X-ray method is based on a comparison of diffraction photographs of deformed and undeformed metals. Deformation of the metal causes distortions of the crystal lattices which alter the diffraction pattern. One of the uses of X-ray diffraction is the measurement of residual stresses in engineering components. The technique at present remains specialised and there are many difficulties of interpretation.[68]

A technique known as acoustoelasticity is based on a stress-acoustic effect similar to the familiar photoelastic optical effect. The passage of polarised ultrasonic waves through a solid body is analogous to that of polarised light through a photoelastic model. The technique is by no means established. Some early experiments in acoustoelasticity were reported in 1959,[69] but there seems to have been no publication of further developments since then.

Certain plastics, for example vinyl polychloride, change colour in regions of plastic deformation. A method named chromoplasticity, based on this phenomenon, has been developed[70] for the study of the formation of plastic hinges in structures, the collapse mechanism and loading capacity of the system. The method is essentially qualitative.

Application to design

During recent years some progress has been made in the presentation of stress analysis results in data sheet or tabular form. The best known books covering such presentations are perhaps Roark[71] and Peterson,[72] and these two probably form the *Genesis* and *Exodus* of many drawing office bibles on stress analysis. To supplement these there is the series of data sheets on stress concentration factors issued by the Royal Aeronautical Society.[75] Savin[7] has attempted to present results (mainly from the Russian School) on stress concentrations around holes in easy-to-use forms.

Scientific design relies to a great extent on determining the stresses in parts under operating conditions and for this purpose a knowledge of stress concentrations is essential. But little quantitative information is available, for example, on the stress distributions in shafts with loaded keyways—a fundamental engineering coupling system. In general, a stress concentration will be reduced if the radius of curvature of the discontinuity can be increased at the position of high stress. A typical example of this is the well-known stopping of cracks by drilling holes at their extremities.

A less obvious method of reducing the stress concentration is the use of stress-relieving holes or grooves. The combined effect of several stress raisers together can be less severe than the effect of a single stress raiser; Fig. 9 shows that a single notch can be more severe than a row of similar notches.[74] Edge reinforcement of a hole can also be used to reduce the stress concentration,[75] but a hazard here is that unsuitable reinforcement can lead to a stress concentration more severe than that of the original hole.[76]

Current stress analysis work is very much concerned with assisting the designer directly[77-79] and providing information for data sheets.[80] On the other hand, problems, for example of impact stresses[81] are not sufficiently well understood for rigid design rules to be proposed; research work here may influence the trend of thought on a particular problem and at least make the designer aware of a possible difficulty.

Photoelastic tests[82] on beams with a single edge notch of semi-elliptical root (with major and minor axes 2a and 2b, respectively) under pure bending, have shown that, for a given combination of notch-width/plate-width and notch-depth/plate-width, there is a particular elliptical notch for minimum stress concentration. Fig. 10 compares this optimum elliptical notch with that of the same geometry but with a semi-circular root.

In the past, the analysis and design of components and structures have, on the whole, been based on the linear theory of elasticity. Design based on the elastic limit fails to take advantage of the work-hardening and ductility of some materials, which lead to a redistribution of stress during plastic flow which enables the structure to carry a load higher than the elastic limit. The modern trend towards such plastic design methods has led to an increased interest in plasticity theory[83-85]. For example, the collapse load of a framed structure can be determined by means of the concept of plastic hinges. The ultimate strength of a pressure vessel under a single load application would be estimated much more accurately from a plasticity analysis[84] than from the corresponding elastic analysis. In some instances a plasticity analysis may be easier.

Effect on materials science

Although the role of stress analysis in the design of every-day highly-stressed components is generally well-known, its role in the field of materials science is perhaps not so obvious, yet this latter field is demanding increasing attention. Mechanical properties of materials are very much related to the specimen shape and the type of loading. Tests range from the simple tensile test—not so simple if one has to consider necking!—to the notch impact test and the determination of fatigue strength. To understand the behaviour of materials in these tests it is necessary to learn more about the history of the stress and strain distributions during the test.

Although many types of specimen have been employed in shear tests, none is considered entirely satisfactory because of the non-uniformity and uncertainty of the stress distribution. Iosipescu[85] illustrated this by a comparative study of various types of specimen by photoelastic methods. He then proposed a new type of specimen and demonstrated the uniform distribution of shear stress in the test specimen under

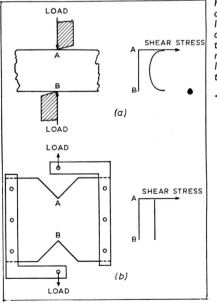

Fig. 11. Shear stress distributions in a loaded test specimen obtained with the conventional method (above) and Iosipescu's new type of specimen

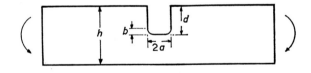

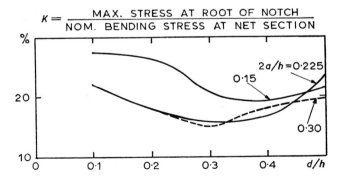

Fig. 10. There is an optimum elliptical notch for minimum stress concentration: the graphs show percentage rise in the stress concentration factor of semi-circular notches for various ratios of slot/plate widths

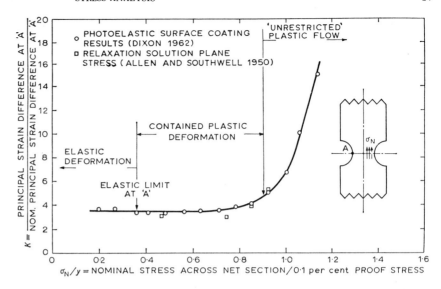

Fig. 12. Even when local plastic flow occurs, the strain concentration factor remains at the elastic value until the specimen is in a state of general yielding. The graph shows variation of K at the root of the notch with applied stress

load. Fig. 11 compares the shear stress distribution across the test section by a conventional method and by Iosipescu.

Non-linear aspects play an important part in materials testing. Many engineering designs must be based wholly or partly on creep behaviour; a review of stress analysis in the presence of creep has been given by Finnie.[86]

Knowledge of plastic flow is also necessary in understanding the behaviour of materials in the presence of stress concentrations. A slip-line plastic analysis of specimens used in impact testing with notched bars has been carried out by Alexander and Komoly,[87] who considered the relative severity of various loading systems as regards producing high tensile stresses at the root of the notch. Dixon[88] investigated the effect of plastic flow on the strain concentration in flat aluminium alloy tensile specimens containing either a central hole or edge notches. Fig. 12 shows the variation of the strain concentration factor K at the root of a semi-circular notch with applied stress; some values from a relaxation solution[19] of the same problem, for a non-work-hardening material, are also included for comparison. In this case, even when local plastic flow has occurred, K remains virtually the same as the elastic value up to a state of general yielding.

During recent years there has been an increasing interest in what is known as fracture mechanics. One of its aims is to establish satisfactory relationships for materials between fracture in service and laboratory tests. A branch of fracture mechanics, dealing with fast fracture of the more brittle materials, is often referred to as the ASTM[89] or Irwin method, and has been extensively developed to describe the behaviour of specimens of different sizes, shapes and loading conditions, in the presence of defects and cracks, but in the absence of any substantial yielding effects. The fracture toughness of the material is related to the elastic stress field near the tip of a loaded crack, and this has led to an intensive study of elastic stress distributions round cracks.[90] The method is also being extended to cover the more ductile materials which fracture under conditions of more general yielding.[91]

The limitations of elastic analysis when applied to real materials has stimulated studies on the effect of local plastic deformation at the tip of a crack, both by dislocation theory[92] and the more conventional continuum mechanics.[93]

As a step in the direction of accounting for the microstructure of a material in a continuum theory, the Cosserat (1909) theory which takes into account couple-stresses, has recently become a very active field of research. The couple-stress theory considers that, within an elastic body, the surfaces of each material element are subjected not only to normal and tangential forces (as in classical elasticity) but also to forces per unit area called 'couple-stresses'. Such an assumption seems appropriate for materials with granular or crystalline structure, where the interaction between adjacent elements may introduce internal moments.

A characteristic length, l, is introduced, which is a material property, presumably related to the substructure of the material and on which the influence of couple-stresses depends strongly. If the ratio of the smallest dimensions of a body to l is large, the theory shows that the effect of couple-stresses is negligible. However, when there are strain gradients and a dimension of the body approaches l, couple-stresses may produce effects of appreciable magnitude.

Mindlin[94] considered the effect of couple-stresses on the stress distribution round a circular hole in an infinite region under plane strain conditions, for a uniaxial stress, a biaxial stress and a pure shear stress at infinity. The couple-stresses had no effect for the biaxial stress field, but resulted in lower values of the stress concentration factor K than given by usual elastic theory for the uniaxial and pure shear stress.

Materials cannot always be treated as isotropic, homogeneous media, even for engineering purposes, and the structure of the material itself can cause complex stress problems.[95] Inclusions in a material may be stress raisers. The composite materials, for example, glass reinforced resins and their metal counterparts such as steel wire or aluminium oxide fibres in aluminium, being composed of two or more dissimilar materials, have complicated internal stress fields on a macroscopic scale. A knowledge of the associated stress distributions is necessary for an understanding of the mechanics of load transfer within the material and its fracture characteristics.

An exact evaluation of the stress distribution in a complex array of fibres within a matrix would be extremely difficult and attention has been directed to evaluating stresses for idealised fibre shapes and without the complicating effect of surrounding fibres.[96] If the fibres are discontinuous, then the attainment of high strength will depend on an efficient load transfer from the matrix to the fibre; it is of considerable interest to know the stress distribution near the end of the fibre.[97]

Future developments

The design of any stressed component is based on a knowledge or estimate of the stresses and deflections so that stress analysis must continue to play a role of increasing importance since there is increasing demand for economic and efficient design. The use of the electronic computer will permit more realistic boundary conditions and less restrictive assumptions in the mathematical analysis and numerical methods appropriate to stress problems will be developed further.

Strain gauges, especially semi-conductor and associated circuitry, will be improved. Developments in photoelasticity and brittle lacquer techniques should take place, possibly through the availability of more suitable materials. The use of the moiré method of strain measurement may become much more popular.

In general, as design methods become more sophisticated, the non-linear behaviour and time dependence of materials in different environments will become more important. It is not surprising, therefore, that many stress analysts are devoting time to studies outside the field of linear elasticity and the current literature already reflects the interest in this class of problem.

The anticipated use of composite materials may well open up new avenues of research.

Acknowledgements

This article is published by permission of the Director of the National Engineering Laboratory, Ministry of Technology. It is Crown copyright and is reproduced by permission of the Controller of HM Stationery Office.

REFERENCES

1. TIMOSHENKO, S. P., *History of the Strength of Materials*. 1953. New York: McGraw Hill.
2. CLARK, E., *The Britannia and Conway Tubular Bridges*, vol. 1, p. 166. 1850. London: Day & Son; John Weale.
3. STEPHENSON, A. C., 'Some Boundary Problems of Two-dimensional Elasticity'. 1943. *Phil. Mag.*, vol. 34(238), p. 766–793.
4. MUSKHELISHVILI, N. I., *Some Basic Problems of the Theory of Elasticity*. 1953. Groningen, Holland: P. Noordhoff Ltd.
5. MILNE-THOMSON, L. M., *Plane Elastic Systems*. 1960. Berlin: Springer-Verlag.
6. MILNE-THOMSON. L. M., *Antiplane Elastic Systems*. 1962. Berlin: Springer-Verlag.
7. SAVIN, G. N., *Stress Concentrations Around Holes*. 1961. Oxford: Pergamon Press.
8. RADOK, J. R. M., 'On the solution of problems of dynamic plane elasticity'. 1956, *Q. Appl. Math.*, Vol. 14(3), p. 289–298.
9. MITRA, M., 'On the Solution of Problems of Dynamic Plane Elasticity for Anisotropic Media'. 1960, *Q. Jl Mech. appl. Math.*, vol. 13(3), p. 369–373.
10. SOBEY, A. J., 'The Estimation of Stresses around Unreinforced Holes in Infinite Elastic Sheets'. *Structures Report* No 283. 1962. Farnborough: Royal Aircraft Establishment.
11. DOLLIMORE, J., 'The Transformation of an arbitrary hole in an Infinite Plate of Elastic Material'. 1963, *Q. Jl Mech. appl. Math.*, vol. 16(2), p. 149–162.
12. MORLEY, L. S. D., 'A Variational Method of Solution for Problems in Plane Elasticity'. *Structures Report* No 293. 1964. Farnborough: Royal Aircraft Establishment.
13. NEUBER, H., *Kerbspannungslehre* (Theory of Notch Stresses), 2nd edn. 1958. Berlin: Springer-Verlag.
14. STERNBERG, E., 'Three-dimensional Stress Concentrations in the Theory of Elasticity'. 1958, *Appl. Mech. Rev.*, vol. 11(1), p. 1–4.
15. REISS, E. L., 'Extension of an Infinite Plate with a Circular Hole'. 1963, *J. Soc. ind. appl. Math.*, vol. 11(4), p. 840–854.
16. *Symposium on Shell Research*, Delft 1961. Proceedings. 1962. New York: J. Wiley & Sons, Inc.
17. *Proc. 11th Internat. Congress Appl. Mechs.* Munich, 1964. Berlin: Springer-Verlag. To be published.
18. FLUGGE, W., *Stresses in Shells*. 1960. Berlin: Springer-Verlag.
19. SOUTHWELL, R. V., *Relaxation Methods in Theoretical Physics*. 1946. Oxford: Clarendon Press.
20. GRIFFIN, D. S. and VARGA, R. S., 'Numerical Solution of Plane Elasticity Problems'. 1963, *J. Soc. ind. appl. Math*, vol 11(4), p. 1046–1062.
21. SCHLACK, A. L. and LITTLE, R. W., 'Elastostatic Problem of a Perforated Square Plate. 1964, *Proc. Am. Soc. civ. Engrs* (J. Eng Mechanics Div.), 10(EMS): pt. 1, Paper 4089, p. 171–188.
22. HOWLAND, R. C. J., 'On the Stresses in the Neighbourhood of a Circular Hole in a Strip in Tension'. 1930, *Phil. Trans. R. Soc.*, vol. 229A, p. 49–86.
23. KOITER, W. T., 'An Elementary Solution of Two Stress Concentration Problems in the Neighbourhood of a Hole'. 1957, *Q. appl. Math.*, vol. 15(3), p. 303–308.
24. SPENCER, A. J. M., 'The Solution of Plane Elastic/Plastic Problems by Relaxation Methods'. 1964, *Appl. scient. Res., Section A.* vol. 12(4–5), p. 391–406).
25. REDSHAW, S. C. and RUSHTON, K. R., 'A Study of the Various Boundary Conditions for Electrical Analogue Solutions of the Extension and Flexure of Flat Plates. 1961, *Aeronaut. Q.*, vol. 12(3), p. 275–282.
26. RUSHTON, K. R., 'Elastic Stress Concentrations for the Torsion of Hollow Shouldered Shafts Determined by an Electrical Analogue'. 1964, *Aeronaut. Q.*, vol. 15(1), p. 83–96.
27. ROSS, D. S. and QURESHI, I. H., 'Boundary Value Problems in Two-dimensional Elasticity by Conducting Paper Analogue'. 1963, *J. scient. Instrum.*, vol. 40(11), p. 513–517.
28. CLOUGH, R. W. and RASHID, Y., 'Finite Element Analysis of Axisymmetric Solids'. 1965, *Proc. Am. Soc. civ. Engrs* (J. Eng. Mechanics Div.), vol. 91(EM1), pt. 1, Paper 9229, p. 71–85.

Dr J. R. Dixon *went to Goole Grammar School and joined the RAF in 1944. From 1948 to 1952 he attended University College, Hull, to take an honours degree in Mathematics and then went to Rolls-Royce Aero-engine Division, as a graduate apprentice in 1952. A year later he became an aerodynamicist with Bristol Aircraft Ltd. After spending one year on photoelasticity research at University College, London, he returned to the Company as Section Leader in the photoelastic laboratory. In 1958 he joined the NEL at East Kilbride, where he is now Principal Scientific Officer in charge of the Stress Analysis Section. He was awarded a PhD in 1967 for work on deformation around cracks. He has published a number of papers on photoelasticity, finite element analysis and fracture mechanics.*

29. DE VEUBEKE, B. F. (Ed.), *Matrix Methods in Structural Analysis*. 1964. Oxford: Pergamon Press.
30. PERRY, C. C. and LISSNER, H. R., *The Strain Gauge Primer*. 1955. New York: McGraw Hill.
31. MURRAY, W. M. and STEIN, P. K., *Strain Gauge Techniques*. 1958. Massachusetts: MIT.
32. STEIN, P. K., *Advanced Strain Gauge Technique*. 1962. Phoenix, Arizona: Stein Engineering Services, Inc.
33. WEYMOUTH, L. J., 'Strain Measurement in Hostile Environment'. 1965, *Appl. Mech. Rev.*, vol. 18(1), p. 1.
34. BERTODO, R., 'Resistance Strain Gauges for the Measurement of Steady Strains at High Temperatures'. 1963–64, *Proc. Inst. Mech. E.*, vol. 178, pt 1, Paper No. 34, p. 907.
35. HIGSON, G. R., 'Recent Advances in Strain Gauges'. 1964, *J. scient. Instrum.*, vol. 4(7), p. 405–114.
36. SANCHEZ, J. C., 'The Semiconductor Strain Gage—A New Tool for Experimental Stress Analysis'. *In* ROSSI, B. E. (Ed.), *Experimental Mechanics, Proc. International Cong.*, p. 256–274. 1963. Oxford: Pergamon Press.
37. HICKSON, V. M., 'Some New Techniques in Strain Measurement— A Survey'. ZIENIEWICZ, O. C. and HOLISTER, G. S., *Stress Analysis*, p. 239–263. 1965. London: John Wiley and Sons Ltd.
38. HICKSON, V. M., 'Special Techniques in Experimental Stress Analysis'. *In* ROSSI, B. E. (Ed.), *Experimental Mechanics, Proc. International Cong.*, p. 221–236. 1963. Oxford: Pergamon Press.
39. LINGE, J. R., 'Some Developments and Applications of Brittle Lacquers'. 1958, *Aircr. Engng* vol. 30(350), p. 94–100; vol. 30(351), p. 142–148; vol. 30(352), p. 173–179.
40. DURELLI, A. J., PHILLIPS, E. A. and TSAO, C. H., *Introduction to the Theoretical and Experimental Analysis of Stress and Strain*. 1958. New York: McGraw-Hill.
41. JESSOP, H. T. and HARRIS, F. C., *Photoelasticity* (Principles and Methods). 1949. London: Cleaver-Hume Press.
42. FROCHT, M. M., *Photoelasticity*, Vols. I and II. 1941: 1948. London: Chapman & Hall Ltd.
43. COKER, E. G. and FILON, L. N. G., *A Treatise on Photoelasticity* 2nd edn. 1957. Cambridge: University Press.

44. LEVEN, M. M., 'Epoxy Resins for Photoelastic Use'. *In* FROCHT, M. M. (Ed.), *Proc. Symp. on Photoelasticity*, p. 145–165. 1963. Oxford: Pergamon Press.

45. DURELLI, A. J. and PARKS, V. J., 'Photoelasticity Methods to Determine Stresses in Propellant Grain Models. 1965, *Exp. Mech.*, vol. 5(2), p. 33–46.

46. DIXON, J. R., 'Problem of Materials for Composite Models in Photoelastic Frozen-stress Work'. 1962, *Brit. J. appl. Phys.*, vol. 13(2), p. 64–67.

47. KUFNER, M., 'Investigation of the Strength of Laminated Materials with the Help of Photoelasticity' (in German). *Internat. Symp. Photoelasticity.* 1962. Berlin: Akademie-Verlag. English transl: *RAE Trans* No. 1070. 1964. Farnborough: Royal Aircraft Establishment.

48. ZANDMAN, F., *Photostress* (Principles and Applications). 1959. Phoenexville: Budd Co., Box 245.

49. FESSLER, H. and HAINES, D. J., 'A Photoelastic Technique for Strain Measurement on Flat Aluminium Alloy Surfaces'. 1958, *Br. J. appl. Phys.*, vol. 9(7), p. 282–287.

50. LINGE, L. R., 'Photoelastic Measurement of Surface Strain'. 1960. *Aircr. Engng.* vol. 32(378), p. 216–221; vol 32(379), p. 261–270; vol. 32(380), p. 296–298.

51. DIXON, J. R. and VISSER, W., 'An Investigation of the Elastic–Plastic Strain Distribution Around Cracks in Various Sheet Materials'. *In* FROCHT, M. M., *Proc. International Symposium on Photoelasticity*, Chicago, 1961, p. 231–250. 1963. London: Pergamon Press.

52. HOLISTER, G. S., 'Recent developments in Photoelastic Coating Techniques'. 1961, *Jl R. aeronaut. Soc.*, vol. 65(610), p. 661–669.

53. ZANDMAN, F., REDNER, S. S. and POST, D., 'Photoelastic Coatings Analysis in Thermal Fields'. 1963, *Exp. Mech.*, vol. 3(9), p. 215–221.

54. ZANDMAN, F., REDNER, S. S. and RIEGER, E. J., 'Reinforcing Effects of Birefringent Coatings'. 1962, *Exp. Mech.*, vol. 2(2), p. 55–64

55. DUFFY, J., 'Effects of the Thickness of Birefringent Coatings'. 1961, *Exp. Mech.*, vol. 1(3), p. 74–82.

56. HOLISTER, G. S., 'Photoelastic Strain Gauges'. 1963. *Applied Materials Research*, vol. 2(1), p. 20–30.

57. RILEY, W. F., 'Photoelastic Study of the Interaction Between a Plane Stress Wave and a Rigid Circular Inclusion'. 1964, *J. Mech. Eng. Sci.*, vol. 6(4), p. 311–317.

58. AUSTIN, A. L., 'Measurements of Thermally Induced Stress Waves in a Thin Rod Using Birefringent Coatings'. 1965, *Exp. Mech.*, vol. 5(1), p. 1–10.

59. GERARD, G., 'Progress in Photothermoelasticity'. *In* FROCHT, M. M., *Proc. Internat. Symp. Photoelasticity*, Chicago, 1961, p. 81–94. 1963. Oxford: Pergamon Press.

60. MONCH, E. and LORECK, R., 'A Study of the Accuracy and Limits of Application of Plane Photoelastic Experiments'. *In* FROCHT, M. M., *Proc. Internat. Symp. Photoelasticity*, Chicago, 1961, p. 169–184. 1963. Oxford: Pergamon Press.

61. FROCHT, M. M. and CHENG, Y. F., 'An Experimental Study of the Laws of Double Refraction in the Plastic State in Cellulose Nitrate—Foundations for Three-dimensional Photoelasticity'. *In* FROCHT, M. M., *Proc. Internat. Symp. Photoelasticity*, Chicago, 1961, p. 195–216. 1963. Oxford: Pergamon Press.

62. HOLISTER, G. S., 'The Photoelastic Method Applied to Materials Research. 1962, *Appl. Mat. Res.*, vol. 1(3), p. 149–159.

63. DUNCAN, J. P., 'Grid and Moiré Methods of Stress Analysis'. *In* ZIENIEWICZ, O. C. and HOLISTER, G. S., *Stress Analysis*, p. 314–345. 1965. London: John Wiley & Sons Ltd.

64. LOW, I. A. B. and BRAY, J. W., 'Strain Analysis Using Moiré Fringes'. 1962. *Engineer, Lond.*, vol. 213(5540), p. 566–569.

65. VINCKIER, A. and DECHAENE, R., 'Use of Moiré Effect to Measure Plastic Strains'. 1960, *J. bas. Engng*, vol. 82(2), p. 426–434.

66. SCIAMMARELLA, C. A. and Ross, B. E., 'Thermal Stresses in Cylinders by the Moiré Method'. 1964, *Exp. Mech.*, vol. 4(10), 289–296.

67. RILEY, W. E. and DURELLI, A. J., 'Application of Moiré Methods to the Determination of Transient Stress and Strain Distributions'. 1962, *J. appl. Mech.*, vol. 29(1), p. 23–29.

68. GARROD, R. J. and HAWKES, G. A., 'X-Ray Stress Analysis on Plastically Deformed Metals'. 1963, *Br. J. appl. Phys.*, vol. 14(7), p. 422–428.

69. BENSON, R. W. and RAELSON, V. J., 'Acoustoelasticity'. 1959, *Product Engineering*, vol. 30(29), p. 56–59.

70. BALAN, S., RAUTU, S. and PETCU, V., *Chromoplasticitatea.* 1963. Romine: Editura Academiei Republicii Populare.

71. ROARK, R. J., *Formulas for Stress and Strain.* 3rd edn. 1954. London: McGraw-Hill Book Co.

72. PETERSON, R. E., *Stress Concentration Design Factors.* 1953. London: Chapman & Hall Ltd.

73. *Fatigue Data Sheets.* London: Royal Aeronautical Society.

74. LIPSON, C. and JUVINALL, R. C., *Handbook of Stress and Strength.* 1963. New York: The Macmillan Co.

75. HICKS, R., 'Stress Concentrations Around Holes in Plates and Shells'. *Inst. Mech. Engrs.*, *Appl. Mech. Convention*, Newcastle-upon-Tyne, *Preprint Paper* 14. 1964. London: Institution of Mechanical Engineers.

76. WITTRICK, W. H., 'On Axisymmetric Stress Concentration at an Eccentrically Reinforced Circular Hole in a Plate.' 1965, *Aeronaut. Q.*, vol. 16(1), p. 15.

77. EDMUNDS, H. G., 'The Contribution of a Mechanical Engineering Laboratory to Prime Mover Development'. 1964, *The English Electric Journal*, vol. 19(5), p. 28.

78. FESSLER, H. and LEWIN, B. H., 'Stresses in Branched Pipes Under Internal Pressure'. 1962, *Proc. Inst. mech. Engrs.*, vol. 176(29), p. 771–788.

79. GREEN, W. A., HOOPER, G. T. J. and HETHERINGTON, R., 'Stress Distribution in Rotating Discs with Non-central Holes'. 1964, *Aeronaut. Q.*, vol. 15(2), p. 107–121.

80. ALLISON, I. M., 'The Prediction of Three-dimensional Stress Concentration Factors'. 1959, *Jl. Aeronaut. Soc.*, vol. 63(585) p. 549–551.

81. DALLY, J. W. and HALBLEIB, W. F., 'Dynamic Stress Concentrations at Circular Holes in Struts'. 1965, *J. mech. Engng Sci.*, vol. 7(1), p. 23–27.

82. TSAO, C. H., CHING, A. and OKUBO, S., 'Stress-concentration Factors for Semielliptical Notches in Beams Under Pure Bending'. 1965, *Exp. Mech.*, vol. 5(3), p. 19A–23A.

83. HODGE, P. G., *Plastic Analysis of Structures.* 1959. London: McGraw Hill Book Co.

84. SHIELD, R. T. and DRUCKER, D. C., 'Design of Thin-walled Torispherical and Toriconical Pressure Vessel Heads'. 1961, *J. appl. Mech.* vol. 28(2), 292–297.

85. IOSIPESCU. N., 'Photoelastic Investigations on an Accurate Procedure for the Pure Shear Testing of Materials'. 1963, *Rev. Mecan. Appl.*, vol. 8(1), p. 145–164.

86. FINNIE, I., 'Stress Analysis in the Presence of Creep'. 1960, *Appl. Mech. Rev.*, 13(10), p. 705.

87. ALEXANDER, J. M. and KOMOLY, J. T., 'On the Yielding of a Rigid/Plastic Bar With an Izod Notch'. 1962, *J. Mech. Phys. Solids*, vol. 10, p. 265–275.

88. DIXON, J. R., 'Elastic–Plastic Strain Distribution in Flat Bars Containing Holes or Notches. 1962, *J. Mech. Phys. Solids*, vol. 10, 253–263.

89. 'Fracture Testing of High Strength Sheet Materials'. *ASTM Bulletin*, No. 243, p. 29–40. 1960. Philadelphia: ASTM.

90. SIH, G. C., PARIS, P. C. and ERDOGAN, F., 'Crack-tip Stress Intensity Factors for Plane Extension and Plate Bending Problems' 1962, J. Appl. Mech.,vol. 29(2), p. 306–213.

91. WELLS, A. A., 'Notched Bar Tests, Fracture Mechanics and the Brittle Strength of Welded Structures'. 1965, *Br. Weld. J.*, vol. 12, p. 2–13.

92. BILBY, B. A., COTTRELL, A. H. and SWINDEN, K. H., 'The Spread of Plasticity From a Notch'. 1963, *Proc. R. Soc.*, Series A, vol. 272, p. 304–314.

93. DIXON, J. R., 'Stress and Strain Distributions Around Cracks in Sheet Materials Having Various Work-hardening Characteristics'. 1965 *International Journal of Fracture Mechanics*, vol. 1(3), p. 224–244.

94. MINDLIN, R. D., 'Influence of Couple-Stresses on Stress Concentrations'. 1963. *Proc. Soc. exp. Stress Analysis,* vol. 20 (1), p. 1.

95. ANDREWS, E. H. 'Report on Conference on the Mechanics of Anisotropic and Inhomogeneous Materials'. Manchester, April 1964. 1964, *Br. J. appl. Phys.*, vol. 15, p. 777–782.

96. STOWELL, E. Z. and LIU, T. S., 'On the Mechanical Behaviour of Fibre-reinforced Crystalline Materials'. 1961, *J. Mech. Phys. Solids,* vol. 9, p. 242–260.

97. TYSON, W. R. and DAVIES, G. J., 'A Photoelastic Study of the Shear Stresses Associated with the Transfer of Stress During Fibre Reinforcement'. 1965, *British Journal of Applied Physics*, vol. 16(2). p. 199–205.

Friction and Lubrication

by Geoffrey W. Rowe, MA, PhD, CEng, MIMechE

Wherever there is mechanical engineering there is friction and usually lubrication problems arise. During the last few years considerable work has been done on new types of lubricant and methods of lubrication, while a new attempt at a comprehensive theory of friction has been put forward. Tribology as a whole has received large-scale Government attention.

Friction appears everywhere in engineering. It is commonly condemned as a source of power loss, but even the humble nut and bolt would need to be redesigned if friction were always low. All engineers are interested in both high and low friction, if only in their own cars. Great care is taken to provide enough good clean oil for car engines,[1] but disaster can follow if oil reaches the brake surfaces. The clutch demands a delicate balance of friction,[2] to provide smooth engagement, yet positive drive, even during furious acceleration. Lubrication of vehicles played an important part in one of the world's most ambitious journeys, the crossing of Antarctica.[3] We may expect to hear about lubrication when the full stories of Gagarin's and Shepard's journeys come to be told.

For low friction, it is usually best, whenever possible, to design the moving parts so that a thick film of lubricant may be built up by hydrodynamic forces.[4] Wear is then also reduced to a very low level. Heavy loads, slow speeds and reciprocating motion make this very difficult in many instances and it is then necessary to rely on lubricant films only a few molecular layers thick. Fortunately such boundary lubricants do exist[5, 6] and have a remarkable tenacity. If no fluid at all is allowed, as in some textile machinery, in vacuum, or at high temperatures, solid lubricants may be used.[7] Really dry sliding of unprotected metal should always be avoided.

This review is concerned primarily with friction experiments of general interest which have been reported since the Conference on Lubrication and Wear[8] sponsored by the Institution in 1957. More extensive bibliographies for the specialist may be found in the Digests of Lubrication literature published by the American Society of Mechanical Engineers for 1958 and 1959–60.[9] A bibliography of research on lubricants has been published by Verein Deutsche Ingenieure.[10]

Air bearings

A remarkable development in the past few years has been the revival of interest in air as a lubricant, first suggested by Hirn[11] over a century ago. Although successful experiments were conducted by Kingsbury[12] in 1897, it is only recently that air has been seriously considered as a bearing fluid.

[By courtesy of Ford, Harris and Pantall,[13]]

Fig. 1. An experimental apparatus incorporating 7 in. diameter journal bearings lubricated by air. The shaft weighs 300 lbs and runs at speeds between 100 and 3000 r.p.m. Such bearings are expected to carry loads up to one ton at 3000 r.p.m.

Like normal liquid lubricants, the air can be used to provide the necessary load-carrying pressure either by utilizing its viscosity in a high-speed bearing, or by supplying air under pressure to orifices in the bearing wall. The aerodynamic method has of course the advantage that no compressor is required. On the other hand it may be necessary to provide aerostatic lift initially to allow the rotor to reach operating speed, since gases in general do not have boundary lubricating properties. It is possible that this disability could be overcome either by including a mildly reactive gas in the air supply or by first coating the surfaces with a lamellar solid such as molybdenum disulphide. The coatings should last well, since under normal operating conditions they are subject to no wear. For large slowly-moving surfaces such as lathe saddles, the static method is the only alternative to conventional boundary lubrication.

Recent experiments[13, 14] have shown that large bearings can be operated at 3 000 r.p.m. with a load of a ton (Fig. 1), while smaller sizes give good performance at speeds of 60 000 r.p.m. or even 90 000 r.p.m. The bearings tend to be large because the viscous drag, and hence the load-bearing capacity, for a gas is very much less than for a liquid. They are not however limited by cavitation, volatilization or decomposition in the fluid. Indeed the viscosity of a gas increases with temperature, so that their performance actually improves as the temperature is raised. Air bearings have been operated satisfactorily at 400–500°C.

Both aerostatic and aerodynamic bearings have been discussed in several papers at the 1960 ASME–ASLE Conference[15] and even higher speeds (160 000 r.p.m.) have been mentioned.[16] Apart from these, various types of hovercraft operate, of course, on the same principle and it has been suggested that this could be extended to linear motion in overhead railways, using an aerodynamic slipper bearing.

Liquid lubricants behave as solids

Many experiments[17] have shown that under sufficient pressure, usually tens of tons per square inch, most lubricants solidify. It has also been recognized that at sufficiently high rates of shear they may behave as solids. This may be of importance, for example in gears, where the individual teeth approach and recede at considerable speed. Barlow and Lamb[18] have examined the behaviour of selected lubricants when subjected to ultrasonic vibration. At relatively low rates of shear (low ultrasonic frequency) the normal viscous properties are found but, as the frequency is increased beyond a certain value, the response is more like that of an elastic solid.

The fluid can be represented in an elementary way by a mechanical analogy with a spring and dashpot in series. If the movement is fast enough the response is that of the spring. Actually, of course, the systems are much more complex than this when large molecules are involved and, for non-Newtonian fluids, the viscosity itself is a function of speed. Each lubricant has one or more characteristic relaxation times. For many, the relaxation time is of the order of a few microseconds, which is comparable with the measured time of contact of high-speed gear teeth. The liquids may thus behave as solids and so support much greater tooth loading than would be predicted on the basis of their viscosity.[19] Experiments are continuing, to extend these ultrasonic investigations to high-pressure conditions, since the local pressures in gear teeth are usually very high.

Nuclear radiation can also lead to thickening of lubricants. For use in an atomic pile this is unsatisfactory and considerable attention has been given to developing lubricants unaffected by radiation.[20, 21]

Squeeze films

The experiments with air bearings and the ultrasonic studies of gear lubricants show that speed can be an ally of the lubrication engineer. At the other end of the scale, some nominally moving parts do often have to remain at rest for long periods. Every car owner knows that start-up wear may shorten the life of his engine because the lubricant has drained away but he may not realize that the same feature is responsible for his personal discomfort after sitting in the car for a long period.

It has often been observed that the static friction of a soft metal is high because the metal flows slowly under load and the contact area increases. Recent experiments have shown this effect for waxes[22] and even the friction of polytetrafluoroethylene may increase by a factor of 2 if the sliders are left in contact for an appreciable time.[23] Apart from this effect, Denny[24] has found that the friction of rubber increases with standing time as a result of the lubricant slowly flowing out of the interfacial capilliary channels. In some experiments the full friction, corresponding to dry conditions, was found

after the lapse of about 20 hours. When a solid lubricant was used, the effect disappeared; static and dynamic friction were equal.

Lewis and McCutchen[25] have examined the behaviour of animal joints, which show a remarkably low friction, $\mu = 0.01$ to 0.02, according to Charnley.[26] This value is characteristic of hydrodynamic lubrication but the relative speeds and frequent reversal of small movements makes it impossible to maintain a hydrodynamic film by viscous forces. On the other hand, boundary lubrication is characterized by coefficients of friction of about $\mu = 0.05 - 0.10$. The explanation given, and borne out by experiments with cellular rubber and with sausage skins,[27] is that the articular cartilege surface of bone resembles a stiff sponge with closed pores, and contains synovial fluid which can be slowly extruded under pressure. The bone joint is thus an internally pressurized hydrostatic bearing. The experiments show that when all the fluid has seeped out after half an hour or so, the friction rises to a much higher value. The authors state that this may be associated with the observation that horses, which sleep standing up, change their position at least twice an hour.

Boundary lubrication

The simple theory of friction proposed by Bowden and Tabor[5] attributes frictional resistance to the shear strength of junctions formed between plastically deformed asperities on the surfaces under load. This interpretation has been widely accepted for most conditions of boundary and dry friction. Briefly, it is stated that because engineering surfaces are not molecularly flat, the load is borne by a number of highly localized contacts. These deform plastically until their total area A is sufficient to support the load W at the average yield pressure p. Then, if the average shear strength of the junctions so formed is s, the frictional resistance F is given by

$$F = As = \frac{W}{p} s$$

Thus a coefficient of friction μ may be defined

$$\mu = \frac{F}{W} = \frac{s}{p}$$

If a boundary lubricant is present, it will support some fraction x of the normal load so that the area of metallic contact is reduced to

$$(1 - x)\frac{W}{p}.$$

Then if the metal has a shear strength s_m and the boundary film an effective shear s_1

$$\mu = \frac{F^1}{W} = [xs_1 + (1 - x)s_m]/p.$$

This simple formulation accords with experimental results for a wide range of conditions, but it does not account for the high friction exhibited by some soft metals, and by all metals when thoroughly cleaned in high vacuum. McFarlane and Tabor[28] have shown that this failure arises from a basic fallacy in the theory. Because the asperities are deforming plastically, the normal and tangential stresses s and p may not be separated.

It is necessary to introduce a yield criterion relating s and p to the flow stress, thus:

$$p^2 + s^2 = k^2$$

If the junctions are just yielding at a pressure p_o under the normal load alone, the application of a very small tangential force will suffice to cause further plastic flow. This increases the contact area and thereby decreases p until a new equilibrium is reached. Experiments show this increase in area to occur with indium, a very soft metal, in air[28] and with harder metals in a high vacuum.[29]

Recently Courtney-Pratt and Eisner[30] have studied the early stages of sliding in great detail, using a multiple-beam interference technique to detect very small movements, of the order of a few microns (1 micron $\approx$ 40 micro-inches). In their experiments, a pair of gold or platinum sliders was carefully loaded and then an independent tangential force was applied. They found that, as the force was slowly increased, the sliders were progressively displaced laterally (Fig. 2). The electrical resistance fell steadily, showing that the metallic area of contact was increasing. A striking feature of these results is that the early stage of junction growth is identical, whether the lubricant is present or not. However, after a finite growth, the lubricated junction sheared and steady sliding ensued, whereas the clean junction continued to grow. This implies that the primary function of the lubricant is to restrict junction growth, in contrast to the assumption of the simple theory that the lubricant primarily reduces the original metallic contact under normal load.

Tabor[31] has examined this apparently basic difference and in so doing has evolved a comprehensive theory which explains the very high friction of denuded metal and the very marked influence of small quantities of surface contaminant on this value, as well as giving a reason for the validity of the simple theory in many applications despite its basic fallacy.

Comprehensive theory

The experiments just quoted show that the initial stage of junction growth is independent of lubricant and is governed by the plastic yield criterion, but that junction growth ceases at an early stage when the lubricant is present. Tabor[31] suggests that if the metal carries a surface film of shear strength s_i, the normal and tangential stresses p and s can be transmitted through the film, provided s does not exceed s_i, to produce plastic flow according to the criterion

$$p^2 + \alpha s^2 = k^2 = \alpha s_m^2$$

Junction growth thus proceeds as before, but as soon as the tangential force F reaches a value $A^1 s_i$, where A^1 is the area of contact at the instant when $s = s_i$, the junction will shear and sliding will occur.

This limiting condition occurs when

$$p^2 + \alpha s_i^2 = \alpha s_m^2$$

If now the shear strength s_i of the interface be described as some fraction β (less than unity) of the shear strength of the metal

$$p^2 = \alpha(\beta^{-2} - 1)s_i^2$$

$$\mu = \frac{F}{W} = \frac{As_i}{Ap} = \sqrt{\frac{1}{(\alpha\beta^{-2} - 1)}}.$$

There are three important features of this expression. When β is unity, the condition for denuded metals, the coefficient of friction is very high indeed. Even a slight reduction in β

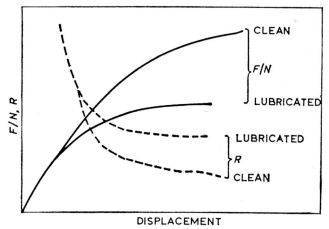

[By courtesy of Tabor,[28] from results by Courtney Pratt and Eisner,[27] and the Royal Society]

Fig. 2. Electrical contact resistance measurements R and tangential force coefficients $\phi = F/N$ corresponding to tangential micro-displacements, for platinum surfaces, under load W = 920 g. The area of contact grows as tangential force is applied, but the growth is limited when a lubricant is present

has however a considerable influence on μ. If for example $\beta = 0.95$ and α is taken to be 9 (the exact value of α may vary from one material to another, but has only small effect on μ), then the coefficient of friction is reduced to about 1.0. This corresponds to values commonly found in laboratory experiments with clean metal surfaces in air.

When the surface film is readily sheared, as we may expect under boundary lubricated conditions, the equation simplifies. If β is less than about 0.2,

$$\mu = \sqrt{\frac{\beta^2}{\alpha(1 - \beta^2)}} \approx \sqrt{\frac{\beta^2}{\alpha}}$$

In terms of the mechanical properties this may be written

$$\mu = \sqrt{\frac{\beta^2}{\alpha}} = \sqrt{\frac{\beta^2 s_m^2}{\alpha s_m^2}} = \frac{s_i}{p_o}$$

$$= \frac{\text{critical shear stress of the interface}}{\text{plastic yield pressure of the underlying metal}}$$

The simple equation for boundary friction is thus a special case of the more general equation, and is valid for all practical lubrication conditions. Only when the shear strength of the interface is greater than about a quarter of that of the metal is it necessary to apply the more rigorous theory allowing for junction growth.

Apart from the theoretical importance of this analysis, it places emphasis upon the function of the lubricant in achieving separation of the metallic surfaces. If the surface film has a finite shear strength, junction growth may occur and coefficients of friction of the order of $\mu = 1.0$ be recorded without metallic contact. The borderline between this state and a rapidly rising friction with inadequate lubrication is dangerously thin. In many practical processes, avoidance of metal transfer is of much greater importance than the precise value of the coefficient of friction.

Surface effects on strength

While it is undoubtedly true, as postulated by the early adhesion theory of friction, that welds resulting in metallic transfer can, and often do, form between sliding surfaces, it is not always necessary to assume actual welds, as the above theory indicates. Recently Cameron[29] has calculated approximate values of the shear strength of hydrocarbon films adsorbed to metal surfaces. His results depend upon a questionable assumption, but indicate at least that the observed frictional force can be entirely accounted for by the calculable van der Waals interaction between the adsorbed molecules. This result is confirmed by direct measurement of the shear strength of adsorbed monomolecular layers of calcium stearate.[33]

Two other effects may also be important. It is widely accepted in the U.S.S.R. that the strength of a metal may be impaired, as Rebinder suggested,[34] by a 'surface plasticization' resulting from absorption of a boundary lubricant. On the other hand, hard surface layers are found to increase the strength.

The Rebinder plasticization was proposed to explain the results of tensile tests on a tin specimen coated with a layer

Fig. 3. Microsections (× 70) from a 0·05% C steel cut with a tungsten carbide tool with rake angle 6° at slow speed. CCl₄ lubricant (top) has a marked influence on the chip and built-up edge formation, compared with dry machining (bottom)

of oleic acid in cetyl alcohol. It was found that the flow stress was less than that of a similar dry specimen and that the spacing of the active slip planes in the deformed metal was also reduced, so that the slip on each plane was less. Rebinder suggested that the oleic acid penetrated small, invisible, surface cracks initially present on the material. The adsorbed material lowered the strength. This hypothesis has been used to explain a number of friction phenomena[35] and has recently been applied[36, 37] to machining. It is found that carbon tetrachloride is a remarkably effective cutting fluid on a previously machined surface, but if the surface is polished before cutting, thereby presumably closing the microcracks, the fluid is ineffective.

There is good evidence that the strength of metals may be increased by, for example, an oxide film.[38] It has been suggested that the plasticization referred to above is attributable to removal of the normal oxide film rather than to any other specific action. This question remains open.

Roscoe[39] found that an oxide film only 20 atoms in thickness could increase the critical shear stress for slip in cadmium crystals by 5%. This has recently been explained in terms of the pile-up of dislocations at the interface. Brame and Evans[40] have examined this type of behaviour in some detail, using oriented films of gold and other metals evaporated on to single crystals of silver. Their experiments show that the gold films can extend a considerable amount (30% elongation) without cracking, because dislocations can be injected into the gold film from the silver substrate. Rhodium films, on the other hand, crack extensively with very little slip. Few dislocations can apparently be injected into the rhodium film, which consequently is unable to accommodate the applied strain.

Further experiments with gold on a non-crystallographic substrate showed cracking. Thus it is concluded that the thin films are inherently brittle and their mode of deformation on a substrate depends on the ease with which dislocations from the substrate can be injected into the film. This in turn is determined by the degree of misfit and the relative elastic moduli.

The consequences in friction have not been explored, but it seems possible that an effect of this sort may explain for example why hard oxide skin can restrict junction growth and metal transfer, while showing a relatively high friction. It may be still more relevant to the type of frictional behaviour examined by I-Ming Feng[41]. He showed that under some circumstances the slip lines on one surface become impressed upon another, forming a mechanically interlocking junction of saw-tooth form, on a very fine scale. This leads to production of small wear particles on sliding, or to local deformation producing heat, softening the surfaces and encouraging surface welding.

Solid-film lubricants

Apart from the possible frictional effect of surface films which modify the behaviour of the substrate metal, there are of course well-known solid lubricants which have inherent low-friction properties.[7] Probably the most familiar of these is graphite, yet despite its familiarity there is still dispute about the exact reason for the good lubrication normally associated with it. The low friction cannot simply be attributed to the layer structure of graphite permitting easy shear like a pack of cards. Graphite itself shows a relatively high friction about $\mu = 0.5$, when all adsorbed gases have been removed by careful heating in high vacuum.

Deacon and Goodman[42] have experimented with layers of graphite, and its kindred layer-lattice substance boron nitride, deposited in various ways on platinum. They conclude that the high-temperature behaviour in air can best be explained in terms of the interaction between the edges of small crystallites composed of many layers rather than the forces between the layers themselves. They infer that the major action of adsorbed gases is to reduce the attractive forces between the edges of the crystallites. Rowe,[43] on the other hand, concludes from experiments on a range of lamellar solids in high vacuum that the inherent low friction of materials like MoS_2 and TiI_2, and the dependance on absorbates of graphite and boron nitride can be explained in detail in terms of the crystallographic forces between the layers. He claims also to explain on this basis why practically all gases reduce the friction of outgassed graphite while only organic vapours reduce that of boron nitride; yet graphite and boron nitride are jointly unique in showing a marked fall in friction with increasing temperature. Electron diffraction studies by Midgley and Teer[44] appear to support the conclusions of Goodman and Deacon. Electron microscope observations introduce another interesting possibility: Bollmann and Spreadborough[45] have noticed rolls of graphite among the wear debris, similar to the rolls found by James[46] on PVC, and indeed to the rolls formed from an ordinary eraser on paper.

Whatever the detailed mechanism of action of these solid lubricants may be, there is no doubt that they are of considerable value to industry. Recent experience confirms that they can be used in high vacuum[47] or inert gas atmospheres; they are increasingly encountered in engineering[48] as well as in space flight. Solid lubricants are applicable at both high and low temperatures[49] and are little affected by nuclear radiation. Naturally, recent advances have been made in all these directions with fluid lubricants also.[10, 84]

Vapour lubrication

Closely associated with solid lubricants is the relatively new interest in vapour lubrication, which depends on the formation of solid surface films having controlled friction properties, and should not be confused with aerodynamic lubrication described above.

There was some discussion of vapour lubrication at high temperatures during the 1957 Lubrication and Wear Conference,[50] in which it was shown that certain reactive gases could be used to form lamellar surface coatings actually during sliding. The action may be considered analogous to that of extreme-pressure additives in conventional oil technology. Experiments at N.A.S.A.[51] have shown that it is not necessary to use highly-reactive gases. They have used dicholoro-difluoromethane (CCl_2F_2) and similar innocuous gases, commonly found in domestic refrigerators, which react only when the metal surface is exposed at high temperatures. This represents a considerable advance, since these gases are neither corrosive nor toxic. Bearings have been operated in such gaseous environments for short periods at temperatures up to 1500°F. The higher the temperature, of course, the more corrosion occurs, but very recent experiments in the same laboratory are aimed at inhibiting gross corrosion.[52] At the same conference Baldwin and Rowe[53] showed that vapour-deposited coatings are effective on deforming metal under loads of 10 tons, and indeed will lubricate under these conditions to 500°C, whereas a conventional E.P. chlorinated oil breaks down well below 400°C.

Even oxide films may provide effective high-temperature lubrication.[54]

Friction at high speeds

Temperature has usually been regarded as an obstacle to very high speed sliding. As the speed increases and the temperature rises, metal sliders become soft and friction increases. Recent experiments have shown, however, that at very high speeds the temperature rises sufficiently to melt the surface, providing a liquid film of very low shear strength which acts as a lubricant. Bowden and Freitag[55] and Bowden and Persson[56] have studied this in detail with a ball suspended magnetically in a high vacuum and rotated at surface speeds up to 2000 miles an hour (176 000 ft/min). The friction is measured by timing the deceleration through a small speed range when the ball is brought into light circumferential contact with three symmetrically spaced plane surfaces.

The same effect is used industrially in the fusion bandsaw, which has no teeth, but relies on the heat generated by friction between the metal to be cut and a band travelling at about 15 000 ft/min.

It is possible that rocket guide-rails could be lubricated in this way, giving positive guidance after the rockets have reached high velocities.

Salomon[57] suggested, as a result of a theoretical study in 1931, that in metal cutting the heat generated should reach a maximum with increasing speed and then fall off. Recent American experiments[58] have been conducted on metal cutting at speeds up to 360 000 ft/min (4000 m.p.h), and have confirmed this prediction. Projectiles fired from a 20 mm cannon flew past a rigidly-mounted cutting tool and it was found that very high metal removal rates, about 6000 in.³/min compared with conventional values of say 25–200 in.³/min, could be attained. A good surface finish was obtained on steel, accompanied by improved tool life. Minimum tool wear was found at about 120 000 ft/min (1360 m.p.h) when cutting carbon steel. Rocket sledges have also been used to carry the workpiece. Russian work in the field of high-speed machining has been briefly reported in English.[59]

While the shapes which can be formed in this way are very elementary the experiments suggest that the limitation is in machine design rather than friction.

Explosive forming naturally involves high speed and has

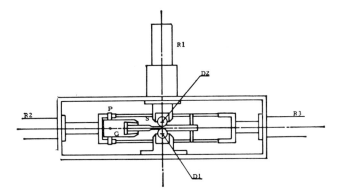

Fig. 4. *A laboratory apparatus, contained in a vacuum tank, for drawing rectangular bars to 55% reduction of cross-sectional area to examine boundary lubrication under severe deformation*

Table 1. Effect of surface preparation on physical properties

Mechanical prepara-tion of bar surface	Max. reduction of area in one pass		Limita-tion	Surface roughness (μ in. CLA)	
	Mild steel	Stain-less		un-drawn	drawn
Polished . .	13	7	Pick-up	2	—
Abraded 600 emery	19	17	Pick-up	10	—
Grit blasted (180 mesh grit) . .	56	58	Tensile failure	50	10

been extensively developed in the past decade[60, 61] but the friction in the process does not appear to have received attention. Some work has been done on lubrication in cold extrusion[62] but this also has not been studied in detail.

Conventional machining

Friction plays a large part in machining with conventional tools, but its influence is complex and not properly understood. Until fairly recently it had been accepted that the chip produced by a cutting tool flows as an entity over the tool face so that its behaviour is dominated by the friction at the interface between tool and chip.

However, it is known that a built-up edge of workpiece metal is often found adhering strongly to the tool nose. Trent[63] has found by microscopic examination of polished tungsten carbide tools that the build-up on the tool surface may remain essentially static, severe shear occurring within the chip itself. On the other hand, Shaw[36] has found that the application of carbon tetrachloride to the back surface of a chip, not the face in contact with the tool, will facilitate flow. He supposes that the fluid penetrates minute cracks left by a preceding cut and plasticizes the metal, after the manner of Rebinder's effect. Recent experiments in the author's laboratory also show the effect of CCl_4 (Fig. 3) and confirm the observation of Shaw that the influence of CCl_4 may persist through several machined layers, each a few thousandths of an inch deep. However, the behaviour is also influenced by the metallurgical condition of the workpiece. Fully work-hardened steel is practically uninfluenced by the presence of CCl_4 and indeed behaves in the dry state in very much the same way as the annealed metal does when lubricated with CCl_4.

Trent[63] has further shown that the static built-up edge occurs at certain speeds and feeds which can be fairly sharply divided from the range where true sliding occurs. Attention is being given to these unresolved questions in several laboratories. The significance of strain hardening and the size of the built-up edge have been discussed in two papers presented at the Institution.[65]

Metal forming

The frictional problem in machining is concerned with high local pressures, very clean surfaces, and high speeds. More generally in metal working, the pressures may be high over considerable areas, producing plastic flow of the whole

of one sliding member. The resultant surface extension produces fresh metal surface, and there seems to be little correlation in general between the performance of lubricants under these conditions and under the light loads of conventional laboratory experiments.

Lancaster and Rowe[66] have attempted to span the range with the apparatus shown in Fig. 4, which is capable of drawing short bars to 60% reduction of area under carefully controlled conditions. One striking result is that in a high vacuum ordinary soap completely fails to lubricate while in a normal atmosphere good draws can be made. As the severity of the normal draw is increased, however, the quantity of lubricant passing through the die, measured by a radio-tracer technique[67], falls rapidly. For a pass giving 15% reduction of area to the bar the soap film is barely sufficient to cover the surface roughness and any further reduction may permit metallic contact, leading to pick-up from the steel bar on the steel die. Recent unpublished experiments[68] show the choice of die material to be important; if, for example, alumina dies are used, the range of satisfactory drawing may be extended to 35%. In contrast, when aluminium is drawn, steel dies give better results than alumina. It is also found[69] that a fine degree of roughening on the dies will reduce the tendency to pick-up. It appears that if the sliding process is regularly interrupted at short intervals, junction growth is restricted and gross pick-up is much less likely to occur. The very high polish normally given to dies may thus be unnecessary though, on the other hand, a coarse die surface may 'coin' into the drawing metal, leaving a poor finish.

A much better method of avoiding pick-up is to provide an adequate film of lubricant. This can be done by encouraging the formation of a thick layer by viscous drag in a low-angle die. Relatively high speeds may be needed, and starting may be a difficulty on short runs. Lubricant may also be trapped in surface pockets, which can and should be much larger than the interruptions of die surface mentioned above, since their purpose is to carry lubricant into the die, where it is compressed by the reduction in size of the pocket as the bar draws. It is found that lubricant eventually bursts out of the trailing edge of such a pocket and lubricates oncoming metal for a considerable distance. Blasting with fine grit affords a convenient method of providing such pockets and a striking improvement is obtained, as shown in Table 1, when steel bars are grit blasted before being drawn with soap lubricant through steel dies.

Similar improvement can be obtained by a commercial phosphating process, and it is probable that a major function of the crystalline phosphate coating serves to provide surface traps for lubricant. The friction of the phosphate itself is high, but it does help to prevent pick-up in the event of lubricant failure.

Apart from its industrial value as a dry process requiring no subsequent chemical removal, the grit-blasting technique of lubricant trapping provides a more fundamental approach to the study of lubricants under heavy deformation. Many lubricants are effective when trapped in this way and it is found that their relative efficiency is then determined by their boundary properties, as found on a light-load apparatus, while viscosity is unimportant.

Thus it appears that the critical difference between friction in small-scale experiments and in operations involving severe deformation is that the lubricant tends to be excluded from the interface in metalworking. If a supply of lubricant is

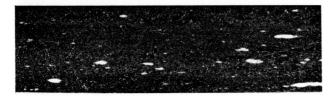

Fig. 5. An auto-radiograph print (above) and photograph of a titanium bar drawn in the apparatus shown in Fig. 4, using a lubricant containing radioactive chlorine. Local reaction has occurred in the regions appearing white, and pick-up has been avoided

provided by surface trapping, there is a close correlation in frictional behaviour, and even in coefficients of friction,[70] provided temperature equivalence is maintained. Butler[71] has made a study of the shapes of deforming surfaces, showing how readily lubricant may be trapped.[72]

Temperature is, however, of considerable importance. Most organic lubricants break down if the temperature exceeds 200°C. Nevertheless, high temperature may be turned to advantage in some instances. There is no boundary lubricant for titanium but certain halogenated compounds will react with a warm titanium surface. Fig. 5 shows an autoradiograph print obtained with a lubricant containing radioactive chlorine. Local reaction has occurred and bars of titanium can be drawn to moderate reductions with this lubricant.[73]

Wallace[74] has suggested close control of temperature to obtain good lubrication in stretch forming. One of the main problems in this operation is to produce a uniform, preferably low, friction over a large area. In his experiments, the wooden former was chilled a little below 0°C and sprayed with water so that a thin film of ice was produced. The metal sheet, at a few degrees above zero, was placed in contact with the ice. The ice surface melted uniformly and gave a low friction over the whole sheet.

Melting glass is widely used to lubricate extrusion of steel[75] and its behaviour can conveniently be examined by a radio-tracer method[76] as shown in Fig. 6.

An interesting example of the use of high temperatures generated in friction is the welding of studs on to plate, and other axially symmetrical applications, where the stud is rotated at high speed, brought into contact with the plate and then stopped.[77, 78]

Metal transfer and wear

The friction process, associated with the interaction of surface asperities, usually involves some transfer of metal from one surface to the other, and is often accompanied by production of loose wear debris. Economically these two factors are of greater importance in engineering than the actual coefficient of friction.

Unfortunately, although many wear experiments have

been reported, most of the results relate to specific circumstances, and there is a lack of general correlation. Some basic experiments at the A.E.I. laboratories have shown that the wear process may be divided into two phases.[79] First, metal is transferred from one sliding surface to the other, where it adheres strongly and builds up an equilibrium layer. If the load is light, sliding proceeds in a steady manner and the transferred film oxidizes. In the second phase, loose wear fragments which are predominantly oxide become detached. Under heavy loads the same pattern is observed, but the detached fragments are larger and consist mainly of metal.[80] Nosovskii[81] has examined the influence of temperature and of atmosphere; he finds that in oxygen, oxidation is the major factor, but in argon local metallic seizure is more important. In normal air the oxidizing form of wear is found at low speeds, changing to thermal wear at high speeds. Lancaster[82] has shown that atmospheric oxidation may form a protective film, provided the load is not excessive.

Water vapour also plays an important part, even when the surfaces are properly lubricated. For example, the wear of tungsten carbide on steel, lubricated with oleic acid, can be reduced by nearly a half if the atmosphere is completely dried.[83]

Significance for engineering

These are exciting days for engineering. Not only is it invading the realm of that purest of sciences, astronomy; it is also being stimulated by other sciences to examine its own traditions, sometimes to destroy, more often to vindicate, but finally to understand them. New conditions are calling for new lubricants, even for new ways of lubrication while, on the other hand, study of the principles of friction is increasingly explaining why current processes work as well as they do. Both these approaches have flourished in the last few years. Even at the time of the last Institution Conference on Lubrication, held only in 1957, hardly any attention was paid to unconventional sliding conditions; yet last year a whole session of the ASME/ASLE Conference was devoted to friction in unusual environments, with similar papers spilling over into at least two other sessions. On the other hand, very little work had given more than passing attention to the details of friction in metal working and cutting processes, though thousands of gallons of carefully-developed fluids, bewildering in their variety, are used for these processes.

Many of the experiments quoted above are of obvious importance to engineers. The air bearing is clearly directly usable. It gives exceedingly low friction and practically no wear, can be used at extremely high speeds, and in much higher ambient temperatures than are permissible with liquid lubricants. In principle, of course, it resembles the hovercraft and even a certain domestic carpet cleaner. Vapour lubrication offers another possibility of high-temperature operation. The principles of lubricant trapping and squeeze-film lubrication can be applied to specific problems where heavy deformation is encountered, as in drawing and press operations. Vacuum friction problems can usually be solved by one or other of the solid lubricants now available. In this category we should perhaps mention the lubricants developed expressly for resistance to nuclear radiation.

The principle of surface melting and high-speed friction appears at first to have little practical importance, but it is in fact already utilized in the lubrication of stretch-forming with ice, and in high-speed cutting.

The second group of experiments does not immediately foreshadow new processes; here the emphasis has been upon understanding lubrication in existing processes. Whether this would lead to improvements it is not possible to say until the process is in fact completely understood. The drawing experiments have shown why thin lubricants did not previously give good results, and how they could be made to do so. The ultrasonic experiments have shown why gear lubricants can support more load than would be expected from their viscosities, and may offer a new basis for selection of gear oils.

Finally, there are experiments which have no immediate application to engineering but which influence the basic interpretation on which further research and development work is designed. Such are the patient experimental studies leading to a more comprehensive theory of friction. This has focused attention on junction growth behaviour in dry and boundary-lubricated sliding and, in the latter condition, also on the inherent properties of the adsorbed layer. The importance of at least some boundary films in determining the plastic flow of the metal itself is becoming recognized as a significant factor in frictional behaviour. The electron microscope offers a new and potentially valuable tool for studying this as well as the influence of dislocation on friction, at present a completely open field, but one which offers real possibilities for metallurgically improved friction materials.

Future trends

An author is severely tempted to present his own research programme as a general trend. Alternatively, he may look at the trend of research over the past years and extrapolate it. Or, again, it is possible to try to envisage the technological advances generally and to think about the special frictional problems they will introduce. Probably a combination of the latter two will give the best chance of accurate prediction.

Higher speeds and higher temperatures seem to be increasingly demanded and their repercussions on friction will probably be met by melting solid lubricants, of which ice and glass alone have technological significance at present. Solid lubricants themselves, apart from graphite, are comparative newcomers and will undoubtedly play an increasing part in special applications. How they work is still in dispute and there will undoubtedly be more arguments before the full theory is evolved. Naturally, space flight will introduce quite new problems of lubrication, which will probably have to be solved with solid lubricants.

The theory of dry friction, almost everywhere accepted for almost two decades, has recently undergone a major revision, confirming the simple concept of asperity welding for most practical conditions but drastically modifying the general view.

The newer theory has yet to be tested, probably best in conjunction with metal transfer experiments. It may be that some light will be thrown on the basic questions of why metallic welds form in the first place (as they appear to do). Do metallurgical factors of crystal structure and work-hardening affect the bonding? Or are the bonds envisaged by the newer theory not really metallic bonds at all, except under special circumstances of cleanliness? Why do the lubricants, in their turn, have such a very remarkable ability to prevent welding or at least to restrict it?

Apart from developments in direct friction experiments we may expect a continuation and extension of the application of friction theory. Until a few years ago there had been relatively little interaction between heavy engineering industry and laboratory friction experiments. It may be hoped that a cordial endeavour to speak the same language may help the practising engineer to use the products of boundary and hydrodynamic lubrication theory and, on the other hand, help the research worker to discover which frictional properties are in fact significant in the industrial processes.

Acknowledgment

I am indebted to my colleague Mr A. T. Male for reading the first draft, and for his helpful suggestions.

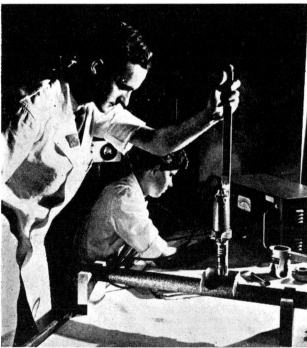

[By courtesy of Tube Investments Research Labs. Photo by W. Nurnberg]

Fig. 6. A Geiger counter being used to measure the thickness of radioactive glass lubricant coating on a stainless steel extrusion

REFERENCES

1. January 1959 *Sci. Lub.*, vol. 11, pp. 9–10, 'Motor Oils 1948–58'.
2. SPRAG, S. R. and CUNNINGHAM, R. G. 1958 *Amer. Chem. Soc.* (*Petrol. Chem. Division*), 'Symposium on chemistry of friction wear'. SPRAG, S. R. and CUNNINGHAM, R. G. 1959 *Industr. Engng Chem.*, vol. 51, p. 1047, 'Friction characteristics of lubricants'.
3. PRATT, D. L. 1959 *Sci. Lub.*, vol. 11, p. 26, 'Performance in Antarctica'. PRATT, D. L. 1958–59 *Proc. Instn mech. Engrs, Lond.* (*Automobile Division*), 'Performance of vehicles under transantarctic conditions'.
4. September 1959 *Machine Design*, p. 178, 'Full film lubrication'.
5. BOWDEN, F. P. and TABOR, D. 1956 *Friction and Lubrication*, Methuen.
6. September 1959 *Machine Design*, pp. 197–200, 'Boundary and mixed film lubrication'.
7. ROWE, G. W. April 1961 *Research Lond.*, vol. 14, pp. 137–142, 'Solid lubricants'.
8. 1957 *Instn mech. Engrs Lond.*, Conference on Lubrication and Wear. November 1957 *Sci. Lub.*, Synopsis of Instn mech. Engrs Conference on Lubrication and Wear.

9. October 1959 *Mech. Engng*, vol. 81, p. 56, 'Digest of lubrication literature 1958'. April 1961 *Mech. Engng*, vol. 83, pp. 53–59, 'Digest of lubrication literature 1959–60'.
10. MOOS, J. 1960 *Z. Ver. dtsch. Ing.*, vol. 102 (1) pp. 32–42 'Schmierstoffe'.
11. HIRN, G. A. 1854 *Bull. Soc. industr. Mulhouse*, vol. 26, p. 195.
12. KINGSBURY, A. 1897 *J. Amer. Soc. nav. Engrs*, vol. 9, p. 267, 'Experiments on air lubricated bearings'.
13. FORD, G. W. K., HARRIS, D. M. and PANTALL, D. 1957 *Instn mech. Engrs Lond.*, vol. 171, pp. 93–128, 'Principles and applications of hydrodynamic-type gas bearings'.
14. 1958 *Trans. A.S.M.E.*, vol. 80, no. 4, pp. 865–878, 'Theoretical and experimental analysis of hydrodynamic gas bearings'.
15. *A.S.M.E./A.S.L.E.*, Lubrication Conference, held in Boston, U.S.A.
16. 1961 *Engineering*, vol. 191, p. 255, 'Running on gas as a lubricant'.
17. 1953 *A.S.M.E.* Pressure-viscosity Report.
18. LAMB, L. and BARLOW, A. J. 1959 *Proc. roy. Soc.*, Ser. A 153, p. 52, 'Viscoelastic behaviour of lubricating oils'.
19. February 1959 *Mech. Engng*, vol. 81, p. 63, 'Viscoelasticity in lubrication'.
20. SCARLETT, N. A. and CLIFFE, J. O. *Nuc. Eng.*, vol. 5, no. 44, p. 23, 'Radiation-resistant greases'.
21. 1958 *Nucleonics*, vol. 16, p. 112, 'Effect of gamma radiation on organic fluids and lubricants'.
22. MILNE, A. A. 1958–59 *Wear*, vol. 2, p. 28, 'Friction experiments with a soft solid'.
23. ROWE, G. W. Unpublished.
24. DENNY, D. F. 1958–59 *Wear*, vol. 2, p. 264, 'Time effects in static friction of lubricated rubber'.
25. LEWIS, P. R. and McCUTCHEON, C. W. 1959 *Nature, Lond.*, vol. 184 p. 1285, 'Weeping lubrication in mammalian joints'.
26. CHARNLEY, J. 1959 *New Scientist*, vol. 6, no. 138, p. 61.
27. McCUTCHEON, C. W. 1959 *Nature, Lond.*, vol. 184, p. 1284, 'Mechanism of animal joints'.
28. McFARLANE, J. S. and TABOR, D. 1950 *Proc. roy. Soc.*, Ser. A 202, pp. 244–253, 'Relation between adhesion and friction'.
29. BOWDEN, F. P. and ROWE, G. W. 1956 *Proc. roy. Soc.*, Ser. A 233, pp. 429–442, 'Adhesion of clean metals'.
30. COURTNEY-PRATT, J. S. and EISNER, E. 1957 *Proc. roy. Soc.*, Ser. A 238, p. 529, 'Effect of tangential force on the contact of metallic bodies'.
31. TABOR, D. 1959 *Proc. roy. Soc.*, Ser. A 251, pp. 378–393, Junction growth in metallic friction'.
32. CAMERON, A. 1960 *Trans. A.S.L.E.*, p. 195, 'A theory of boundary lubrication'.
33. BAILEY, A. I. and COURTNEY-PRATT, J. S. 1955 *Proc. roy. Soc.*, Ser. A 227, p. 500, 'Shear strength of monomolecular films'.
34. REBINDER, P. 1947 *Nature, Lond.*, vol. 159, p. 866, 'New physicochemical phenomena in the deformation and mechanical treatment of solids'.
35. KUZNETSOV, V. D. 1957 *Surface Energy of Solids*, H.M.S.O.
36. SHAW, M. C. 1958–59 *Wear*, vol. 2, p. 217, Action of cutting fluids at low speeds'.
37. PLETENEVA, N. A. and EPIFANOV, G. I. 1956 *Dokl. Akad. Nauk.*, SSSR 77, p. 1051, 'Carbon tetrachloride reduces tool life'.
38. ANDRADE, E. N. da C. and RANDALL, R. F. V. 1952 *Proc. phys. Soc. Lond.*, Ser. B 65, p. 445.
39. ROSCOE, R. 1934 *Nature, Lond.*, vol. 133, p. 912.
40. BRAME, D. R. and EVANS, T. 1958 *Phil. Mag.*, vol. 3 (33), p. 971, Deformation of thin films on solid substrates'.
41. FENG, I-MING 1955 *J. appl. Phys.*, vol. 26 (1), pp. 24–27, 'Metal transfer and wear'.
42. DEACON, R. F. and GOODMAN, J. F. 1958 *Proc. roy. Soc.*, Ser. A 243, pp. 464–472, 'Lubrication by lamellar solids'.
43. ROWE, G. W. 1960 *Wear*, no. 2, pp. 274–285, 'Observations on friction and wear of boron nitride and graphite'.
44. MIDGELEY, J. W. and TEER, D. G. 1961 *Nature, Lond.*, vol. 189, p. 735, Orientation of graphite during sliding'.
45. BOLLMAN, W. and SPREADBOROUGH, J. 1960 *Nature, Lond.*, vol. 186, 'Action of graphite as lubricant'.
46. JAMES, D. I. 1952 *Wear*, no. 2, p. 183, Surface damage caused by PVC sliding on steel'.
47. 1960 *Product. Engng*, vol. 31, no. 16, p. 74, Vacuum bearings use solid lube'.
48. 1960 *Metal Progr.*, p. 65, 'Refractory metals worked in argon-filled room'.
49. 1960 *Machine Design*, p. 30, Naval Air Materials Centre, 'Research on solid lubricants'.
50. ROWE, G. W. 1957 Inst. mech. Engrs Conference on Lubrication and Wear. 'Vapour lubrication'.
51. BUCKLEY, D. H. and JOHNSON, R. L. 1958 *Wear*, no. 2, p. 70, 'Friction and wear in reactive gases up to 1200°F'.
52. BUCKLEY, D. H. and JOHNSON, R. L. 1960 *A.S.M.E./A.S.L.E.* Lubrication Conference. 'Lubrication with halogenated gases up to 1500°F'.
53. BALDWIN, D. J. and ROWE, G. W. 1960 *A.S.M.E./A.S.L.E.* Lubrication Conference. 'Lubrication at high temperatures with vapour-deposited coatings'.
54. JOHNSON, R. L. and SLINEY 1959 *Lub. Eng.*, vol. 15, p. 487, High temperature friction and wear with bonded PbO films'.
55. BOWDEN, F. P. and FREITAG, E. 1958 *Proc. roy. Soc.*, Ser. A 248, p. 350, 'Friction of solids at very high speeds'.
56. BOWDEN, F. P. and PERSSON, P. A. 1961 *Proc. roy. Soc.*, Ser. A 260, p. 433, 'Deformation, heating and melting in high speed friction'.
57. SALOMON 1931 April 27th, Patent.
58. May 1960 *Metalworking Production*, vol. 104, 'U.S. Research in ultrahigh speed machining'.
59. 6th May 1960 *Engineering*, p. 622, Machining metal at ultrahigh speeds'.
6th February 1961 *New Scientist*, vol. 9 (222), p. 408.
60. HOLLIS, W. S. 1961 *Machinery*, vol. 98, p. 267, 'Developments in explosive forming'.
61. PEARSON, J. 1960 *J. Metals*, vol. 12, p. 673, 'Metal working with explosives'.
62. JAMES, D. 1961 *Machinery*, vol. 98, p. 84, 'Phosphate and lubricant for cold extrusion'.
62a. PUGH, H. LL. 1960 *Metal Treatment and Drop Forging*, vol. 27, p. 189, 'Cold deformation of metals'.
62b. BERESNOV 1959 *Bull. Acad. Sci. U.R.S.S.*, vol. 1, p. 128, 'Liquid pressure extrusion'.
63. TRENT, E. M. 1959 *J. Instn Prod. Engrs*, vol. 38, p. 105, 'Tool wear and Machinability'.
64. SPICK, P. T. 1961 Unpublished.
65. OXLEY, P. L. B., HUMPHREYS, A. G. and LARIZADEH, A. 1961 *Proc. Instn mech. Engrs, Lond.*, 'Influence of the rate of strain-hardening in machining'.
HEGINBOTHAM, W. B. and GOGIA, S. L. 1961 *Proc. Instn mech. Engrs, Lond.*, 'Metal cutting and the built-up nose'.
66. LANCASTER, P. R. and ROWE, G. W. 1958–59 *Wear*, vol. 2, p. 428, 'A comparison of boundary lubricants under light and heavy loads'.
67. GOLDEN, J., LANCASTER, P. R. and ROWE, G. W. 1958 *Int. J. App. Radiation and Isotopes*, vol. 4, p. 30, 'Examination of lubrication with soap, using radioactive sodium stearate'.
68. LANCASTER, P. R. and WELCH, D. E. P. 1960 Unpublished.
69. BLADWIN, D. J. and ROWE, G. W. 1960 Unpublished.
70. LANCASTER, P. R. and ROWE, G. W. Unpublished.
71. BUTLER, L. H. April 1960 *Metallurgia*, p. 167, 'Surface conformation of metals under high nominal contact pressure'.
72. BUTLER, L. H. 1960 *J. Inst. Met.*, vol. 89 (4), p. 116, 'Effect of lubricants on growth of plastic contact'.
73. GOLDEN, J., LANCASTER, P. R. and ROWE, G. W. Unpublished.
74. WALLACE, J. F. 1960 *Metal Ind., Lond.*, vol. 97, p. 415, 'Stretch-forming control by phase change lubrication'.
75. SEJOURNET, J. and DELCROIX, J. 1955 *Lubric. Engng*, p. 389, 'Glass lubrication'.
76. GOLDEN, J. and ROWE, G. W. 1961 *Sci. Lub.*, 'Radiotracer study of glass lubrication'.
77. TESMAN, A. B. 1960 *Met. Progr.*, vol. 78 (2), p. 101, 'Friction welding —a lesson from Russia'.
78. August 19th 1960 *Engineering*, p. 250, 'Friction welding—a practical process'.
79. ARCHARD, J. F. and HIRST, W. 1957 *Proc. roy. Soc.*, Ser. A 238, p. 515–518, 'An examination of a mild wear process'.
80. KERRIDGE, M. and LANCASTER, J. K. 1956 *Proc. roy. Soc.*, Ser. A 236, pp. 250–268, 'Stages in a process of severe metallic wear'.
81. NOSOVSKII 1958 *Referat. Zhur. Met.*, p. 10714, 'The Influence of temperature and atmosphere on wear'.
82. LANCASTER, J. K. 1957 *Proc. phys. Soc., Lond.*, vol. 70, p. 112, 'The influence of temperature on metallic wear'.
83. NUNN, T. A. and ROWE, G. W. Unpublished.
84. VAILE, P. E. B. 1961 *Proc. Instn mech. Engrs, Lond.*, 'Lubricants for Nuclear Reactors'.

Dr G. W. Rowe *was educated at Culford School and Emmanuel College, Cambridge, where he took Mechanical Sciences Tripos in 1944 and Natural Sciences Tripos Part II (Physics) in 1950. He wrote a PhD thesis on adhesion of clean metals in 1953 and was awarded a DSc at Birmingham in 1968 for publications on tribology and metalworking. He has worked at the Telecommunications Research Establishment and the Atomic Energy Research Establishment, and in 1954 was a TI Research Fellow, after which he joined Tube Investments Research Laboratories as leader of the lubrication and wear group. Since 1960 he has been at Birmingham University, where he is now Senior Lecturer in Industrial Metallurgy. Dr Rowe has contributed to three books on tribology, and about 70 papers.*

Heat transfer has long been of great importance to mechanical engineers but only in recent years has it tended to become the critical design limitation in such plant as gas turbines or nuclear reactors. The separate theoretical study of heat transfer is, therefore, a comparatively new discipline but a considerable foundation has now been laid, much of it confirmed by experimental results and of practical use to designers.

Heat Transfer

by F. J. Bayley, DSc, PhD, CEng, MIMechE

Fig. I. The scale of heat exchangers involved in nuclear power stations can be seen in this picture of the erection of one of the twelve boilers at Bradwell

Although mechanical engineers have utilized heat transfer since the earliest times, their interest in its principles is comparatively recent. The subject as such has appeared in mechanical engineering syllabuses to any significant extent only in the last 15 years and, with one or two notable exceptions,[1, 2] textbooks on heat transfer rarely antedate the 1939–45 war.

The Institution's index[3] of papers published since 1937 suggests, however, that problems of heat transfer are becoming of increasing concern. The papers listed under 'Steam Turbines[4]', for example, are uniformly spread throughout the period, whereas of those under 'Heat Transmission', most have appeared since 1950.

This increasing preoccupation with heat transfer is a consequence of the inevitable advance of technology. It is only in recent times, due to the increasing tendency to operate thermal plant near the temperature and pressure limits of available materials, that heat transfer processes have often become controlling factors in design. Most recent developments in mechanical engineering, certainly in the power and process industries, have required the solution of heat transfer problems.

Nuclear power

The nuclear power industry did not exist in 1950 and yet today it supplies a significant, if controversial, part of Britain's energy; it also provides an excellent example of the role of heat transfer in modern engineering. The British decision to concentrate upon the gas-cooled, graphite-moderated reactor immediately created a need for heat transfer surfaces of a very high efficiency, measured in terms of the ratio of heat transfer to fluid pressure loss, if the gas pumping power and the size of the plant were not to be prohibitive.

Studies of the forced convection heat transfer process within the reactor led to many novel and interesting configurations of heat transfer surface; but the so-called 'polyzonal' nuclear fuel can has now been almost universally adopted.

The nuclear fuel, in which energy is liberated by the fission process, is contained within the can, on the outer surface of which there are a number of spiral axial fins, integral with the base tube.

Part of the reactor coolant flows between the fins, but unless arrangements are made to replenish these narrow passages with coolant from the mainstream in the annulus around the element, the gas in them rapidly approaches the metal temperature with a consequent reduction in the overall heat transfer rate.

This is the function of the 'splitters', the four straight fins which run the full length of the fuel can and extend radially from the base tube surface almost to the periphery of the circular coolant passage in the moderator matrix. By blocking the spiral passages at intervals along the length of the fuel can, the splitters cause the fluid flowing between the fins to be periodically flung out and mixed with the mainstream. The consequent flow system is inevitably very complex and it is found experimentally to result in a much higher rate of heat transfer than can be obtained in a straight, smooth passage,[5] without the disproportionate rise in pressure loss which has usually to be accepted.

The irregular nature of the flow, alternating between the fin passages and the surrounding mainstream, leads, as might be expected, to irregular heat transfer rates, and thus to variations in surface temperature along the fuel can. These variations are often quite unsystematic; they are seriously affected by minor changes in passage geometry and are therefore a source of grave concern to the reactor designer who is required to guarantee, within very close limits, the maximum metal temperature in the reactor.

Thus, while the 'strategic' problem of developing a surface with good overall heat transfer might be considered solved, the detailed 'tactical' understanding of the process still presents difficulties. As we shall see, this is often the situation in the application of heat transfer principles.

The problems posed in transferring the heat from the reactor coolant to the working fluid of the associated steam turbine mainly arise from the scale of nuclear power projects, which can be judged from Fig. 1.

In most designs of nuclear boilers, heat transfer surfaces of quite conventional form have been used, usually radially finned tubes in cross-flow, about the performance of which a good deal has been known for some time. However, while the effects of small changes in surface geometry—in fin or tube pitching, say—were known approximately, the degree of approximation is quite critical in the nuclear power programme. For example, the first nuclear power stations each used heat transfer surface in the boilers valued at around £1 million, so that a ten per cent uncertainty in performance— a usually accepted margin—represented a very sizeable financial commitment.

For this reason a wealth of very accurate data on the heat transfer performance of conventional surfaces[6] has emerged from nuclear power research, as well as new arrangements like the polyzonal fuel can.

Boiling water

In America and, to a lesser extent, in Europe, the emphasis in nuclear power production has been placed on reactors cooled and moderated by liquids.[7] The heat transfer efficiency of forced convection in terms of pumping power is much higher with liquids than with gases, so that very high heat fluxes, and thus low reactor volumes, become economic. This, indeed, is one of the principal advantages of liquid cooling. Consequently, heat transfer in liquid-cooled systems has been studied less with a view to obtaining accurate data for design at normal operating conditions, but with the object of determining the maximum heat flux which can be accommodated at overload.

This is usually the controlling design feature with liquids. As the heat flux to a liquid is raised, the normal mechanism of convective heat transfer rapidly gives way to boiling and many reactors have been designed to take advantage of the very high heat fluxes which can then be obtained. For example, 100 000 Btu per ft²h is typical, compared with 2000 in a gas-cooled system. Ultimately, however, the vapour formed coalesces to become a continuous film on the hot surface which, compared with the liquid, offers very high resistance to heat flow.

This film boiling condition, in which an increase in temperature is associated with a reduction in rate of heat flow (as shown in Fig. 2) is inherently unstable. Unless an almost instantaneous reduction in power can be achieved when it begins, a catastrophic rise in temperature, or 'burn-out', of the heating surface is inevitable.

The problem is, therefore, to take advantage of the high heat transfer rates associated with boiling without the disastrous consequences of burn-out and a vast amount of research has been carried out towards this object, particularly in the U.S.A.,[8] but also in Britain[9] and the U.S.S.R.[10] However, there is still no really basic and general explanation of the mechanism of heat transfer in boiling, and hence no universally accepted method of correlating experi-

mental data, comparable with, for example, Osborne Reynolds' classically simple account of turbulent forced convection and its analogy to momentum transfer.[11] The boiling process is, of course, much more complex, and is critically affected by small factors like the condition of the surface and the purity of the liquid. This makes for very severe difficulties in analytical and experimental studies.

Water is the liquid most commonly considered as a cooling medium. Fortunately it has an almost ideal specification as a heat transfer fluid—a very high latent heat of evaporation, high thermal conductivity, low viscosity, lack of toxicity and high chemical stability. However, it has certain shortcomings, most notably its low critical temperature, 704°F, which means that the very high heat transfer rates associated with the evaporation of water are only available below this temperature.

Attention has therefore been turned to other liquids as alternative heat transfer media. Non-Newtonian fluids— those in which shear stress is not a linear function of the velocity gradient—are beginning to be studied[12] in this connection but probably the most important alternatives to water are the liquid metals.

It can be shown that the rate of heat transfer, q, through a fluid in turbulent flow (the most common engineering condition) is given by

$$q = \left[\frac{\nu}{Pr} + \epsilon \right] \frac{d\theta}{dy}$$

where $d\theta/dy$ is the temperature gradient and ν is the kinematic viscosity. ϵ is known as the 'eddy diffusivity' and represents the viscosity in turbulent flow, just as ν represents resistance to shear in the fluid in laminar or streamline flow. It is the basis of Reynolds' analogy,[11] that the mechanism by which momentum is transferred in turbulent flow is the same as that by which heat is transferred, so that if ϵ is determined from friction or pressure drop measurements, then the rate of heat transfer may be computed for similar flow conditions. In turbulent flow ϵ is very much greater than ν, which means that the molecular transport processes, which cause non-zero values of ν and thus of shear stress in laminar flow, are

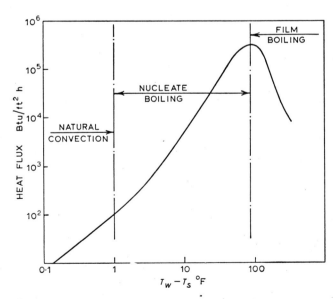

Fig. 2. Variation of heat flux with difference between surface temperature, T_w and liquid saturation temperature, T_s, in boiling heat transfer

negligible compared with the random macroscopic scale motion across the mainstream of turbulent flow. Whether or not ν/Pr is negligible, compared with ϵ, and thus whether the effects on heat transfer of molecular-scale motion are significant in turbulent flow, clearly depends upon the value of Pr, the Prandtl number of the fluid concerned.

The dimensionless Prandtl number is defined as

$$Pr = \frac{\rho \, \nu \, Cp}{k}$$

where ρ, Cp and k are, respectively, the density, specific heat and thermal conductivity of the fluid concerned. Typically, oils have strongly temperature-dependent values of Pr, but ranging up to 1000; for water, Pr varies between 10 at normal temperatures and unity at the highest occurring in practice; and gases have nearly constant Prandtl numbers of just less than unity over a very wide range of temperature. Thus, in the heat transfer equation for turbulent flow, the eddy diffusivity remains the predominant term for all these fluids, and data upon their heat transfer characteristics are broadly similar over a wide range of conditions.

The liquid metals, however, have very high thermal conductivities and relatively low viscosities, and values of Pr ranging down to 0·01. As a consequence, the eddy diffusivity term does not necessarily predominate in the heat transfer equation, and their heat transfer characteristics are accordingly quite different from those of other fluids and have been extensively studied in recent years, both analytically[13, 14] and experimentally.[15, 16] Their high conductivities mean that large rates of heat transfer are possible without the need for evaporative processes, and their acceptable nuclear properties have made them important heat transfer media for this purpose.[17]

The gas turbine

The development of the gas turbine stimulated the study of heat transfer processes at least as much as did the application of nuclear power. In contrast to the intermittent combustion of the piston engine, the gas turbine subjects its combustion system and high pressure stages continuously to the maximum operating temperature. Upon this depends the power and thermal efficiency of the cycle.

Thus there is a powerful incentive to develop effective cooling methods. For long life, cooling of the combustion system has always been necessary but due to the relatively low stresses in this component, simple methods have been adequate in most applications.[18] It is in the turbine that the real problem lies and acceptable cooling systems for the rotor blades in particular are becoming essential in aircraft engines as flight speeds are pushed higher into the supersonic regions, and, indeed, are already essential if the gas turbine is to play a significant part in industrial power generation.

The rotor blades of the most advanced gas turbines in current use are cooled by forced convection: air is forced through small holes in the blade section to mix with the main flow at the tips.[19] A typical blade section is shown in Fig. 3. This is the simplest possible cooling system from the heat transfer viewpoint and the problems posed are mainly of a 'tactical' nature. They arise from the need to know, in addition to the mean blade temperature, the detailed variation of the heat transfer rates from the hot gas around the section. From this data local temperatures of the metal can be determined and hence the magnitude of any induced thermal stresses which can become significant in highly loaded aircraft engines.

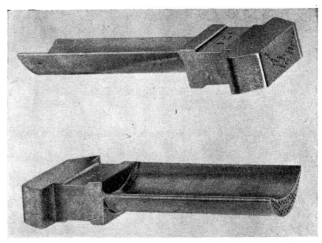
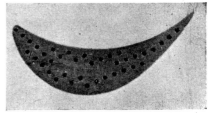

Fig. 3. Air is forced through small holes in the turbine blade-section to mix with the main flow at the tips

This problem has aroused a great deal of interest in the detailed behaviour of thermal boundary layers in the complex flows over turbine blading. These have been extensively studied, both theoretically[20] and experimentally.[21]

Although, from the 'plumbing' viewpoint, air has clear advantages for turbine cooling, it has the disadvantages typical of gases—low thermal conductivity and specific heat—where maximum cooling is required. These are overcome to some extent by the 'sweat' or 'effusion' method in which the surface to be cooled is made of a porous material and the air forced through it.[22] A double cooling effect is obtained—very high rates of heat extraction in the pores and a protective blanket of coolant over the outer surface, which partially insulates it from convective heat transfer. There are, however, many problems in developing suitable porous materials and, for operation at much higher temperatures than are currently used, liquid cooling may be necessary.

Natural convection

Simple forced convection liquid cooling systems for turbines have been studied[23] but the buoyancy obtained in heated liquids subjected to the very high centrifugal acceleration in the rotors, make natural convection very attractive. This rather prosaic mode of heat transfer, more usually associated with pipes and radiators, has been the subject of much research since the earliest proposal for a natural convection liquid-cooled turbine, shown in Fig. 4, was put forward.[24] Water fed into the rotor is flung outwards to replace the heated fluid which moves radially inwards over the hollow blade walls, as a result of its buoyancy in the centrifugal field. This has become known as the 'thermosyphon' and poses some interesting hydrodynamic problems because of the boundary which must form in the fluid between the outward and inward moving streams in the blade passage.

The turbine shown in Fig. 4 never ran successfully, for the blade cooling passages were ideal collecting points for impurities and rapidly became blocked up. A natural development of the original proposal is the so-called closed thermo-syphon shown in Fig. 5, in which a quantity of liquid is sealed within the hollow blade to act merely as a conveyor of heat by natural convection from the blade span to the root. The root itself can be cooled by a secondary gas or liquid. This arrangement permits the use of fluids like the liquid metals which are quite impracticable in the open system. By allowing condensation and evaporation to occur to and from a thin film of liquid,[25] very effective cooling can be obtained with small quantities of sealed coolant.

A more mundane problem is associated with the design of heat exchangers to recover heat from the turbine exhaust by preheating the compressed air on its way to the combustion system. The basic requirement is for high heat transfer, low pressure loss and a minimum of bulk and weight. The first two of these conditions mean that a very large surface is required, and if this is to be contained within a reasonable volume, the smallest possible diameter of flow passage must be used. The practicability of construction sets a lower limit of about $\frac{3}{8}$ in to tube diameter for conventional shell and tube heat exchangers. Even for a moderate size of gas turbine the number of end fixings is measured in tens of thousands and the heat exchanger dwarfs the rest of the engine, as may be seen in Fig. 6.

The equivalent passage diameter may be reduced to about $\frac{1}{8}$ in by the use of matrices constructed of alternate layers of plane and corrugated plate to form the gas passages, and the heat transfer characteristics of many novel surfaces following this general arrangement have been studied.[26]

More striking reductions in bulk are, however, possible by the use of passage sizes down to 0·010 in in the rotary re-generator, which works on the 'capacitance' principle.[27] A disc or drum of some material through which the air and gas streams may flow with the minimum of resistance is rotated so that each element of the matrix passes alternately through each stream, to be heated by the one and cooled by the other.

It has been stated[28] that the theories of the regenera-tor are among the most difficult and involved that are encountered in engineering. With the aid of computers, however, the performance of the regenerator has now been completely analysed. Under gas turbine conditions in which

the period of the thermal cycle is very short, the theory fortunately corresponds closely to that of the equivalent recuperator—i.e., the more usual 'conductance' type of heat exchanger.

Other problems of heat transfer continue to arise as the gas turbine is developed, especially in the aircraft industry. The turbine disc itself is beginning to present a cooling prob-lem as temperatures get higher, and the heat transfer characteristics of the complex flow patterns between rotating and stationary discs are therefore becoming of interest.[29] Also, as a direct consequence of the great success of the aircraft gas turbine, the ultimate heat transfer problem is now beginning to face the aeronautical industry; that of determin-ing the steady and transient temperatures of the aircraft structure itself in very high-speed flight. This, too, will call for further intensive research into problems of heat transfer, both in simulating operating conditions on the ground, and solving the problems in the air.

Steam plant

Although the development of nuclear power and the gas turbine have provided much of the stimulus for the immense amount of heat transfer research undertaken in recent years, a good deal of attention has been given to similar problems in the more traditional industries. The internal combustion reciprocating engine, which continues to dominate the automotive field, was once in the position in which the gas turbine is now—of requiring the solution of a cooling problem before being universally accepted. However, heat transfer is no longer a critical feature in the development of the piston engine and very little work appears to be in progress in this field.

The steam turbine, which supplies something over 98 per cent of this country's electrical power, has made tremendous technical advances in recent years. However, it seems that, temporarily at least, the limit of temperature has been reached because of difficulties in the design of the super-heater tubes in the boilers. A significant step forward will prob-ably require im-proved materials

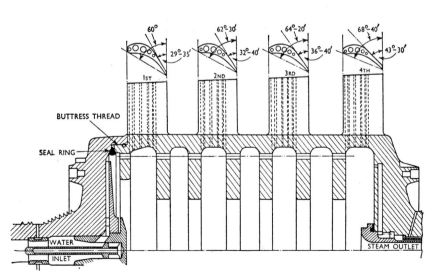

Fig. 4. *The earliest proposal for a natural convection liquid-cooled turbine. Water is flung outwards by the centrifugal field to replace the heated coolant in the blade passages*

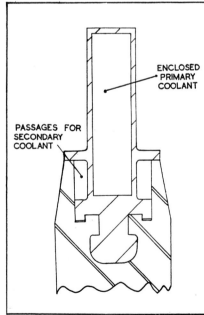

Fig. 5. *In the closed thermosyphon, liquid is sealed into the blade to cool by convection*

[Courtesy: British Ship Research Assn]

Fig. 6. If conventional tube and shell plant is used for preheating gas turbine air, the heat exchanger may dwarf the turbine

but methods of increasing heat transfer rates on the steam side could ease the problem by making the metal temperatures approach more nearly that of the steam, and this would repay further study.

Attention is already turning to the disc temperature problem, mentioned above in connection with the gas turbine. The condenser is one of the largest single items in the steam plant and improvements in its thermal performance are continuously being sought. One of the most promising is to promote dropwise condensation on the tubes. Normally vapours condense into a continuous film of liquid on the cooled surface which provides the principal resistance to heat flow on the vapour side. If the film can be broken up into discrete drops, as shown in Fig. 7, this resistance to heat transfer very largely disappears, and much less condensing surface is required for a given duty.[30]

Dropwise condensation can be maintained for very long periods, if not continuously, by the addition of chemical coatings to the steam side of the tubes, but there seems to be some uncertainty about the effect of these upon the boilers and high-pressure feed systems of modern large power plant. The latter, comprising seven or eight stages of regenerative feed heaters, are examples of plant in which heat transfer, though vital to its operation, is not the controlling design factor as, for instance, is the mechanical design of the high-pressure units.

Direct conversion

The power industry is turning its attention to methods of eliminating the heavy plant that has so far been necessary for the conversion of chemical energy into electrical power.[31] Direct conversion, while probably still a generation or more away from large-scale application, is now becoming a practical proposition. Here again heat transfer problems are very much to the fore and are probably the controlling factors.

The direct conversion principle, in its simplest terms, involves moving a conducting fluid through a transverse magnetic field, like the armature of a conventional alternator. For a worthwhile output, a high fluid velocity and a very intense magnetic field are essential. The high velocity is obtained by the expansion of a gas through a high pressure ratio in a suitably designed nozzle. To make the gas conducting it is 'seeded' with particles of one of the alkali metals.

Nevertheless, temperatures in the region of 2000°C remain essential. Thus, there is the immediate need to provide effective cooling of the convergent-divergent expansion nozzle, and the consequent requirement for accurate forced convection heat transfer data under these extreme conditions of flow. The only comparable conditions are found in the nozzles of rocket motors where, however, operating lives are measured in minutes, compared with the thousands of hours of trouble-free operation required for base-load power generation.

To promote the intense magnetic field required it has been suggested that the phenomenon of super-conductivity of fluids near absolute zero should be utilized. Here indeed we find the ultimate in the range of heat transfer problems.

Process industries

All but an infinitesimal proportion of the total energy of Great Britain is derived from heat engines, so that the increasing preoccupation of power engineers with heat transfer problems is inevitable. Equally, however, the process industries often present heat transfer problems on a very large scale. Heat exchangers, an ever-present part of process plant, are generally easier to design, for adequate margins for uncertainties here do not carry the same financial penalties as we have seen, for example, in the nuclear power industry.

Perhaps one of the most interesting recent developments in the process industry is the use of fluidized beds for heat and mass transfer. If a gas or liquid is blown through a mass of solid particles, the latter become suspended at a rate of flow at which their resistance to the flow becomes comparable with their weight. The whole bed of particles then behaves like a fluid. Under these conditions the rates of heat and mass transfer are found to be very much greater than through the undisturbed bed of particles. This phenomenon has been extensively used, for example in drying towers,[32] and is being considered as a means of obtaining high combustion intensities, which depend upon rapid rates of heat and mass transfer, in furnaces fired by solid fuels.[33]

However, as new processes are utilized, new gaps in our knowledge of heat transfer are revealed. For example, air-cooled condensers have begun to show certain advantages in the chemical industry and their use is being considered for steam power plant. It is convenient to arrange for the air to be blown across the outside of the tubes, which may carry external fins to increase the effective heat transfer surface area. In water-cooled condensers, the velocity of the vapour over the tubes is quite insignificant, but inside the tubes of air-cooled condensers the effect of the vapour velocity upon the condensation process cannot be disregarded, and there is an increasing preoccupation with this effect in the literature.[34]

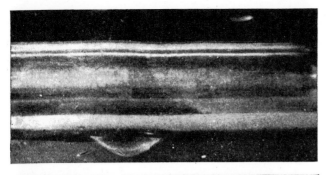

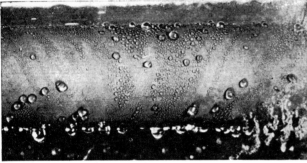

Fig. 7. Dropwise condensation (lower picture) can give heat transfer rates up to 10 times those with an unbroken film of condensate which acts as an insulator

Even traditional heat transfer devices begin to present problems with the advance of process technology. For example, the stressing of the tube plates of shell and tube heat exchangers has only recently progressed to the stage where allowance is made for the support given by the tubes themselves,[35] and thermal stresses appear to be entirely ignored. Indeed, it seems to be the case that a tube plate is stressed on the assumption that, at the hot end, it reaches the maximum fluid temperature. This is clearly absurd in view of the heat transfer processes going on behind it, but there seems to be a complete absence of information which would enable the designer to assess even the average metal temperature, to say nothing of the temperature gradients which will undoubtedly exist. Here again we have an excellent example of a heat transfer problem which it has up to now been possible to ignore, but which, with the inevitable pressure for more severe operating conditions, may well become quite critical.

Fundamental studies of convection

In this article I have tried to associate the recent very rapid increase in the mechanical engineer's concern with heat transfer with those developments to which this relatively new discipline has been intimately and obviously related, and where its immediate application is clear. Perhaps because of the often urgent need for rapid answers to practical problems, the subject is sometimes held to lack the fundamental background of academic research which reinforces other parts of the mechanical engineer's repertoire. This is in fact not the case and basic research has placed much of our knowledge of heat transfer on the soundest foundations.

Forced convection is the commonest mode of heat transfer consciously applied in engineering. Laminar flow is fully understood and theoretical solutions associated with the

flow of real fluids are possible for many practical arrangements, largely as a result of Prandtl's classical boundary layer concept[36] and its effect in simplifying the rigorous equations of motion. Complications arise in theoretical studies, as, indeed, in experimental research, through the variation of the physical properties of fluids—conductivity, viscosity, density and specific heat—with temperature. And, by the very nature of the heat transfer process, temperature variations are bound to occur.

Nevertheless, analytical methods of coping with this problem have now been established.[37]

The flow of fluids in engineering practice is less likely to be laminar than turbulent, and our understanding of the nature of this type of motion is much less than complete. Statistical methods of representing turbulence seem not to have been successful in accounting for observed phenomena, and semi-empirical methods, based on Reynolds' analogy between fluid friction and heat transfer, are the usual basis for analysis. The modifications to Reynolds' original concept which are necessary to take account of the inevitable regions of laminar flow near the boundaries of even violently turbulent systems, are well understood. This is partly through much basic experimental work upon velocity distributions in turbulent flow,[38] and also because many fluids in common engineering practice have Prandtl numbers near unity and for this single value the analogy continues to apply even in laminar flow.

Thus, providing the frictional pressure loss of a turbulent flow is known (and the effects of form drag discounted), the rate of heat transfer by forced convection can often be quite accurately predicted.

The concept of eddy diffusivity has already been used in describing the particular problems of the liquid metals. An interesting paradox is associated with this parameter, in that despite Reynolds' universally accepted hypothesis concerning the similarity of the mechanism of momentum and heat transfer, values of ϵ differ according to whether they are determined from thermal or friction measurements. This is a fundamental problem, fit to occupy the most academically-minded worker in the heat transfer field.

The remarks concerning forced convection apply equally to natural convection, in which the pumping power of the former is replaced by the buoyancy forces arising from the density changes in a heated or cooled fluid. Natural convection has been the subject of a number of quite basic studies, both experimental and theoretical.[39]

The results have been applied almost universally in the day-to-day calculation of heat losses from engineering structures of all kinds.

As was noted in discussing gas turbines, studies of natural convection have received a recent stimulus through the extension of the range of centrifugal accelerations to 30 000g, (compared with the 1 g of most heat loss calculations). In addition, a number of novel regimes of flow were predicted theoretically[40] to occur in the thermosyphon and their existence was very largely confirmed by experiment.[41]

Conduction and radiation

Heat transfer by conduction has been the happy hunting ground of applied mathematicians since the days of Fourier and most engineering problems of interest have been studied in some detail, certainly for conditions of equilibrium heat flow. Transient conduction problems, which are often important in determining thermal stresses, lend themselves very well to numerical analysis[42] and, with the aid of the

Professor F. J. Bayley *was educated at King's College, University of Durham and received his practical training at Fraser and Chalmers Engineering Works. From 1948-51 he worked at the National Gas Turbine Establishment, Farnborough, and from 1951-53 at PAMETRADA, Wallsend-on-Tyne, at both establishments on heat transfer problems in gas turbine design. In 1953 he was awarded a James Clayton Fellowship by the Institution and returned to King's College to undertake heat transfer research. In 1955 he was appointed lecturer in the Department of Mechanical Engineering at the University in Newcastle and to a personal Readership in 1959. In 1966 he was appointed to the first Chair of Mechanical Engineering at the University of Sussex. Professor Bayley has acted as consultant to industry and Government Departments and has published numerous papers in the Proceedings of the Institution and elsewhere.*

modern high-speed digital computer, accurate temperature-time plots can, in most cases, be obtained for even the most complex components.

In conduction calculations it is often necessary to know the resistance to heat flow across the interface between dissimilar bodies, usually lagging and the lagged surface. Uncertainty about the value of this resistance makes calculation quite meaningless without experimental data obtained under identical conditions; but recently attempts have been made to correlate, by statistical methods, the experimental values of these very important resistances.[43]

Energy transfer by radiation has been the subject of some of the most fundamental research and has thrown much light upon the nature of matter but this is the concern of the physicist rather than the engineer.

Most radiation calculations involve the exchange of heat between gases and solids. Here the grey body concept is quite inadmissible and well-established empirical methods[44] must be relied on. The radiation from non-luminous gases (of which only the polyatomic need be considered) is usually a fairly small percentage of the convective heat flow, and uncertainties about the applicability of the experimental data on the method of calculation are rarely critical, although some caution is necessary at high pressures.

The principal problems arise with luminous gas radiation, that is, where flames are present and rates of heat transfer by radiation become much greater. An extensive programme of experimental research[45] has been undertaken to determine heat transfer rates from luminous flames which is intended to obviate some of the wilder assumptions about radiation in furnaces.

Although, as we have seen, empiricism is still essential in many engineering heat transfer calculations, much more is now known about the fundamental processes. Experimental research will always have an essential part to play in providing heat transfer data for the engineering designer but analytical methods of predicting rates of heat exchange are much more significant than was the case until very recently in the history of mechanical engineering.

It is no longer true to say of heat transfer, as was once said of fluid mechanics, that it consists of observations which no-one could explain, and explanations which no-one could observe.

REFERENCES

1. FISHENDEN, M., and SAUNDERS, O. A., *The calculation of heat transfer.* 1932 H.M.S.O.
2. MCADAMS, W., *Heat Transmission* 1st edition, 1933 McGraw-Hill
3. *Brief subject and author index of papers published by the Institution 1937–62* 1963 I. Mech. E.
4. HIRST, A. W. C., 'Review of progress—Steam Turbines'. *The Chartered Mechanical Engineer.* Dec. 1963.
5. BOWDEN, A. T., 'Some engineering developments in the nuclear power field'. 1959 Trans. N.E.C.I.E.S. vol. 75.
6. *Symposium on the use of secondary surfaces for heat transfer with clean gases.* 1960 I. Mech. E.
7. ISBIN, H. S., 'Catalogue of nuclear reactors'. 1958 *Proc. Second U.N. Conf., Geneva.*
8. ROHSENOW, W. M., and GRIFFITH, P., 'Correlation of maximum heat flux data for boiling of saturated liquids'. 1955 *A.I. ChemE-A.S.M.E. Heat transfer Symposium.*
9. IVEY, H. J., 'Acceleration and the critical heat flux in pool boiling heat transfer'. 1962 *Proc. I. mech. E., Lond.*
10. KUTATELADZE, S. S., *Fundamentals of heat transfer.* Chap. XVIII. 1963 Arnold.
11. REYNOLDS, O., *Proc. Manchester Lit. Phil. Soc.* 1874 Vol. 8.
12. CLAPP, R. M., *Turbulent heat transfer in non-Newtonian fluids of the pseudoplastic type.* 1959 U.K.A.E.A. report X/M 186.
13. MARTINELLI, R. C., 'Heat transfer to molten metals'. *Trans A.S.M.E.* 1947 Vol. 69.
14. BAYLEY, F. J. 'An analysis of turbulent free convection heat transfer'. 1955 *Proc. I. mech. E., Lond.,* vol. 169, no. 20.
15. TREFETHEN, L. M. 'Liquid metal heat transfer in circular tubes and annuli'. *Proc. Gen. Discussion on Heat Transfer.* 1951 I. Mech. E.
16. BAYLEY, F. J., et al., 'Heat transfer by free convection in a liquid metal'. 1961 *Proc. Roy. Soc. A.,* vol. 265.
17. DIAMOND, J., and HALL, W. B., 'Heat removal from nuclear-power reactors'. 1956 *Symposium on Nuclear Energy.* I. Mech. E.
18. BAYLEY, F. J., *Air cooling methods for gas turbine combustion systems.* 1959 A.R.C.R. and M. No. 3110, H.M.S.O.
19. AINLEY, D. G., et al., 'An experimental single-stage air-cooled turbine'. 1953 *Proc. I. mech. E., Lond.,* vol. 167 (A) no. 4.
20. SQUIRE, H. B., *Heat transfer calculations for aerofoils.* 1942 A.R.C.R. and M. No. 1986, H.M.S.O.
21. WILSON, D. G., and POPE, J. A., 'Convective heat transfer to gas turbine blade surfaces'. 1954 *Proc. I. mech. E., Lond.,* vol. 168, no. 36.
22. GROOTENHUIS, P., 'The mechanism and application of effusion cooling'. 1959 *Jnl. Roy. Aero Soc.,* vol. 63, no. 578.
23. ELLERBROCK, H., 'Some N.A.C.A. investigations of heat transfer of cooled gas turbine blades'. *Proc. Gen. Discussion on Heat Transfer,* p. 410. 1951 I. Mech. E.
24. SCHMIDT, E., 'Heat transfer by natural convection in water-cooled gas turbine blades'. *Proc. Gen. Discussion on Heat Transfer,* p. 361. 1951 I. Mech. E.
25. COHEN, H., and BAYLEY, F. J., 'Heat transfer problems of liquid-cooled gas turbine blades'. 1955 *Proc. I. mech. E.,* vol. 169, no. 53.
26. KAYS, W. M., and LONDON, A. L., *Compact heat exchangers* 1958 McGraw-Hill.
27. COX, M., and STEVENS, R. K. P., 'Regenerative heat exchangers for gas turbine power plant'. 1950 *Proc. I. Mech. E., Lond.,* vol. 163.
28. JAKOB, M., *Heat transfer,* Vol. II. 1957 Wiley.
29. RICHARDSON, P. D., and SAUNDERS, O. A., 'Studies of flow and heat transfer associated with a rotating disc'. *J.M.E.S.* Dec. 1963.
30. HAMPSON, H., and OZISEK, N., 'An investigation into the condensation of steam'. 1952–53. *Proc. I. mech. E., Lond.,* vol. 1B, no. 7.
31. LINDLEY, B. C., 'Large-scale power by direct conversion'. 1960 *Engineering* vol. 189, p. 567.
32. COX, M., 'A fluidised absorbent air-drying plant'. 1958 *Trans. Inst. Chem. Engrs.,* vol. 36.
33. JOLLEY, L. J., and STANTON, J. E., 'Fluidisation in beds of coal and coke particles'. 1952. *Jnl. of App. Chemistry,* vol. 2.
34. ROHSENOW, W. M., et al., 'Effect of vapour velocity on laminar and turbulent-film condensation'. 1956 *Trans. A.S.M.E.,* vol. 78.
35. MILLER, K. A. G., 'The design of tube plates in heat exchangers'. 1952 *Proc. I. mech. E., Lond.,* vol. 1B, no. 6.
36. PRANDTL, L., *3rd Int. Mathematiker-Kongresses,* 1904 Heidelberg.
37. PICKERING, A. R., *Heat transfer to fluids with variable physical properties—a review.* 1964 U.K.A.E.A. report AEEW R290.
38. VAN DER HEGGE ZIJNAN, B. G., 'Measurements of velocity distribution in the boundary layer along a plane surface'. 1928 *Proc. Acad. Science, Amsterdam,* vol. 31.
39. SAUNDERS, O. A., 'Natural convection in liquids'. 1939 *Proc. Roy. Soc. A.,* vol. 172.
40. LIGHTHILL, M. J., 'Theoretical considerations on free convection in tubes'. 1953 *Quart. Jnl. of Mech. and App. Math.,* Vol. VI.
41. MARTIN, B. W., 'Free convection in an open thermosyphon'. 1955 *Proc. Roy. Soc. A.,* vol. 230.
42. DUSINBERRE, G. M., *Numerical analysis of heat flow.* 1949 McG. Hill.
43. FENECH, H., and ROHSENOW, W., 'Prediction of thermal conductance of metal surfaces in contact'. Sept. 1962 *Trans. A.S.M.E.*
44. HOTTEL, H. C., 'Radiant-heat transmission'. 1954 Chap. 4, in *Heat transmission* 3rd Edn. W. McAdams, McGraw-Hill.
45. BROEZE, J. J., et al., 'Flame radiation research joint committee reports'. 1951 *Journal of the Inst. of Fuel,* vol. 24.

Pneumatic Power and Control

by J. Hodge, MA, CEng, FRAeS, FIMechE

Air has long been used as a medium for transmitting power and in many applications, such as mining and contracting plant, it is still outstandingly successful for this purpose. At the same time low pressure air has become an increasingly popular means of transmitting signals for process measurement and control; and this in turn has led to the design of pneumatic actuators and computing elements. Due to better compressors and prime movers, an entirely new field has recently been opened up in anti-friction devices such as fluidized beds, air bearings and hovercraft.

In ancient Egypt the priests made use of compressed air to perform various 'miracles'; Aristotle used it in his diving bell. In 1653 Denys Papin was the first to try transmission of power to a distant point by this means. Among the earliest, and still one of the most important, practical uses was rock drilling, the first successful instance being in the 7·5 mile Mont Cenis tunnel under the Alps in 1861.

Since then compressed air has become an important means of power transmission in engineering; new uses for it are being developed constantly and there are few sections of industry that can do without it.

It is difficult to draw the line between pneumatics and other applications of air which are more properly described as aerodynamics but all depend on an initial supply of compressed air in the form of a fan or compressor.

Rotary compressors

Most equipment using compressed air is operated within a range of 60 lb/in² to 150 lb/in²; however, demand is arising for pressures up to 500, and in a few cases up to 10 000 lb/in². Usually high pressures are required only when more power has to go into a confined space, or for certain chemical processes. The efficiency of power converting machines tends to decrease with increasing operating pressures, because of the difficulty of using those pressures expansively.

Recent years have also seen a growing demand for slightly compressed air at up to 20 lb/in² for such purposes as fluidization,[1] pneumatic conveying of powders and granular solids, and for signal transmission in control and measuring systems.[2,3]

The efficiencies of compressors supplying pneumatic equipment are usually expressed not as isentropic or isothermal values but in terms of cubic feet of free air per minute per horsepower, or hp per 100 ft³/min at a given delivery pressure. Fig. 1 shows how hp varies with delivery pressure for both isentropic and isothermal compression. At 100 lb/in²g delivery pressure isentropic compression gives 5·6 ft³/min/hp and isothermal 7·6 ft³/min/hp.

For many years the reciprocating compressor was the only practical means of producing air in the quantities and pressures required and it still has an important part to play in fixed installations and high-pressure work. It seems unlikely that there will be any startling improvements in this already reliable and efficient machine. Rotary compressors work either on aerodynamic principles (axial or centrifugal) or by positive displacement (screw or vane). The former, as used, for example, in an aircraft gas turbine, are really only suitable when high flow velocities (and hence large mass flows) can be used, and have found little favour for general industrial use, though in specialized applications, such as blast furnace blow-

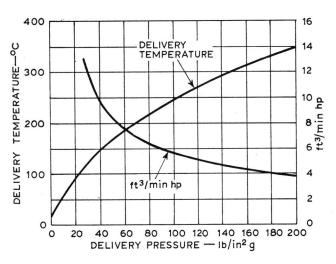

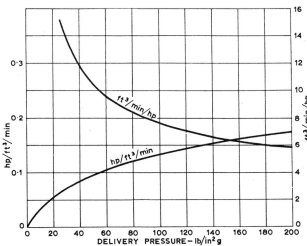

Fig. 1. How compressor power varies with delivery pressure in isentropic (left) and isothermal compression. Inlet conditions 14·7 lb/in²abs at 15°C

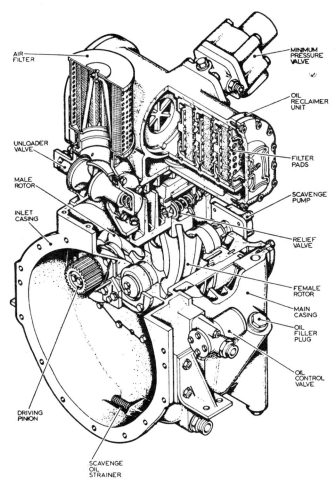

AIR FILTER

MINIMUM PRESSURE VALVE

OIL RECLAIMER UNIT

UNLOADER VALVE

FILTER PADS

MALE ROTOR

INLET CASING

SCAVENGE PUMP

RELIEF VALVE

FEMALE ROTOR

MAIN CASING

OIL FILLER PLUG

OIL CONTROL VALVE

DRIVING PINION

SCAVENGE OIL STRAINER

Fig. 2. No phasing gear is required with the Lysholm screw compressor and oil injection is necessary for sealing, cooling and lubrication

ing or natural gas pumping, they have proved very successful.

There has been a marked change from the reciprocating to the rotary types in recent years, especially for portable machines, because of lighter weight and reduced maintenance costs which are due to fewer moving parts. Efficiencies are approximately equal so that power consumption is not an important factor in the change.

A comparison of British exports[4] in 1956 and 1961 shows that, whereas in 1956 reciprocating portables were valued at three and a half times as much as rotaries, in 1961 the rotaries were worth twice as much as the reciprocators. Stationary machines were not affected to anything like the same extent, largely because of the efficiency of the reciprocator at low speeds.

The first kind of rotary compressor to be developed to a useful stage was the vane type, in which a cylindrical rotor with axial slots is mounted eccentrically in a cylindrical casing. Vanes are placed in the slots and slide radially in and out as the rotor turns, altering the volumes trapped between adjacent vanes. With suitable positioning of the ports, the air is compressed between intake and outlet.

Oil injection

In the modern version of this design which initiated the widespread change to rotaries for portable use, large

quantities of oil are injected into the compressor casing; this oil serves three purposes—it absorbs much of the heat generated in the air during compression, it lubricates the moving parts and it acts as a seal between high and low pressure regions.

The oil entrained with the delivered air must of course be separated from it, cooled and recirculated. This is achieved in two stages, one, macroscopic and the other microscopic. In the first, gravitational forces are used in various ways by means of baffles, vortices, etc., to remove the bulk of the oil. In the second stage the remaining oil is removed by passing the air through a series of filter packs, usually of wool or other fibres. This is very effective and oil losses from the system can be kept down to one gallon per five million cubic feet of free air or better. Sometimes the two stages are housed in separate vessels, but often they are combined within the same casing.

The other type of rotary compressor now in common use is the Lysholm screw type, shown in Fig. 2. This consists of a pair of intermeshing rotors, with male and female helical lobes.[5]

Like the vane type, the Lysholm only became practicable for portable use as the result of oil injection which here too has the functions of sealing, cooling and lubricating. As a result, a pressure ratio of at least 10 is possible in a single stage, with a reasonable efficiency. However, an additional, and very important, advantage results from the oil injection: it can be shown that, with certain lobe forms, which fortunately coincide closely with those required for best efficiency, nearly all the power input is absorbed by the male rotor. The female rotor takes only little more than is accounted for by mechanical friction—and acts simply as a rotary valve. Therefore, if the drive is imparted to the male rotor, either directly or through step-up gearing, no phasing gears are required as the small transfer of power to the female rotor can take place through the oil film.

In the engine-driven rotary sets, it is usual for the oil to be cooled by air in a pressurized radiator, placed in series with the engine radiator, both being served by a common over-sized engine-driven fan. Since the introduction of the rotaries, the upper output limit of portable sets has been increased from about 350 ft³/min to 1200 ft³/min (though the largest built in this country so far has an output of 900 ft³/min.)

The early screw compressors did not use oil flooding, and this type is most useful where the need for oil-free air is combined with moderate pressure ratios (up to 3 or 4 per stage), or where the gas being compressed is incompatible with oils. In this case it is essential to maintain accurate clearance between male and female rotors by synchronizing gears. Vane compressors can be operated 'dry' when used in low pressure applications and so can various other machines. For higher pressures and small quantities of oil free air, such as are generally used for control and instrumentation, food-stuffs or pharmaceuticals, unlubricated reciprocating compressors are used. These are usually of the crosshead type and have graphite or P.T.F.E. piston rings.[6,7]

The future of compressors

It is difficult to forecast the next step in compressor development. The slow speed reciprocator should hold its place for flows in the range from 1000 to 5000 ft³/min and the oil-flooded rotaries will continue to dominate the market for portables, extending also into the field of small stationaries.

The most interesting new possibility is that one of the rotary internal combustion engine designs now being developed (such as the Wankel[8] or Renault[9] types), will prove

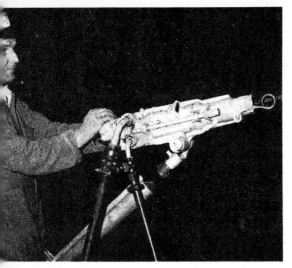

Fig. 4. A modern rock drill in use: the basic principle has remained the same for many years. In this machine the integral airleg support thrusts against a sprag forced into the ground

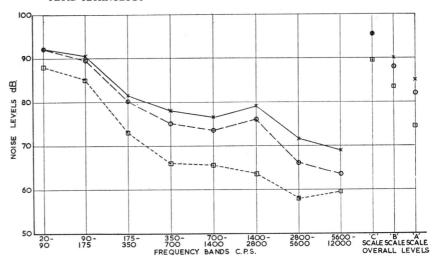

Fig. 3. Much effort has recently been devoted to reducing compressor noise and some success has been achieved, particularly at the obnoxious upper end of the frequency spectrum

————— standard machine   — — — ditto inside 'Pfister tent'   - - - - - special silenced machine

to have advantages as a combined compressor and engine. The gas turbine may come to the fore again in the larger portable sizes or in specialized stationary applications where waste heat usage is practical. The first set specifically designed for production of compressed air at 100 lb/in² and for long industrial life was the Holman T.100.[10] This has not so far been put into production but, as with other small gas turbine projects, a good, cheap heat exchanger and low cost production methods could transform the prospects overnight.

Another far-out possibility, closely related to the gas turbine is the pressure exchanger.[11] This is a device which accomplishes compression and expansion of gases in a single rotor. It consists in its simplest form of a rotor made up of two concentric cylindrical tubes with radial vanes between them, forming axial cells. The rotor runs at a comparatively low speed between end plates in which are inlet and outlet ports. These are arranged so that pressure waves, caused by their opening and closing as the rotor cells pass them, produce compression or expansion. Added to this is the effect of constant volume heating. Generally speaking, operating cycles are similar to those obtainable with a gas turbine and output is in the form of compressed air or gas, or cold air.

This design could conceivably be made into a cheap, compact, combined compressor and prime mover, perhaps suitable for small outputs down to 100 ft³/min.

Recently much effort has been devoted to the reduction of noise in portable compressor sets and the tools which they power. Some idea of the success achieved in a typical small machine is given in Fig. 3.

These results have been achieved by careful attention to all the points at which noise can escape from the machine. Sound absorbent baffles are placed wherever this is possible without restricting the air flow. Anti-vibration mountings are used and a glass-fibre canopy, which is lined with sound absorbent foam, and improved engine silencers are fitted.

Rock drills

One of the earliest and still one of the most important applications of compressed air is in rock drilling. Fig. 4

shows a modern rock drill in use at a mine face. It is supported by a pneumatic feed cylinder or airleg. The basic principle of the rock drill has remained unaltered for many years, although improvements in detail have now produced an extremely effective tool. Until comparatively recently nearly all practical machines were of the conventional percussive type with rotation of the bit superimposed to enable the cutting edge to strike a fresh piece of rock at each impact.

The machine in Fig. 4 has the airleg integral with the drill and the controls for both are combined. The end thrust of the airleg support is taken by a sprag forced into soft ground or by a chain ladder attached to the rock face. This mechanical configuration offends the purist since there is only one position in which there is static balance between airleg thrust, rock drill weight and feed pressure. However, this device is very simple—a most important feature in mining—and has relieved the operators of a great deal of the hard manual work involved in supporting and feeding the drill.

The large machines, known as drifters, are mounted on various forms of carriage. Cylinder bores are normally up to 4¼ in but there are machines with 6 in bore, weighing up to 600 lb. Feeds may be by pneumatically driven screws or chains. There is an increasing tendency for drifters to be mounted on rigs which can position them by hydraulic means because of the rigidity required. These may have a number of rams giving great freedom of motion and a whole rig can be operated by one man.[12] Fig. 5 shows such a rig, which is track mounted and propelled by air motors, powered by the towed compressor.

More than one drill may be mounted on a single carriage, and one can conceive of the drilling process being made completely automatic. However, the problems are tremendous.

Another recent development in rock drilling is the 'down-the-hole' machine. Here the percussion tool is fed down with the drill bit attached directly to it. The rotary component is imparted to the percussive hammer by a series of large diameter steel tubes. The holes produced so far have a minimum diameter of about 2½ in and a maximum of about

10 in. This type of machine is particularly valuable for deep drilling in quarries (100–200 ft) since the efficiency does not fall at depth anything like as rapidly as with a conventional drill where the percussive energy is transmitted down the hole through a series of coupled steel rods. Another advantage is that the whole of the exhaust air is used to clear the hole of debris, forcing it to the surface for disposal.

Much larger holes have been drilled for special purposes, using clusters of small down-the-hole drills.[13] Because the percussive mechanism is removed from the operator, the process is much less noisy. Higher working pressures are possible—one model works at 350 lb/in².

The use of higher air pressures not only has the obvious effect of increasing the energy imparted to the percussive piston, but also the quantity of air passed through the machine. This makes for more effective removal of chippings from the hole so that the energy available can be used to cut fresh rock, rather than smashing up already cut material.[14, 15]

Noise

Noise is a serious problem with percussive tools; in a small rock tunnel it can reach 120 dB which is near the threshold of pain. Considerable efforts have been made to reduce the noise level but so far without much success except at the expense of cumbersome additions which are not generally acceptable.

The noise comes mainly from two sources: the exhaust and the impact of piston upon drill steel. It is much easier to suppress the former than the latter; a simple exhaust hose, for example, can reduce it by 10 dB. The impact noise can be reduced only slightly by enclosing the drill in a sound absorbent casing. But it is not the rock drill as much as another member of the same family, the road breaker, which most afflicts the public ear.

A variety of different absorbent covers are now on the market, varying considerably in bulk, weight, effectiveness and cost. The best of them do achieve a worth-while result with little or no sacrifice of performance. In many countries the past year or two have seen much greater attention to noise problems and hazards, resulting in most cases in legislation limiting noise levels, both for the public and the operator. There is no doubt that progress in silencing will be speeded up as a result.

The use of compressed air in the mining industry extends far beyond the drilling process. The purpose of drilling holes in the rock face is, of course, to enable explosive charges to be placed in the holes, breaking up the rock to the depth of the hole. This is often dangerous and always expensive but, until recently, no other technique has been available. High pressure air is now used on a considerable scale to replace explosives in breaking coal, which is of course much softer than most rocks.[16, 17] This eliminates the risk of pieces of 'live' explosive remaining in the coal as, instead of the explosive charge, a steel shell is inserted into the shot hole and charged with high pressure air. Discharge is triggered by fracture of a shear pin, and the rapid increase in pressure bursts the coal from the face.

The pressure needed is 12 000 lb/in², which introduces a new range of problems in compression and transmission. So far air blasting has been used only for coal mining and it seems unlikely that it will prove possible to use it in harder measures.

Control devices

One of the most important steps forward in the use of pneumatics and, indeed, in the progress of industry generally towards automation has been the development of widely applicable control components. These take the form of a great variety of valves, whose operation can be initiated or 'piloted' by many different means—manual, mechanical, pneumatic or electrical. The output from the valves is eventually used to operate some mechanical device—generally by means of a pneumatic cylinder or motor which can push, pull, position, lock or rotate something.

Though many of the operations and devices are very simple indeed (as for instance, the opening and closing of doors) some applications involve highly complex circuits, built up mainly from simple pneumatic valves and cylinders but also incorporating electrical and hydraulic elements where appropriate. Many modern transfer lines and process plants rely on this type of actuating and control mechanism. It is safe in explosive atmospheres and circuits can be made to fail safe and provided with ample interlocks. It is inexpensive and is not sensitive to vibration or moderate amounts of heat. Fault finding and correction are usually straightforward, power consumption small and work outputs—both in terms of force and distance—widely variable.[18]

So far most valves have been of substantial size—even pilot valves seldom using pipelines smaller than about 1/8 in bore. The valves themselves have either used 'O' ring or similar seals to prevent leakage between adjacent parts or, in some cases, have relied on metal-to-metal sliding fits of very high accuracy and fine finish—comparable to those of diesel fuel pump pistons. These latter valves, assuming reasonably clean air supplies, while more expensive than those using elastomeric seals, are more reliable and less subject to wear. They are also capable of withstanding higher operating temperatures and, if necessary, radio-active environments.

A tendency which has recently been accelerating is the use of plastic parts. In several cases almost the whole of a valve, and even a cylinder, have been made out of synthetic materials. The reasons for this are low cost and weight and the ability to operate with dry air. Temperature limits are of course more restricted than in the case of all-metal components but not greatly different from those for normal rubber seals.

Using control elements of the kind described, it is possible to devise simple feedback mechanisms. However, these usually only operate with on/off response. Other limitations are imposed by the size of the valves and the velocity of sound in air, both of which limit the speed of response and the ratio of signal to actuating power.

Fig. 5. Large drills are mounted on carriages and even self-contained hydraulically positioned rigs. The compressor is in the trailer

In many ways these limitations are analogous to those existing in electrical and mechanical control gear though not in electronics. The most fascinating prospect in pneumatic control is the development of components or devices which, in effect, can be regarded as the pneumatic equivalent of transistors. These are referred to by various workers as 'pneumatic logic elements', 'solid state pneumatics' and so on.

Pneumatic logic

The first step in this direction was simply the miniaturization of valves, so that signal circuits could operate with very small powers and forces and could act more quickly because of smaller fluid or solid inertias. This kind of thinking led to devices such as the shuttle or ball type valves developed both in the U.S.A. and Russia. For example, the Kearfott ball valve, shown diagrammatically in Fig. 6, can be made in very small sizes; it has been claimed that some thousands can be assembled in a cubic inch. Response times can be reduced to a point where several hundred operations per second become feasible.

It seems, however, that this type of valve is likely to be only an intermediate step towards the ultimate in which there are no moving parts at all. With suitable choice of materials, such 'solid state' pneumatics obviously gives tremendous advantages by way of freedom to operate at high temperatures and in conditions of severe vibration, with the possibility of extremely good reliability as well as the advantages of miniaturization. It can combine the virtues of the transistor with those of the printed circuit.

Most of the known development in this field has been carried out in the United States, where already there are several devices on the market and in use on a small scale. Similar work has also been carried out in Russia.

It is difficult to describe all possible fluid switching and amplification devices, as there are so many different ideas being developed. However, Fig. 7 shows several in principle. The first is perhaps the simplest type of jet interaction device; a control jet is used to divert a power or input flow to one or other of two output channels. It is possible to obtain some amplification of the control signal, but not to any great extent.

A more sophisticated form of bistable amplifier-cum-switch is shown top right. It relies on the fact that, in the absence of any flow from the control passages, the main flow will attach itself to one wall due to a boundary layer re-attachment effect. This can be nullified by the injection of a small control jet into the passage adjacent to the wall, which causes the flow to 'lose its grip' on the wall and to jump across to the opposite wall, thus diverting the output to the other channel, where it will remain, whatever happens to the control jet. It will only be dislodged by a second control jet. By proper aerodynamic design a very small control jet can be made to divert a much larger main stream.

In the turbulence amplifier (Fig. 7, bottom) the undisturbed stream is laminar and crosses the gap between the input and the output orifices without appreciable spread, so that all the input appears in the output. When the very small control jet is applied, it injects sufficient turbulence to spread the power stream considerably, so that little of it enters the output channel. Again, because the control acts only as a trigger, high degrees of amplification are possible.

These are simple devices; by the use of various forms of boundary layer control, Coanda effect, turbulence control and so on, more complex devices can be built up, with feedback built in to give stability or 'memory' which enables them to carry out a variety of logic functions. There is a large and rapidly growing literature on the subject.[19-28]

The uses of such elements as these may be expected to expand rapidly when they are more widely available and their capabilities become more appreciated. At the simplest they can be used merely as miniature replacements for current pneumatic or electric pilot devices—limit switches, go/not-go gauges, and so on. At the other end of the scale they can be built up into full-scale pneumatic computers. Obviously these will not supplant electronic computers, as they cannot have such high speeds or such large memories. However, they can work at surprisingly high speeds and where their other attributes of reliability, freedom from the effects of heat, radiation and vibration are important, they may find applications.

Between these extremes there are many control functions in which the ability to perform comparatively simple logic and computation operations without changing from one type of power to another, and back again, will be invaluable, especially if costs can be brought down well below those of electrical equivalents, as seems possible.

Such elements can be very readily controlled by means of pneumatic signals generated by punched tape or card readers, giving a simple and robust system throughout. An interesting example is a pneumatic typewriter developed by I.B.M.[29] The elements do not have to be very small but it seems likely that most will be quite tiny, with nozzle sizes down to a few thousandths of an inch. They will have to be accurate to achieve reliability in systems containing a large number of elements.

Already several completely new manufacturing processes designed to achieve this have been described. Apart from lost wax precision casting methods, the most interesting involve the use of light-sensitive plastic, ceramic or glass, comparable to printed circuit technique in electronics. The form required is first drawn to a large scale. This is then projected with ultra-violet light on to the prepared material, reduced to the required size. The material's properties are affected by the ultra-violet light in such a way that it is selectively attacked by an acid or alkali, so that the desired shape is chemically etched. Fig. 8 shows in actual size a pneumatic digital logic circuit—a decimal ring counter.

Analogue devices

The control devices described so far, to which probably most attention has been devoted, are all basically digital—that is they can be used as on/off switches and for counting. It is equally possible also to use pneumatic devices to perform analogue functions for which a linear response to an input signal is generally required. To achieve this requires much more fundamental investigation of flow patterns and attention to all the many factors, such as Reynolds and Mach numbers, accuracy of manufacture, and so on, which can affect the output characteristics.

A simple form of analogue device is similar to the bistable one in Fig. 7, except that the stability or memory action is deliberately suppressed by the use of gaps in the side walls where the boundary layer reattachment would otherwise operate. The output can be made responsive to the input, so that, over a fair range of input pressure difference, the output signal responds linearly and can then be further amplified by similar stages or used to operate positioning or controlling mechanisms.

Another analogue device takes the form of a controlled vortex. In this the pressure drop between inlet and outlet is

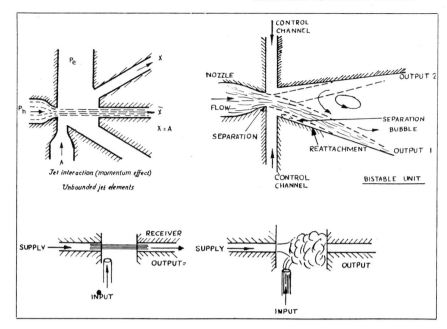

Fig. 7. Three types of 'solid state' fluid control device, all based on the use of small jets of air. The interaction type (top left) works due to the displacement of the input jet into one of two output channels. The bistable unit (top right) relies on the boundary layer attachment to the wall; and the turbulence amplifier at the bottom reduces output flow because of the turbulence created by a small control jet at right angles

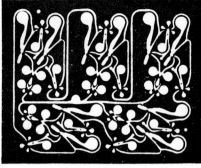

[Courtesy: Bowls Engineering Corp.]

Fig. 8. Actual size of a pneumatic digital logic circuit for a counter

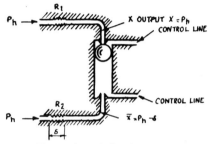

Fig. 6. Thousands of ball valves such as this could be assembled in one cubic inch

determined by the amount of angular momentum injected by the control flow.

The application of pneumatics to process control seems certain to make big strides in the next few years, with the introduction of both digital and analogue elements and developments based upon them.[30-36]

Widely used are air gauges, in which the pressure drop of a controlled air supply through a small orifice, positioned close to the article being measured, indicates its size. This is especially useful where large quantities of parts have to be inspected to a high degree of accuracy, as it gives a very easily interpreted reading, is not susceptible to wear or damage and can be used by much less skilled personnel than conventional gauges.

The system is particularly suitable for easily deformable materials as no physical contact is necessary. Obviously air gauges lend themselves very readily to combination with the control devices described above, either to act as 'go/no-go' gauges or to give proportional control for positioning or machining. High sensitivity is obtainable because of the accuracy and the very small forces involved.[37]

Other uses

There are, of course, a host of different tools—percussive and rotary—which are operated by compressed air. Probably the greatest advantages over other forms of power are achieved by the percussive types—concrete breakers, chippers, rivetters, spaders, and so on. A number of newer types have been introduced recently, such as nibblers, staplers, tools for assisting assembly of small electrical components, etc. These will doubtless be added to as the need arises. The advantages of the compressed air tool—effectiveness, safety and robust-

ness—are enhanced by the reduced temptation to theft—most home workshops are not equipped with a compressed air supply!

Small expansion turbines are now quite widely used—in high-speed hand tools, mainly grinders, dentist's drills and, probably most important, in cryogenic applications, where they provide an essential step in the cooling processes and where their ability to operate with high expansion efficiency and without lubrication at very low temperatures is of primary importance.[38, 39]

Air is used directly to achieve a desired effect in a number of ways. Of great potential importance is the prevention of freezing of waterways by bubbling air up from the bottom to stir the water and bring up some of the warmer water near the bottom. Surprisingly small amounts of air are remarkably effective in this way—figures under 100 ft³/min per linear mile of pipe are being quoted.[41]

By passing air upwards through a bed of powdered or granular material the load can be 'fluidized' (i.e. made to behave in a manner similar to a fluid). It will then flow without jamming and articles can be dipped into it just as they could into a liquid bath. These properties make possible a number of interesting processes; for example, coating metal objects with plastics by heating them and dipping into a fluidized bed of the powdered plastic.[41] Fast heat transfer between the fluid, the immersed article and the powder bed are possible and may be used for heat treatment and heat storage.[42, 43] Transport of powder from one place to another is, of course, greatly facilitated by fluidization and this process is now very widely used; so is straightforward pneumatic movement of solids along pipelines without true fluidization.

On the other hand, the old idea of using compressed air

to control wave action has not progressed far because of the difficulty of determining quantities of air required. There is little doubt that the system can be made to work but estimates of air quantity vary greatly. Probably the quantities required are too large for widespread use.[44, 45] As an intermediate step, curtains of compressed air bubbles to limit the effects of shock waves from underwater explosions have proved quite effective. Similarly, the use of curtains of air instead of doors at the entrance to department stores and other buildings has proved very successful.

Although entirely different in scale, air cushion vehicles and gas bearings have a common principle. Either could easily be the subject of a separate article, even to summarize the state of the art adequately. The air cushion vehicle has not progressed as fast as many expected since its first appearance a few years ago. There are now many variants but all require the use of large quantities of low pressure air to provide support.

The essence of design is to minimize the power required to sustain the vehicle clear of the ground. This can be achieved by keeping down the dead weight and increasing the efficiency of the compressors. While there are obvious advantages to be gained by the use of air cushion vehicles in some applications, these have only resulted in rather specialized uses. It is by no means clear how widely they will be used eventually. For military purposes the advantages might well be substantial.[46]

Air bearings have obvious advantages where conventional lubricants cannot be used, as in nuclear plant or oxygen separation plant, or where very low friction or high accuracy and stability of shaft location are required. They may be self pumping or they may rely on an external air supply to support the load.[47-49]

REFERENCES

1. LEVA, *Fluidization*. 1959 McGraw Hill.
2. ALLEN and FERRAND, 'Pneumatic Handling of Powdered Materials'. Manchester Association of Engineers.
3. VAN DER LINGEN and KOPPE, 'Moving Granular Materials by Air Line'. 13th Sept., 1963 *Engineering*.
4. CROCKER, 'Dramatic Increase in Compressors'. 2nd January, 1962 *Times Review of Industry*, p. 44.
5. LYSHOLM, A. J. R., 'A New Rotary Compressor'. 1943 *Proc. Instn mech. Engrs, Lond.*, vol. 150, p. 11.
6. DU PONT, 'Rings and Rods with "Teflon" running after 4½ years'. May/June 1964 *Journal of Teflon*, vol. 5, no. 3.
7. 'Oil Free Air for Pneumatic Systems'. June 1964 *Instrument and Control Engineering*.
8. ANSDALE, R. F., 'NSU—Wankel Engine'. May 1960 *Automobile Engineer*, p. 166.
9. 'Some Epitrochoidal Engine developments'. 22nd Nov. 1963 *The Engineer*, p. 846.
10. 'Developed design for compressor and shaft-drive duties'. Jan. 1962 *Oil Engine and Gas Turbine*, p. 339.
11. BARNES and SPALDING, 'The Pressure Exchanger'. Feb. 1958 *Oil Engine and Gas Turbine*.
    DARCHTOLD, 'The Comprex Diesel Supercharger' June, 1958. *Society of Automotive Engineers*.
12. INETT, 'Recent Developments in the Design and Use of Tractor Drills'. Aug. 1962 *Quarry Managers' Journal*, p. 333.
13. JAMES, R., 'Big Holes Fast'. Oct. 1963 *Compressed Air Magazine*, p. 14.
14. MARSHALL, 'Progress in Down the Hole Drilling'. Aug. 1962 *Quarry Managers' Journal*.
15. PFLEIDER and LACABANNE, 'High Air Pressures for down the hole percussive drills'. October/November 1961 *Mine and Quarry Engineering*.
16. 'Experience in Compressed Air Blasting'. 13th May 1961 *Mining Journal*.
17. 'Air Blasting in Mining'. Jan. 1962 *Compressed Air and Hydraulics*.
18. 'Circuit Design'. Maxam Power Ltd. 1964 Camborne.
19. MITCHELL, A. E., GLAETTLI, H. H. and MUELLER, H. R., 'Fluid Logic Devices and Circuits'. June 1963 *Trans Soc. Instrument Technology*.

20. *Symposium on Fluid Jet Control Devices*. Nov. 1961 A.S.M.E.
21. Diamond Ordnance Fuze Laboratory, 'Fluid Amplification Symposium'. Washington Oct. 1962. Pub.: Office of Technical Services, U.S. Dept. of Commerce.
22. MITCHELL, A. E., MULLER, H. R. and ZINGG, 'Some Recent developments in the design of Fluid Switching Devices and Circuits'. London 1964 *Fluid Power International Conference*.
23. AUGER, R., 'Pneumatics in computers'. Dec. 1960 *Automatic Control*.
24. DEXTER, 'Vortex Valve Development'. April 1961 Soc. Automatic Engineers.
25. Novoi v pneumo-grdaulitcheschoi Automatika, Institut Avtomatiki y Telemechaniki (Isdatelstvo Akademia Nauk U.S.S.R.) Moskva 1962 (New Pneumatic and Hydraulic Automation).
26. ZALMANZON, L., 'Pneumatic Computing and Control Devices'. April 1964 *Engineering Materials and Design*, p. 228.
27. KOMPASS, 'The Present State of Development of Fluid Amplifiers'. Feb. 1963 *Engineers Digest*, p. 82.
28. MITCHELL, A. E., 'Calculating with Jets'. 7th March 1963 *New Scientist*, no. 329.
29. 'Pneumatic Typewriter'. Dec. 1963 *Compressed Air Magazine*, p. 12.
30. *Pneumatic Components and Computing Devices for Control Systems*. 1963, Butterworths.
31. 'Centralised pneumatic control for ships', 1963 *Control*, vol. 7, p. 39.
32. PAMELY-EVANS, 'Analogue data handling in pneumatic form'. 1961 *Process Control and Automation*, vol. 8, p. 2810.
33. 'Automatic control for Lurgi plant and high pressure gas grid'. *Ibid*, p. 353.
34. 'Automatic inspection of engine valves'. 1962 *Ibid*, vol. 9, p. 574.
35. YOUNG, L., 'The control of boilers fired by solid fuels in suspension'. 1952 *Proc. Instn mech. Engrs, Lond.*, vol. 166, p. 91.
36. MILLER, J. T., 'Automatic Control Systems: Controller Units and Correcting Units'. June 1964 *Instrument Practice*, p. 617.
37. EVANS, J. C. and MORGAN, I. G., *Principles of Pneumatic Gauging*. D.S.I.R. Notes on Applied Science. No. 34.
38. 'New Liquid Helium Refrigerating Plant using Turbo-Expanders'. *Sulzer Tech. Review* 3/1963. p. 121.
39. 'The Brown Boveri Cryogenic Expander'. 1963 *Brown Boveri Review*, vol. 50, no. 6/7, p. 366.
40. 'Beating Ice Packs in the St. Lawrence'. 15th April 1960 *Engineering*, p. 526.
41. 'Survey of plastic coated metals'. Jan. 1964 *Engineering Materials and Design*, p. 30.
42. 'Fluidised beds: heat treatment as well'. Aug. 7 1964 *Engineering*, p. 177.
43. BOTTERILL, J. S. M., and ELLIOTT, D. E., 'Fluidised beds: answer to peak power'. 31st July 1964, *Ibid*. p. 146.
44. 'Pneumatic Breakwaters to Protect Dredgers'. Jan. 1961 *Proc. A.S.C.E.* (Internal Waterways and Harbour Div.).
45. BULSON, 'Currents Produced by Air Curtains—Deep Water'. May 1961 *The Dock and Harbour Authority*, vol. 42.
46. COULTHARD, W. H., 'Military use of ground effect vehicles'. May 1962 *Journal of Royal Aeronautical Society*, vol. 66, no. 617.
47. 'A Bibliography on Gas Lubricated Bearings—Revised'. Sept. 1959 Franklin Institute Laboratories for R & D, Philadelphia. Interim Report I-A2049-6.
48. 'Gas Bearings and their use in Industry'. *Sulzer Review Research*, no. 1961, p. 65.
49. 'Principles and Applications of Hydrodynamic Type Gas Bearings'. 1957 *Proc. Instn mech. Engrs, Lond.*, vol. 171, no. 2.

Mr James Hodge is Group Chief Engineer of Holman Bros. and Technical Director, Maxam Power. Educated at St Austell and St John's College, Cambridge, where he took the Mechanical Sciences Tripos in 1941, he joined Power Jets Ltd and worked on jet engine design until 1946. After a short period with the Gas Turbine Dept of English Electric he spent 18 months on reactor design at Harwell, then returned to Power Jets (R & D) Ltd as a member of the newly formed consultancy team, and became Chief Engineer. In 1953 and 1954 he was granted leave of absence to become Visiting Professor of Mechanical Engineering at Columbia University, New York. Mr Hodge joined Holman Bros in 1958. He is the author of 'Gas Turbine Cycles and Performance Estimation', published in 1955.

Fluidics — Principles and Practice

by A. Linford, BSc, CEng, MICE

Fluidics is a control technique in many ways analogous to, and in competition with, electronics. Its chief drawbacks are relatively slow response and—pending mass production—high unit costs. However, it has great advantages in reliability, resistance to environmental conditions and utilisation of process fluids. Its versatility is enormous but, as yet, little known to most engineering designers.

Fig. 1. Comparison of size between a model of large Coanda fluidic element with a powerful output and a small one (left) for a logic operation

Fluidics is the term used to describe a new technology dealing with the utilisation of well known phenomena of fluid mechanics for the purpose of controlling fluid flow without the use of mechanical devices. Until the last decade or so, fluid-operated devices for performing logic functions, amplifying signals and carrying out work have employed moving parts, the displacement of which, however small, has been essential to obtain the required output. Typical examples of such devices are industrial pneumatic and hydraulic process control assemblies.

In pure fluidic devices, on the other hand, there are no mechanical moving parts at all, the effects resulting solely from changes in the conditions of the driving fluid which may be a gas or a liquid.

In many respects, fluidics is analogous to electronics; both can be used to carry out the same classes of functions. Both have their advantages and disadvantages. As fluidic techniques develop it is becoming more apparent that the two control techniques complement each other.

A considerable impetus was given to fluidics in 1959 by the work done in the Harry Diamond Laboratories in the USA. Work in the first few years was confined mainly to the fundamentals and the development of such logic devices as AND, OR, NOR and memory elements, to amplification, and to counters, oscillators, timers and sensors. In the last few years the emphasis has been on an ever widening range practical applications.

Depending on its form and function a fluidic element, can range in size from a logic system, with flow passages measured in hundredths of an inch and etched on epoxy resin, to a diverter valve of the order of cubic feet. One of the largest fluidic elements[1] is used to steer a barge. It has a

flow passage with a cross section of 12 ft² and produces a thrust of 2 500 lb. It is claimed that units have been designed capable of producing thrusts up to 35 000 lb. A comparison between a large experimental fluidic element (of the Coanda type) for providing a powerful output and a small one for a logic system is shown in Fig. 1 with an enlargement of the small fluidic element, left.

Basic elements

Although not ideal—since hybrids exist—the most satisfactory classification appears to be on the basis of principle of operation. Thus, we have turbulence amplifiers and elements based on vortex effect, Coanda effect and momentum effect.

In the turbulence amplifier the power jet normally flows through in laminar fashion, its pressure forming the output signal. The control, also a jet of fluid, is applied at an angle to the driving jet which, under the influence of the disturbance, becomes turbulent. The level of the output pressure signal is then much reduced, practically to zero. By using two or more control jets, a NOR element is obtained, as shown in Fig. 2.

In the vortex element (Fig. 3) one form of which is known as the vortex amplifier, the power jet enters a cylindrical chamber axially at one end and leaves it tangentially at the other. When the direction of its flow is reversed, the vortex in the chamber, produced by the tangenial entry of the fluid, greatly increases the resistance to flow. Ratios of pressure loss up to 50 to 1 are claimed.[2],[3] Because of the high resistance to reverse flow these elements are known as 'diodes', other designs of diodes have also been developed.

By means of a control jet a rotational component may be

imposed on the power jet and the pressure drop across the element can be varied in proportion to the value of the control signal; this results in 'proportional' action. However, a drawback is that the control jet pressure must be greater than the power jet pressure. In this form of element, of course, the flow takes place in one direction only.

In the Coanda effect element, shown in Fig. 4, the flow passages are in the form of a 'Y'. The power jet is applied to the bottom and flows through either passage. The control signal is applied to one or other side, just upstream of the bifurcation. The power jet then attaches itself to the opposite wall and flows through the corresponding arm of the 'Y'. This jet attachment is known as the Coanda effect. Since the jet remains attached to the passage wall after the control signal is removed the element constitutes a memory and is therefore frequently referred to as 'flip-flop' or 'bi-stable' element, like its electronic equivalent. For some applications the element may be designed to be (mono) stable in one direction only.

The momentum effect shown in Fig. 5 also uses a 'Y' shape but the passage walls are cut away to prevent attachment so that the jet is deflected and split between the two arms in accordance with the magnitude of the control signal. Thus, a 'proportional' control action is obtained.

These elements are amplifiers because, to a greater or lesser degree, a large output signal can be achieved from a weak control signal. The gain may be in terms of pressure, flow or power. It will be seen that the first and third element described have a digital (on/off) action, while the others have an analogue (proportional) action. Of the four basic types, the Coanda element is the most versatile and highly developed.

Conditions for application

The value of fluidic devices lies in the fact that they can withstand shock and vibration, be used in hazardous atmospheres and remain highly reliable even in extremely adverse conditions. They can have simple inputs and outputs and are suitable for sensing as well as control.

On the other hand, compared with electronic equipment, they are bulky, give a low amplification (gain) and a slow response. Among the factors which determine their speed of operation are inertia. A small gas-operated element has a frequency response of the order of thousands of cycles per second, far higher than a liquid yet, even so, the response of an equivalent electronic switch would be of the order of Mc/s.

Among the unique properties of fluidic elements is that they can be designed to work with any fluid. Thus in process control, for example, it may be possible to use the process fluid to power the control devices. The units can be made from virtually any solid to suit the environment. Some elements, eg, turbulence amplifiers, can be actuated from a distance by pressure pulses.

Applications must justify themselves and the temptation must be resisted to use fluidics simply for the novelty. Perhaps the best method of assessing the progress during the past decade or so, and the future potential, is to analyse the literature. Notable among the publications on the subject are those of three international meetings.[4-6]

An analysis reveals that, to date, the applications for fluidics, in descending order of frequency, are to machine tools; aerospace; computers; process measurement and control; and others.

The first four areas appear to constitute about 80 per cent of the market. The miscellaneous area comprises applica-

tions which cover virtually the whole of the remaining field of technology from power generation and transport to medicine. It may be that in the aerospace and atomic power fluidics is much more advanced than is apparent, because much of the material may be secret; and that the research effort put into machine-tool applications has been excessive. It is, perhaps, not surprising that while, in Western Europe, the main emphasis has been on machine tools, in the USA and the USSR most interest has been show in aerospace applications.

Machine tools

Practically all the expansion in the use of fluidics in this field has taken place since 1964, probably because widespread automation here coincided with the upsurge of interest in fluidics. Fluid pressure impulses can be readily used for switching positioning devices. Digital, ie, pulsed, methods of operation are inherently more accurate than continuous, analogue systems; they are not influenced by drift or zero shift and noise.

Other contributing factors are that high-frequency responses are not required for machine tool control and pneumatic and hydraulic power are usually readily available. However, to ensure reliability, the required speed of operation of the controls should be kept well below the maximum possible, otherwise there is a danger of lost or spurious signals. Finally, other things being equal, the mechanical engineer will avoid electrics when this is possible.

Perhaps one of the most interesting developments in this field are the fluidic numerical position control systems utilising wall attachment elements. A tape reader scans the information on a standard tape and feeds this into a fluidic logic system by having air nozzles, each corresponding to a hole position on the tape and sensing the presence of a hole by the lack of back-pressure in the nozzle. Reading speeds up to 25 lines/s have been obtained.

The logic system, after correcting for datum level, tool position, etc, feeds the required tool positions and movements to a comparator. A disc or drum digitiser which uses air holes to encode the degree of rotation is used to detect actual tool position. This information too is fed to the

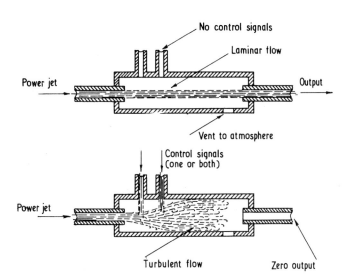

Fig. 2. In a turbulence amplifier the control signal, applied at an angle to the main flow, causes turbulence which reduces the output practically to zero. Two control inputs provide a NOR element

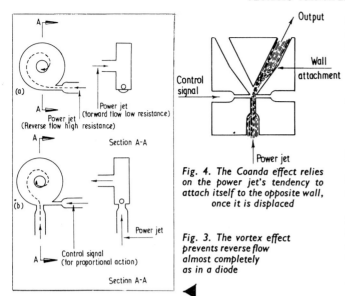

Fig. 4. The Coanda effect relies on the power jet's tendency to attach itself to the opposite wall, once it is displaced

Fig. 3. The vortex effect prevents reverse flow almost completely as in a diode

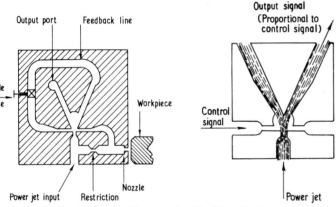

Fig. 6. How the Coanda effect is used in a proximity gauge

Fig. 5 By using the momentum the control can proportion the flow

atmosphere. The pressure on the upstream side of the nozzle increases as the distance between the nozzle and the work-piece decreases. At the predetermined distance, this pressure attains a value which switches the jet to the output port[9].

Turbulence amplifiers also have a place in machine tool control, eg for stator forming machines and shot blasting.[10] and point-to-point positioning of machine tool work tables.[11]

Aerospace

In the aerospace industry the expansion of fluidics has shown a healthy trend. This is not surprising, since reliability and environmental tolerance are factors of paramount importance. Unfortunately much of the work in this field is classified.

However, from published literature it is evident that fluidics has an important role to play in the flight control of aircraft, space craft and missiles. For example, fluidic devices for logic switching, amplification of signals, temperature, pressure and rate sensing, etc, have been used to build aerodynamic controllers, and navigation and vector thrust systems.

The vortex effect element with a proportional action can be made sensitive to changes in direction of motion and therefore plays an important role in various flight applications. It can be used as a fluidic gyroscope, for example, to improve the reliability of stability augmentation systems for helicopters. It requires less maintenance and costs less than electromechanical systems.

In one design of vortex rate sensor, a cylindrical chamber is fitted with a porous partition so that an annular chamber is formed into which the power jet is introduced. The effect of rotating the element is that the fluid emerges from the porous partition into the control chamber with a tangential velocity component, the magnitude of which is determined by the turning rate of the sensor. In accordance with the principle of the conservation of angular momentum, the tangential velocity is greatly increased by the time the control outlet is reached. Therefore, a small rate of change of rotation produces a significant change in pressure at the sensor outlet. It is claimed that the vortex principle can be used to detect rates of deflection of the order of 10^{-2} deg/s.

A completely fluidic automatic pilot has already been developed[12] which gives control in three axes. Vortex sensors are used in conjunction with fluidic amplifiers and, it is claimed, equal the performance of conventional systems.

Another aerospace application is for replacing mechanical diverter valves in V/STOL aircraft.[13] The function of these valves is to divert the hot engine exhaust gases, either to the jet (cruising) or to the fan (hovering). Among the difficulties experienced in this application is that the thrust efficiency is some 10-15 per cent less than with a corresponding mechanical valve. Further work is in progress to eliminate these shortcomings.

Yet another application under test is a fluidic attitude control system for a solar probe which would approach the sun to within a radius equal to 0.3 of the earth's orbit.[14] Fluidic elements would be used in the stabilisation system.

Computers

In the early days of fluidics it was thought that these devices would have a wide use in a number of computing systems, but this has not proved to be the case, because of their bulk, low speed and high cost, in comparison with their modern electronic counterparts. In the USA a very small air-operated, general-purpose digital computer, 21 in × 24 in ×

comparator which applies the necessary correction signals to the machine actuating unit, eg a hydraulic motor, which will then move the tool in the direction to reduce the difference of the two signals to zero. That is to say, the machine automatically follows the instructions stored on the tape.

Tool positioning to within 0·001 in. can now readily be obtained, the fluidic tape readers giving point-to-point control over four decades, ie, 1·000–0·001 in. The repeatability is of the order of 0·00025 in.[7]

Simpler circuits, controlled by fluidic card readers, can be used, for example, for drilling sequence control with positioning in one axis.[8] Fluidic systems of this form could bridge the gap between fully automated machine tool control systems and manual or semi-automatic ones.

The detection of the position of a rapidly moving object without making contact can be achieved with an accuracy better than 0·001 in. as shown in Fig. 6, by a kind of digital air-gauge. A monostable Coanda element is used, modified to incorporate feedback. The power jet through a fixed resistance provides the right-hand control signal of the element and also feeds a nozzle which discharges into the

14in. consists entirely of fluidic logic elements. It can be programmed to perform a few elementary calculations.[15] British computer firms have also done research work in the fluidics field.

Perhaps a reduction in manufacturing costs of the elements, resulting from wider use in other industries, will lead to renewed interest in their use for the smaller machines, eg, calculators and accounting systems. The principle of the calculations is of course the same as in electronics, active and passive bi-stable wall-attachment elements being used as binary counters. The latter detect the direction of the signals.

Somewhat more complicated is a binary/decimal converter circuit. Wall-attachment (Coanda) elements are used throughout, as OR-NOR gates; that is to say, there is an output signal in a certain direction if one OR other of the control signals is present and there is no signal in this direction if neither one NOR the other of the control signals is present.

There is a potential for fluidic techniques in what may be termed peripheral computer equipment. That is to say, in areas where certain logic operations have to be carried out during the relatively long periods taken by mechanical parts to move. A typical example of this is the sequencing and control of computer input and output mechanisms. The most common fluidic elements required for such applications are the OR-NOR gates and bi-stable elements already described.

To use fluidic elements in this manner calls for electro-transducers, a number which have been developed. For example, the fluidic element may be actuated by electromagnetic or piezo-electric means by inserting an obstruction into a flow passage; generating a pressure pulse by electromagnetic or electrostatic means; or disrupting the flow by means of an electric discharge, using loudspeaker input to control channels, and so on. Some methods result in a short pneumatic pulse when the value of the electric function changes, while others cause fluid flow changes which persist until the electric signal returns to its original value.

Among methods for pneumatic/electric conversion is fluid pressure used for straining piezo-electric elements.

To detect the flow as distinct from pressure, a hot body, (filament or thermistor) has its electric resistance changed as soon as it is cooled by the fluid.[16]

Process control

Fluidic systems can be used as sensors for detecting the values of physical variables, can perform logic operations as a result of these measurements and can amplify the control signals so obtained. Finally, 'heavy current' fluidic devices can produce the physical effort required to regulate a plant.

It would seem, therefore, that fluidics should have a very wide appeal in the field of industrial process control. However, changes in the dimensions and geometry of the passages due to corrosion, etc, can cause breakdowns in these precision elements. Preventive maintenance may overcome this problem but there is little information on this new technology in that section of the technical press devoted to instrumentation and automatic control.

When fluidics is used for process control, the tendency is to design the system for the specific application. However,

design problems arise from this: for example, it is not possible to assemble Coanda and other fluidic elements without giving thought to the effects of 'fan out', interaction, instabilities, etc.

For temperature sensing, a self-sustaining oscillator principle can be used which is illustrated in Fig. 7. The element is a bi-stable or flip-flop wall-attachment device with feedback passages; the process gas forms the power jet and the transfer of part of it from one feedback passage to the other switches the main jet from one output to the other. The frequency of switching depends on the speed of travel of pressure waves in the gas and this is inversely proportional to the square root of the gas density. Since the density is inversely proportional to its absolute temperature, the pulse frequency is directly proportional to the square root of temperature.

The output, in digital form, could be converted to analogue signals and then amplified, for example, to a 3–15 lbf/in² air pressure for operating conventional pneumatic control equipment. Very fast response to temperature changes is obtained, eg, in gas turbine control.[17] Alternatively, the oscillating frequency can be detected by a piezo-electric transducer, an electric digital output related to the temperature being obtained. Depending on the size of the device, frequencies of up to about 7 kc/s at 4 000°F are claimed and high resolutions can be obtained by using a beat frequency technique.

This temperature-sensing device, made of ceramic and powered by air, could be used in the steel industry for molten steel, to give a continuous measurement, replacing disposable thermocouples.

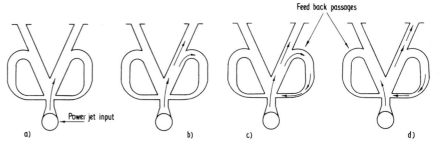

Fig. 8. A self-sustaining fluidic oscillator can be made from this bistable wall-attachment device

Fig. 9. A Coanda diverter valve under test at BHRA: the only limit to size appears to be structural

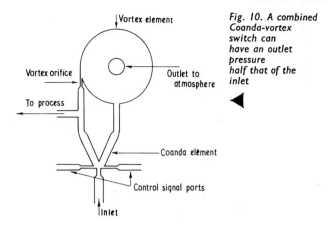

Fig. 10. A combined Coanda-vortex switch can have an outlet pressure half that of the inlet

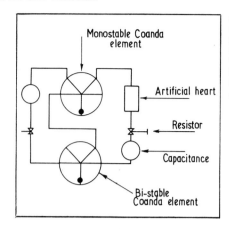

Fig. 12. This fluidic artificial heart pumping system, now being tested on animals, is expected to prove extremely reliable

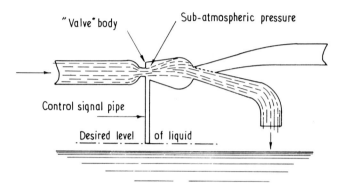

Fig. 11. A Coanda level control valve suitable for slurries

For diverting the flow of difficult fluids such as acids, hot gases, etc, the bi-stable Coanda element appears to be ideal; it should have a useful potential in various industrial processes. An experimental fabricated element, with an outlet bore of $1\frac{1}{2}$ in and discharging liquid into the atmosphere, is shown in plan in Fig. 9. A control pulse applied at C would immediately switch the discharge from outlet A to B with no leakage from A. The only limit to the size of a device of this form is imposed by structural considerations. For example, for switching flows through large culverts, the device could be incorporated into the main structure and made, say, from reinforced concrete. However, until sufficient experience has been gained, some research work will be required to establish the operational parameters for any particular application.

The outlet back pressure of a fluidic switch of the form described above is limited. To overcome this difficulty a combined Coanda-vortex switch, illustrated in Fig. 9, has been designed.[18], [19] It is claimed that, with this device, an outlet pressure of about 50 per cent of the inlet pressure can be achieved on one side, but approximately atmospheric pressure is obtained when switched. When the flow is through the right-hand branch of the Coanda element, the flow through the vortex chamber is radial and the back pressure in the vortex chamber is approximately atmospheric. When the flow is switched to the left-hand branch, the flow through the small-bore tangential orifice produces a vortex which causes a high back-pressure within the vortex chamber. This allows the main flow to overcome the process back-pressure and discharge through the left-hand horizontal passage.

The 'heavy current' Coanda switching valve can be used as an automatic level control system with a step (on/off) action.[20] Such a system would be useful for controlling the level of a slurry, for example, which presents difficulties to conventional methods. In the fluidic level controller, illustrated in Fig. 10, the bi-stable wall-attachment element has only one control signal input, formed by a pipe, the lower open end of which terminates at the desired liquid level. The valve is switched when the liquid reaches the desired level and seals the open end of the pipe.

To illustrate the wider possibilities of fluidic techniques examples from the very different fields of medicine and nuclear power generation must suffice. The circuit illustrated in Fig. 11 is a fluidic pumping system for an artificial heart.[21] The reason for using fluidic elements, of course, is their reliability which is vital here. The system utilises bi-stable and monostable Coanda elements connected so as to set up an oscillation. This action simulates that of a natural heart. By an adjustment of the resistors and capacitors, the frequency of oscillation (pulse rate), pressure amplitude and ratio of increasing/decreasing pressure duration can be varied. Similar circuits have been implanted into animals and experimental evaluation of more advanced fluidic circuits is pending.

In atomic energy reactors it is necessary to monitor the radioactive level of the coolant passing through the fuel cartridges. By this means bursts in fuel elements can be detected. This operation calls for automatic and sequential withdrawal of samples of the coolant which pass through the instruments and are then returned. Because of the radiation hazard, the sampling gear should require no maintenance: a fluidic system is therefore a suitable choice. Three 'heavy current' Coanda type elements, activated by the coolant being sampled, act as valves,[2] one being on while the other two are off. The switching is obtained by applying a control pulse to the power jet inlets of small Coanda elements which carry out the necessary logic operations. This application is a good example of the use of the same type of fluidic elements, both as valves and as logic devices.

Production techniques

Problems in the development of fluidic devices and their applications arise not only from fluid mechanics but also in manufacture. The development of miniature elements and the associated circuitry for logic is done with models very much larger than life-size. The principles of dynamic similarity must be observed in scaling down. That is to say, among the important criteria may be the Reynolds and Mach numbers, as well as surface roughness of the flow passages. Other items which must be taken into account are the lengths of the interconnecting lines between the individual elements and temperature effects. This is true particularly of wall attachment, momentum effect and turbulence amplifiers, since these elements have by far the widest application and therefore demand quantity production.

It is not easy to produce small fluidic elements in quantity with high dimensional accuracy of the critical sections, eg, nozzles. But there are a number of satisfactory methods.

In the *Dycril* process, now mainly used in laboratory work, a layer of photosensitve material on a metal plate is exposed through a mask, as in printing electrical circuits, and the resulting pattern reproduced in depth, on plastic material by means of contact printing under ultraviolet radiation. The chemical properties of the plastic exposed to the radiation are changed while the unexposed plastic is etched away with a chemical solution, leaving behind the passages forming the element or circuit, which is then covered with a Perspex plate. The required accuracy is achieved by starting with a drawing 10 times life size and reducing the size photographically.

A similar method uses photosensitive glass. Metal etching techniques are also used and so are epoxy moulding and replication for small-scale production.

In epoxy replication, a life-size master is produced by copy milling from a large template, say 20 times life size, which is used to produce a pattern by casting. Epoxy castings from the pattern give accurate reproductions of the master.

Fluidic units of this type can also be cast from acrylic material poured into a sandwich mould, sealed by an impervious material and polymerised in a furnace.

A detailed account of the various production methods was given at the 1967 BHRA Conference on Fluidics.[23]

Future trends

There is no doubt that fluidics is moving out of the laboratory stage and is taking its place alongside hydraulics, pneumatics and electronics in all areas of industrial activity. Elements of various types and standard circuits and modules are commercially available which, for relatively simple operations such as amplification, logic and so on, are frequently more economic than electronics.

However, the future significance of fluidics in the UK appears to depend largely on the engineers. It would seem that British engineers do not fully appreciate the versatility of fluidic equipment for heavy-duty work as well as for logic operations, the generation of analogue and digital signals, sensing, etc, as well as the possibility of operating it by means of process fluids.

Cost remains a problem. The cost of fluidic equipment can only be reduced by the rationalisation of circuit design. This would broaden the field for the use of standard modules and eliminate much special testing. This, in its turn, would reduce manufacturing costs and so increase the appeal of fluidic devices. Such development work calls for much capital and industrialists must be convinced that the expenditure will bring suitable rewards. The instrument industry in particular should give serious consideration to investing in fluidics.

It would appear that in the UK fluidics activity has been more pronounced in educational establishments than in industry. In the USA interest is far more intense—20 to 25 corporations have major development programmes.

REFERENCES

1. ANON. Fluidic thruster steers a barge. *Machine Design.* 40, 12, p. 16. (May, 1968).
2. ZOBEL, R. *Experiments on a Hydraulic Reversing Elbow.* Mitt. Hyd. Inst, Munich. 8pp. 1-47. 19 (United Kingdom Atomic Energy Authority, Risley, Trans. 430).
3. HEIM, R. *An investigation of the Thoma Counterflow Brake.* Trans. Hyd. Inst, Munich Tech. Univ, Bult. 3, pp. 13-28. Published by ASME, 1935.
4. First International Conference: *Fluid Logic and Amplification,* Sept. 1965. Organised by BHRA. and the Dept. of Prod. and Ind. Admin, College of Aeronautics, Cranfield.
5. Second Cranfield Fluidics Conference, Cambridge, Jan. 1967. Organised by the BHRA and sponsored by the IMechE, Soc. of Inst. Tech, and the Dept. of Prod. and Ind. Admin, College of Aeronautics, Cranfield.
6. Third Cranfield Fluidic Conference, May 1968. Organised by the BHRA and the Inst. di Technologia Meccanica, Milan.
7. RAMANATHAN, S. and BIDGOOD, R. E. The development of a Pure Fluidic Position Control System. Paper A6. *Proc.* 3rd Cranfield Fluidic Conf, BHRA, May, 1968.
8. ABBATE DAGA, A. 'The AF-Relay—Some Applications to Automatic Control Systems.' RIV-SKF, Sezione Calibri e Strumenti, Torino, Italy. Pape J3, *Proc.* 2nd Cranfield Fluidic Conf., BHRA Jan, 1967.
9. FOSTER, K., MITCHELL, D. G. and RETALLICK, D. A. 'Fluidic Circuits used in a Drilling Sequence Control'. Paper 114, *Proc.* 2nd Cranfield Fluidic Conf, BHRA Jan. 1967.
10. CHARNLEY, C. J. '*Fluidics in Production Engineering*'. Paper presented at IProdE meeting. Nov. 1967 as the George Bray Memorial Lecture.
11. MANN, B. G. 'Fluidics in the Garment Industry'. Paper presented at *Fluidics in Industry* (discussion) organised by IProdE. Session 1, Paper 3, April, 1968.
12. VETTER, L. J. *Now Fluidics Takes Over Automation Flight Control.* SAE Journal. April 1968, No. 76, pp 41-3.
13. CAMPAGNUOLO, CARL J., and HOLMES, ALLEN B. *A Study of Two Experimental Fluidic Gas Diverter Valves.* Paper C1. *Proc.* 3rd Cranfield Fluidic Conf, BHRA May 1968.
14. WALL, D. B., and PATZ, B. W. *Fluidic Control System—Solar Probe.* Paper presented at 5th Space Congress, Cocoa Beach, Florida, March, 1968.
15. ANON. 'Unvac Fluid Logic Computer'. *Machinery and Production Eng.* Vol. 10, No. 2839, p.800. 12th April 1967.
16. HAWGOOD, D. 'Electrical Transducers for Fluidic Systems in Computer Peripherals'. *Proc.* 2nd. Cranfield Fluidic Conf. BHRA Jan. 1962.
17. REILLY, R. J. *Aircraft Gas Turbine Control.* Lecture No. XXIII. AGARD, London, September, 1966.
18. ROYLE, J. V. and HASSAN, M. A. *Characteristics of Vortex Devices,* 2nd Cranfield Fluidic Conf. BHRA, Jan. 1967.
19. HART, R. R. 'Performance of a Coanda Device as a Pressure Maintainer and as a Switch or Selector'. *Proc.* 2nd Cranfield Fluidic Conf. BHRA, Jan. 1967.
20. ADAMS, R. B. 'Application of Fluidic Valves.' *Proc.* Winter Annual Meeting, ASME, Nov. 1965.
21. HARADA, M., OZAKI, S., and HARA, Y. 'Output Characteristics of Wall Attachment Elements'. Paper F2, *Proc.* 3rd Cranfield Fluidic Conf. BHRA, May, 1968.
22. FISHER, M. J. and THOMSON, A. 'The Application of Fluidics to the Detection of Burst Fuel Elements in Nuclear Reactors', Paper K5. *Proc.* 2nd Cranfield Fluidic Conf. BHRA, Jan. 1967.
23. *Fluid Amplifiers, State of the Art,* NASA Report, 1964, Vol 1, Section 8, CR-101.

Mr A. Linford, BSc, a graduate of Kings College, London, has been engaged in the instrument manufacturing industry for over 30 years, and his interests have included not only flow measurement but also all aspects of instrumentation and of automatic control. His experience, obtained with George Kent Ltd, has extended to research and design, and to the manufacture of measuring instruments and automatic controllers. His responsibilities have included the consideration of special problems and applications, and field work in conjunction with industrial users throughout the whole field of industrial undertakings. He left in 1963 to join the British Hydromechanics Research Assn as Principal Research Engineer. He has travelled widely in the course of his career, which has enabled him to gain first-hand experience of an unusual variety of industrial metering requirements. He is well known as the contributor to the technical press of authoritative articles with all aspects of instrumentation.

Hydraulic Power Transmission

by A. E. Bingham, CEng, FIMechE

Apart from servomechanisms, hydraulic theory does not change very rapidly but applications of fluid power are multiplying fast. Electrically operated valves and the use of oil as a medium have been major factors in this expansion. The development of elastomers eases one of the most difficult problems still facing designers and manufacturers.

Fig. 1. Endurance test rig for aircraft hydraulic jacks

Hydraulics or, more properly in these days, fluid power, is expanding very rapidly in many fields, from presses exerting forces of many thousands of tons, to light machinery and small mobile equipment. The combination of hydraulic power transmission with electrically operated valves, using oil as the fluid medium, has been responsible for most of this expansion. Even further broadening has come about by the incorporation of electronics into the control system.

Hydrostatic systems incorporating variable delivery pumps are widely used, especially on mobile equipment.

The use of oil, as opposed to water, is one of the main factors which account for the great increase in the number and scope of applications. The work of the chemists in developing fluids and suitable elastomers for seals is appreciated and they should be encouraged to progress still further; but the work on elastomers must not lag behind that on fluids.

Metallurgists, machine tool designers and metrologists all come into the picture and each specialist is very dependent on the others. Fortunately, co-operation between them is well in evidence.

BS. 2917/1957[1] is now coming into more general use and, once the symbols have been learnt, it greatly assists the understanding of hydraulic circuit diagrams which are frequently quite complicated. Without such a system being used universally, much confusion and misunderstanding can result.

It is to be hoped also, that all absolute viscosities will be expressed in stokes or centistokes, together with temperatures in degrees centigrade. The present confusion in these units is beyond polite comment.

Apart from servomechanisms, hydraulic theory does not change rapidly but the same cannot be said for its applications. Recent developments include machine tools, process control, mining equipment, vibration testing equipment and the control of radar screens and radio telescopes. Endurance testing rigs such as that shown in Fig. 1, are now widely used to ensure long term reliability of much hydraulic equipment.
used to ensure long term reliability of hydraulic equipment.

This review will endeavour to cover the last decade in particular and, in places, the last two decades; hard and fast lines cannot easily be drawn. For a brief review covering the last fifty years there is a masterly work by T. E. Beacham[2] presented at the College of Aeronautics, Cranfield.

Piston pumps

Pumps must obviously form the starting subject and in these, as in many other mechanisms, the increase in speed of rotation is a major criterion, which automatically leads to a reduction of weight and volume. The aircraft industry has probably given the greatest lead because of its need to save space and weight.

Further saving in both these directions is obtained by increasing pressures. For example, at the beginning of the 1939–45 war, pressures in aircraft hydraulic systems were approximately 800 lb/in^2 and the familiar gear pump was able to deal with such pressures quite adequately, without too much refinement.

However, in a very few years the pressures crept up to 1800 lb/in^2 and piston pumps had to be introduced with steel pistons working in bronze bores. Soon it became obvious that more research on materials would be necessary, despite the introduction of chromium bronze. In the case of the *Live Line*, self-limiting pump, the rotor had to be changed to cast iron with modifications to give extra strength geometrically; later spheroidal graphite cast iron was used to permit pressures over 2000 lb/in^2 and speeds above 3000 rev/min.

Fig. 2. This compact pump has a common crank which rotates while the radial cylinders remain stationary

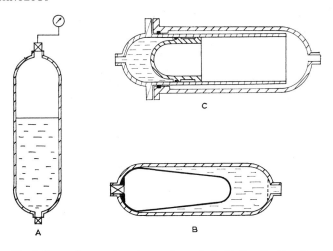

Fig. 3. Types of accumulator in use: A, vertical with free liquid surface; B, a synthetic rubber bag separating gas from liquid; and C, a separator piston in a pressure-balanced thin cylinder

Manufacturing problems had to be overcome by increasing the precision of grinding and subsequent lapping and these, in turn, posed problems for the metrologist who introduced pneumatic instruments to cope with them.

Centreless grinding of pistons gave good dimensional control but, with the clearances becoming finer as the development progressed, trouble still persisted. This was traced to out-of-roundness; though the diameter, as measured by micrometer, was correct. It was noticed that there were 13 high zones around the pistons, revealed by strips parallel to the axis that had been worn to a bright appearance. Therefore, after centreless grinding, a ring lapping process was introduced, followed by very careful inspection.

The close co-operation of the metallurgist is essential to engineers, both to develop improved materials and to control the material quality in accordance with specifications. The checking of mechanical properties is not always sufficient, especially where the nature of the surface is of extreme importance. The depth of surface hardening can often be critical, whether the process be cyanide hardening or nitriding and, with high expansion steels for insertion into light alloy bodies, the grain size can be a criterion. In special cast irons, the percentage of graphite is not the only factor; the form and distribution must be correct for the application.

Having found the right material in the right form, it is necessary to adhere rigidly to the specification. The history of the material and the inspection checks must be recorded throughout the whole of the processing. This applies not only to pumps but to all parts in a modern high-pressure hydraulic system; and permits pressures to be increased to 4000 lb/in² which is now common in aircraft and many other applications.

Fig. 2 shows a pump which is now used in many aircraft and numerous industrial installations. The radial cylinders are stationary and the common crank rotates. Its compact form has great advantages where space is limited. For industrial applications, bodies and cylinder covers are made in steel which gives a longer life.

Rotary pumps

Where the maximum pressure does not exceed 2500 lb/in², the well-known gear pump is still a cheap and robust unit. In the modern types, a high efficiency has been obtained by reducing the number of details to a minimum, by the use of plain bearings instead of ball or roller types, and the pressure balancing of the gears, together with cartridge construction. Hitherto the main source of internal leakage was between the gears and the side plates but by feeding the output pressure to one pair of side plates, or bushes, and proportioning the area over which the pressure acts, the bushes are held in contact with the gears, which limits the clearance to the minimum needed for lubrication.

The plain bearings are supplied with a metered amount of oil from the gear meshing zone so that lubrication is practically independent of the pump operating pressure. A cartridge arrangement can be used to eliminate alignment difficulties. By these developments, volumetric efficiencies of 94 per cent were obtained, even up to 2500 lb/in². The outputs over the complete range are from 0·5 to 50 gallons per minute.

For lower pressures vane-type pumps are still in use and there are many variations. A development during the last few years is the slipper type, used in millions of cars, especially in America.[3, 4] Instead of thin vanes, relatively wide, spring-loaded slippers are used in which the outer surfaces are contoured and lead to the formation of wedge-shaped oil films. The eccentricity between rotor and casing is small so that wear and noise is reduced to a minimum. This was very necessary as these pumps were designed for automatic transmissions.

For very heavy presses, pump pressures of 5000 lb/in² and 6000 lb/in² may be used in order to keep ram diameters and overall sizes of cylinders down to manageable limits but they lead to difficulties in other parts of the system.[5] These higher pressures have also been suggested for aircraft but so far they have not been received with favour. However, it is understood that pressures of 10 000 lb/in² are contemplated for guided weapons and spacecraft. Although such hydraulic systems have to be exceedingly reliable, their working life is so short that comparisons with industrial equipment are rarely practical.

For land-based applications, the most convenient pump

drive is by electric motor, at up to 3000 rev/min. For mobile equipment, the pumps can be driven, via gearboxes if necessary, by internal combustion engines or gas turbines. With the latter, it is often more convenient to run at higher speeds, say 4000 to 5000 rev/min, and on missiles pumps are running at 10 000 rev/min.

Following the pump, the most important item in any hydraulic system is the actuator or jack. Here again, as pressures increase, the effective piston area and bulk decrease and further reductions can be made by using materials of higher tensile strengths. Jacks generally rely on elastomers for sealing both the piston and the piston rod. The surface finishes of the cylinder and piston rod have to be of a high order, say 4 to 8 μ in C.L.A. at 3000 lb/in² and above. Parallelism of rod and cylinder bore is important, especially when the jack has to work at low temperatures, for instance in aircraft and in vehicles operating in cold climates.

Accumulators

In many systems accumulators are necessary and much of what has been said about jacks applies to them. As most modern accumulators are air or gas loaded, a further consideration is necessary, that of safety: a burst hydraulic jack may be a considerable nuisance but it is seldom dangerous; a burst pressure vessel containing air or another gas can be catastrophic, due to the considerable stored energy. The fatigue life of the accumulator must be determined if it cannot be made virtually infinite. Ductile materials must be used and great attention paid to design and manufacture to avoid

Fig. 4. Low weight and bulk combined with first-class fatigue life and endurance are features of aircraft valves like the miniaturized 'Hydel', which is 4 in long and weighs 2·16 lb

stress raisers. Great strides are being made in forming these vessels, thus keeping down the number of joints.

If a separator piston is essential, then one end of the vessel must be open for assembly and machining the bore; but if not, both ends can be manufactured by forming. Modern methods of non-destructive testing have made such forming methods safe. Making the large separator piston both gas-tight and oil-tight presents the seal designer and manufacturer with one of the most difficult problems, especially when a large temperature range is involved.

The reason why the accumulator seal presents such a difficult problem is that on one side there is liquid and on the other gas; therefore the seal has to satisfy two different sets of conditions. A small leak from liquid to gas is not serious but a small leak from gas to liquid is a great nuisance in a system with an open reservoir, and results in inefficient working; in the case of a closed reservoir, the pressure in the low pressure side of the system is raised, possibly to the blow-off pressure of the relief valve. This results in loss of gas, liquid and reserve power in the accumulator.

In order to prevent the expansion, under pressure, of the cylinder in which the separator piston travels, this cylinder can be made comparatively thin and arranged so that it has pressure both inside and outside. In that case it has to be surrounded by an external cylinder barrel to take the pressure. This method has proved very successful in many difficult cases and the extra weight is virtually only that of the thin inner cylinder.

In land or marine systems, the piston and its problems can be eliminated by the bag-type accumulator which has been very greatly improved during the last few years. Its use in aircraft, however, is not yet viewed with great favour because the very wide temperature range has made the choice of a suitable elastomer for the bag an extremely difficult one. Three types of accumulator are shown diagrammatically in Fig. 3.

Valves and pipes

Valves of all kinds cannot be discussed in detail, the subject is far too vast; but, speaking generally, the improvements have been due to research in flow characteristics; concentration on detail design, especially with regard to tolerances and clearances; improved materials and finishes; and endurance and fatigue testing (Fig. 1), including pressure pulse testing to find out weak spots.[6-10]

The zone between converging ports is a source of weakness and ports should enter chambers at right angles. This is sometimes difficult to achieve but is worthy of close study because an acute angle between a port and a chamber is a zone very susceptible to fatigue.

One of the greatest improvements in general system design is the development of the electrically operated valve, either motor or solenoid, which can be placed where it is most effective hydraulically, and remotely controlled from a convenient point. Modern valves of this nature are extremely light for their duty and are widely used on aircraft. Much research and development have been put into this type of valve and its fatigue life and endurance are now first class. Some well-known valves of this nature are being miniaturized for still further weight and bulk reduction but, of course, at some cost in greater pressure drop. Fig. 4 shows a typical miniature valve of this kind.[11]

Solenoid-operated valves are very varied in design. Many incorporate a small servo system in which a solenoid operates a small ball valve which controls the fluid acting on the end, or ends, of a piston valve which, in turn, controls the main fluid flow. The principle is shown in Fig. 5.

Practically all hydraulic systems use pipes to connect the various items in the system. Up to pressures of 3000 lb/in² the well-known types of pipe joints give little trouble but, at 4000 lb/in² and over, special designs may be necessary. During the last decade stainless steel tubing and fittings have been used for the higher pressure aircraft fittings with great advantage. Brazing by torch or induction methods is satisfactory, provided that the joints are adequately inspected, using X-rays or gamma rays and static pressure tests.[12] With the latter it may be necessary to boil the joints in water first, to remove flux which may provide a temporary seal during testing. Welding generally gives a sound joint but can produce internal scale which is often difficult to remove.

Flexible pipes are now manufactured in small sizes for pressures of 4000 lb/in² and over. The improvements during the last few years have been quite remarkable. However, many designers avoid them if possible. They have very strong competition in the swivel coupling, which has been developed to give low friction and high reliability.

Operating fluids and seals

The choice of a fluid[13,14] is governed by many circumstances and Table 1[15] is a rough initial guide. But before chosing one in practice a specialist should be consulted. The table shows that the number of types now available is quite large and the following considerations must be included in the correct choice.

1. Usual running temperature and temperature range;[16,17]
2. Minimum viscosity acceptable for bearings, vanes, pistons or gears;
3. Maximum viscosity acceptable for the suction line;
4. Is there a definite fire risk?
5. Cost;
6. Availability.

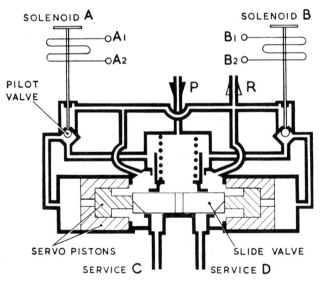

FUNCTIONAL DIAGRAM

Fig. 5. The 'Hydel' valve: with both solenoids de-energized the pressure of supply fluid at P holds the pilot valves against the 'return' seats. Energizing one solenoid opens the return line R to the servo cylinder and allows the central servo piston on the opposite side to move the slide valve to operating position C or D

Having chosen the fluid it is most important to choose the right sealing medium[18,19] and Table 2 shows the many types of elastomer now available and which of them match the fluid.

There have been great developments in elastomers during the last two decades and these are continuing with considerable urgency. The choice of the elastomer has to be made in conjunction with the working fluid, the pressure and the temperature range. Before the technology of synthetic rubber was as advanced as it is today, natural rubber had to be used. This severely limited the nature of the working fluid which, generally, had to be based on castor oil.

Synthetic rubbers can now be made which are compatible with mineral oil and many other types of hydraulic fluid. This marriage between mineral oil and synthetic rubber coincided with the spectacular development of fluid power.

From the fatigue point of view, ever greater attention is

Table 1.—Types and Scope of Operating Fluids

Item	Fluid type	Purpose	Temperature Range °C	Remarks
A	Water + soluble oil	Heavy equipment	2 to 71	Cheap, but not a very good lubricant.
B	Water + soluble oil + inhibitor	Equipment in irradiation zone of nuclear power stations	2 to 71	More expensive than A because distilled water should be used.
C	Water glycol base with inhibitors	Equipment in areas where there is a fire risk e.g. plastic factory	−20 to 80	Reasonable lubricant but fairly expensive. Mixture must be checked chemically periodically.
D1	Hydro-carbon (petroleum base) specially blended e.g. Specn. DTD 585	General for aircraft and light machinery	−60 to 80	Reasonable price, fair lubricant, upper end of temperature range can be raised, by incorporating pressurised reservoir, to 135. World wide availability.
D2	As D1 but higher viscosity and to commercial specifications	General purpose, land and marine	−15 to 90	Fairly cheap, good lubricant, wide choice from all well known oil companies. Good availability. Temperature range required will determine best viscosity.
E	Silicone (viscosity 20 cs @ 25°C) other viscosities available	Small systems where small change of viscosity with temperature is important	−70 to 175	Very wide temperature range, more difficult to seal than D1 or D2, not a good lubricant, expensive. More compressible than D1 or D2.
F	Chlorinated silicone (viscosity 5 cs @ 25°C)	Aircraft when low inflammability required	−60 to 200	Very wide temperature range. Fair lubricant, very expensive.
G1	Phosphate Ester. Viscosity 7000 cs @ −40	Aircraft of limited performance. Land and marine systems (low inflammability)	−20 to 70	Moderately expensive. Lubricating properties good. Butyl rubber seals required. Obsolescent for aircraft.
G2	Phosphate Ester. Viscosity 500 cs @ −40	Civil aircraft. Land system when extra low temperature and low inflammability required	−40 to 90	Slightly more expensive than G1. Lubricating properties good. Resin cured butyl rubber seals required.
H	Silicate Ester	Experimental aircraft with large temperature range required	−40 to 175	Moderately expensive. Susceptible to hydrolysis. Contact with copper or cadmium to be avoided at high temperature.
J	Alkyl Silicate Ester and similar	Experimental aircraft (supersonic) and missiles	Varies with type, from −65 to 300	These types are still in experimental stage.

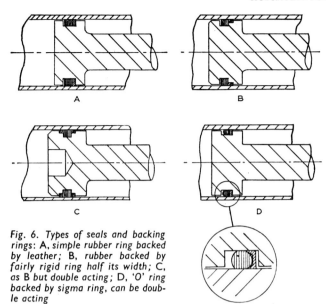

Fig. 6. Types of seals and backing rings: A, simple rubber ring backed by leather; B, rubber backed by fairly rigid ring half its width; C, as B but double acting; D, 'O' ring backed by sigma ring, can be double acting

Table 2.—Elastomer Seals for Various Fluids

Item	Fluid Type	Temperature Range (°C)			Elastomer Base	Hardness BS Deg	Remarks
		Low	High	Short Peak			
A	Water + soluble oil	2	71	80	Butadiene acrylonitrile*	65 to 77	Up to 2000 lb/in²
B	Water + soluble oil + inhibitor	2	71	80	As A	As A	As A
C	Water glycol + inhibitors	−20	80	85	As A	As A	As A
D1 & D2	Hydrocarbon	−60	80	90	As A	As A	As A
E	Silicone	−70	80 150	90 175	As A Viton A†	As A 85 to 95	As A Up to 4000 lb/in²
F	Chlorinated silicone	−60	150	200	Viton A or resin cured butyl	85 to 95	Up to 4000 lb/in²
G1	Phosphate ester	−20	70	80	Butyl**	65 to 77	Up to 3000 lb/in²
G2	Phosphate ester	−40	80	90	Resin cured butyl	80 to 90	Up to 4000 lb/in²
H	Silicate ester	−40	150	175	Butadiene acrylonitrile	65 to 77	Up to 3000 lb/in²
J	Alkyl silicate ester				Viton B or ethylene propylene‡		Still experimental

* There are different compounds on a butadiene-acrylonitrile base.
† The Vitons are co-polymers of vinylidene fluoride and hexafluoropropylene.
** Butyl is a co-polymer of isobutylene and isoprene.
‡ Ethylene propylene is a recent development which is claimed to have much better properties than existing synthetic rubbers.

Table 3—Maximum Permissible Contamination

	Particles measuring more than					
Fibres	100μ	50–100μ	25–50μ	15–25μ	5–15μ	
3	5	60	3000	15 000	45 000	

being paid to detail design of metal parts and the type and form of the seals. To assist the latter, backing rings are employed; some years ago certain types of leather were considered highly satisfactory for this purpose but if temperatures go above 90 degrees C, leather becomes inadequate. Fabric-reinforced laminated plastics have often supplanted the leather but at temperatures above 125 degrees C these are reaching their limit.

Fig. 6 (A) shows a simple rubber ring, backed by a leather ring, the latter being the full width of the rubber. The metal of the cylinder and that of the piston must be chosen carefully to avoid scoring, e.g., a brass or brass-plated piston would be necessary in a steel bore, or the piston could be steel, working in a nitrided bore.

The backing rings in (B) are only about half the radial width of the rubber ring but are made of fairly rigid material, such as fabric-reinforced synthetic resin, and the metal piston has a substantial clearance, say 0·006 in to 0·010 in radially, so that the piston cannot come into direct contact with the cylinder bore. In such cases, both cylinder and piston can be in steel, high tensile if necessary.

If the jack, or actuator, is double-acting, then the arrangement could be as shown in (C). Positive seals, such as square section or 'O' section, should not be duplicated as high pressure can be locked up between the seals, thus causing high friction and rapid wear. 'O' rings are frequently backed up by sigma rings as shown in (D) and, if double acting, the sigma rings can be on each side of the same rubber ring.

In the last few years PTFE backing rings with various types of reinforcement, e.g., glass fabric or random fibres, have proved satisfactory. Considerable development is now in progress and materials such as graphite and molybdenum disulphide are being added. These stand up to temperatures of 200 degrees C, or more, for short periods.[20]

Filtration

The water-based fluids are practically non-inflammable but, if spilt and the water content is allowed to evaporate, the residue is inflammable. The work on fire-resistant fluids is being pursued with considerable vigour but the perfect non-inflammable fluid has not yet been found. The provisional specification, DTD 5507, covering fire-resistant fluids has now been superseded by DTD 5526 (low inflammability, high-temperature hydraulic fluids) which has reached its final draft and should be issued shortly.

Filtration of the working fluid has always been important but with modern hydraulic systems using high speed pumps, complicated valves and, especially, servo-valves, the filtration standard must be very high. Success or failure of such a system often depends on it and experience during the last ten years shows that the contamination should not exceed that given in Table 3 which is based on a 100-ml sample.

Various methods of determining the contamination are now available, some involving quite elaborate apparatus which is only justified if a large number of counts per week are required. A reasonably simple method not requiring very expensive apparatus is to take a 100-ml sample through a specially designed adaptor which excludes any airborne dust and contains a *Millipore* disc about 1½ in diameter, which only passes particles below 0·8 μ. This disc is then microscopically examined in a dust-free room at about 90 diameters magnification. It is ruled in 3·8-mm squares to facilitate counting. The binocular microscope has a graticule so that particle sizes can be estimated reasonably accurately.

There are differences of opinion on whether the sample

tapping point should be situated in the pressure line or the return line. If in the pressure line, which may be at 4000 lb/in², a needle valve control is necessary, the narrow annulus will act as a filter for the larger particles.

This is not necessary in the return line but the turbulence may be less, so that some larger particles might settle out. Wherever the tapping is situated, it must be after the pressure line filter, otherwise the sample is not representative of the fluid with which the mechanisms are working.

Power transmission

The hydrostatic transmission, which incorporates a positive variable displacement pump coupled to a positive displacement hydraulic motor is used in a multitude of stationary and mobile machines.[21] Generally, the pump is of the tilting head type, the angle of which can be varied from full forward position, through neutral, to full reverse. The hydraulic motor is of fixed capacity so that its speed is infinitely variable in each direction. If the system is small, the angle of the tilting head can be hand-controlled, but in larger systems a hydraulic servo jack is incorporated so that only a light hand load is necessary. The system automatically provides braking characteristics for, by reducing the angle of the tilting head, the roles of pump and motor are reversed and the pump drives the motor, or engine. If the angle is reduced to zero, the system becomes locked and the pressure is reduced through the relief valve.

Such systems are employed on tractors, dumpers, mobile cranes, shunting engines and dockside capstans, all of which require a high degree of manoeuvrability and nicety of control. Conventional clutch, gear shifts and reversing mechanisms are eliminated which reduce driver fatigue. The foregoing remarks also apply to similar systems using swash plate pumps, against tilting head drives.

A recent development for field work,[22] both for farming and civil engineering, is the slow speed hydraulic motor, with dual capacity, built directly into the driving wheels; a device under development at the National Institute of Agricultural Engineering.

Infinitely variable drives are being applied to lathes, especially those turning large diameter faces, so that the optimum cutting speed can be maintained as the tool is moved radially across the face. This is a great advantage to the manufacturer, the customer and the turner.

In crawler tractors, the hydrostatic transmission, besides driving the tractor, steers it by setting the differential speeds of the tracks. Similar systems are used on bulldozers.[23]

Aircraft

Originally, the aircraft hydraulic system was introduced to operate the undercarriage but today it carries out many tasks including the operation of fairing and bomb doors, etc; wing flaps; operating winches on helicopters and the flying controls on large or high-speed aircraft where the forces required are too large to be exerted by the pilot. Nose-wheel steering is also hydraulic. The horsepower of hydraulic systems has now increased from about three to well over a thousand on large supersonic bombers.

The system diagram for modern large high-speed aircraft is complicated and is rendered more so by the fact that pumps and many vital services have to be duplicated for reasons of safety.[12] In the case of malfunctioning or failure of one system of flying controls, the automatic switching to the duplicate system must be instantaneous as, at high speed, control over the aircraft must not be lost for a second.

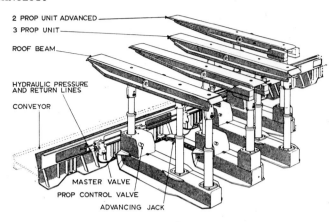

Fig. 7. Walking pit props; by supporting the roof on one pair the other pair can be released and pushed forward by hydraulic jacks

Emergency services for undercarriage and landing flaps are provided by compressed air and the change-over is by hand control, the timing, in this case, being not so critical.

In the case of missiles hydraulic systems carry out similar functions of control power transmission.[24]

Automatic forging

Turning to heavy plant, one of the most spectacular developments during recent years was the application of automatic controls to forging presses. Power-assisted manual controls for forging presses have been in use for a considerable time but the replacement of water by oil as the working fluid permitted the use of much lighter and high-speed valves to be introduced, which has now led to the successful design and operation of fully automatic presses. Speeds of 150 blows per minute have been obtained on a 500 ton press and an inverted 3000 ton press was put into commission in 1961, the moving parts of which weigh 250 tons. Accuracies down to 1/32 in can be obtained. Naturally the control system is highly complicated and costly but the overall forging cost is reduced, together with the wear and tear on tools and press, and there is a significant reduction in furnace time.[25] The same type of control can be extended to the forge manipulator.

To supply the large quantities of high pressure oil it has been necessary to design special pumps and an outstanding development during the past ten years is the 500 hp pump. This is a double axial poppet valve pump and is not, therefore, capable of being used as a motor, but it has the advantage that it will operate at high pressure under arduous conditions such as obtained on forging and extrusion presses.

The largest high-speed precision forging press with automatic thickness control has been installed in Britain. This is a 3000 ton forging press and is powered by six 500 hp pumps. The equipment includes a double swashplate with opposed pumps on either side. The special feature which has made possible a compact pump, reliable at high pressures, is the Fogg thrust washer. In other words, hydrodynamic bearings are used, instead of roller bearings, as no roller type thrust bearings were available which were adequate for the forces involved within so small a space.[26]

Mining applications

Dropping from surface level to below ground, the use of hydraulics in mining has developed at an increasing rate since the introduction of the hydraulic pit prop, just after the 1939–45 war. The use of this permitted the coal cutting machines to be used to their full advantage, for it had not been possible to follow the machine fast enough while erecting wooden props. The individual hydraulic prop has been followed by the 'walking' pit prop which consists of several props attached to a pair of roof and floor beams which, in turn, were coupled to a similar pair. By supporting the roof on one pair, the other pair can be released and pushed forward by hydraulic jacks; and the operation is then repeated with the first pair, as shown in Fig. 7.

This system has now been developed further so that a complete coal cutting face can be covered and electrically controlled from a console in the main gallery. This not only saves manpower but, to a great extent, eliminates risk. The system is still in the development stage but a full-size plant is working well in a dummy mine.

Another underground machine of high importance is the Collins Miner[27, 28] which is a coal cutting machine and transport system. This is also remotely controlled and again uses electronics to guide its way into the coal seam. Both these hydraulically actuated devices, still being developed, enable thin seams to be worked economically and obviate hitherto appalling working conditions.

The coal so won is transferred by belt conveyors to tubs. Tub arrestors are hydraulic devices to retard the vehicles at the end of the rail journey. These brakes have now been developed further to control the movement of full-size railway wagons for the purpose of automatic wagon control in marshalling yards; the system includes speed sensing so that wagons which are travelling above the critical speed of the siding are retarded and those which are below the critical speed are accelerated. The hydraulic devices are triggered by, and work on, the wheel flanges. Finally the wagons are brought to rest quickly and smoothly at any predetermined point.

Future prospects

To prophesy on any matter is a hazardous proceeding; particularly so in a specialized engineering subject. However, a prophecy may promote interest or discussion and therefore be worthwhile.

The combination of hydraulics with electromechanics has already led to great advances and the further combination with electronics is now well on its way. The development of these three skills in combination augurs well for more complicated problems now being tackled. Dare we predict fluid power automation?

The vast efforts now being put into research on both

fluids and elastomers must surely bear fruit in the not too distant future so that the ideal can be more nearly attained in both cases. For special high temperature work, e.g., supersonic aircraft, the metallurgist and hydraulics engineer will have to work in even closer co-operation with machine tool designers and metrologists to produce more accurate geometrical shapes at reasonable prices. This applies to actuator rods, cylinders and metallic seals.

There is still scope for development and application of fluid power in mobile equipment which can be operated by semi-skilled labour.

Except for very special purposes, the general maximum pressure is not likely to exceed 4000 to 5000 lb/in² but for large presses it may gradually rise to 10 000 lb/in².

REFERENCES

1. B.S. 2917/1957. 'Graphical Symbols for use in Diagrams for Fluid Power Transmissions and Control Systems'.
2. BEACHAM, T. E., 'Fifty Years Development in Oil Hydraulics'. British Hydromechanics Research Association. Publication SP. 720.
3. GUTERMAN, J. W., 'Transmission Oil Pumps'. S.A.E. Preprint no. 311c.
4. 'Ninth International Automobile Technical Congress'. Report, part 2. 1962, 25th May The Engineer, vol. 213, p. 901.
5. 'Pumps up to 230 hp at 6000 lb/in²'. Hydraulics and Pneumatics Ltd (Turner Manufacturing Group).
6. MACLELLAN, G. D. S., MITCHELL, A. E., and TURNBULL, D. E., 'Flow Characteristics of Piston-type Control Valves'. 1960 Instn mech Engrs, Proc. Symposium on Recent Mechanical Engineering Developments in Automatic Control, p. 13.
7. MANNAM, J., 'Further Aspects of Hydraulic Lock'. 1959 Proc. Instn mech Engrs, Lond., vol. 173, p. 699.
8. HARDING and UPSHALL. British Hydromechanics Research Association. Publication RR. 704.
9. WHITEMAN, K. J., 'Hydraulic Lock at High Pressure'. 1957, 12th April The Engineer, vol. 203, p. 554.
10. NOTON, G. J., and TURNBULL, D. E., 'Some Factors influencing the Stability of Piston-type Control Valves'. 1958 Proc. Instn mech. Engrs, Lond., vol. 172, pp. 1065–81.
11. 'Dowty Solenoid Valve'. 1962 February Hydraulic Power Transmission, vol. 8, no. 86, p. 118.
12. CONWAY, H. G., ed., Aircraft Hydraulics. Vol. 1, 'Hydraulic Systems' Published under the authority of R. Aero Soc. 1957 Chapman and Hall, London.
13. JACKSON, T. L., 'Some Notes on Mineral Oil Viscosity Requirements of Hydraulic Systems'. Shell International Petroleum, Lubricants Division.
14. HAYWARD, A. T. J., 'Air Bubbles in Oil'. 1962 30th April Product Engineering.
15. Information from Monsanto Chemicals Ltd, Midland Silicones Ltd, I.C.I. Ltd and 'Shell Industrial Lubricants', Shell Petroleum Co., Ltd.
16. DAMASCO, F., and SPAR, C., 'High-temperature Hydraulic Fluids'. Nat. Conf. Industrial Hydraulics, vol. 15, p. 104.
17. HAYWARD, A. T. J., 'Research on Lubricants and Hydraulic Fluids at the National Engineering Laboratory'. 1963 D.S.I.R., National Engineering Laboratory Report No. 75.
18. 'General Review of Recent Seal Research at B.H.R.A.' British Hydromechanics Research Association, publication SP. 723.
19. CANNIZARO, S., LEE, J., and SCHROEDER, R., 'Sealing in Severe Environments'. Nat. Conf. Industrial Hydraulics, vol. 15, p. 187.
20. BINGHAM, A. E., 'Hydraulic Seals for Extremes of Working Temperatures with special reference to Aircraft'. 1962 Proc. Insn mech. Engrs, Lond., vol. 176, p. 409.
21. BOWERS, E. H., 'Hydrostatic Power Transmissions for Vehicles'. 1961 Instn mech. Engrs, Proc. Conf. on Oil Hydraulic Power Transmission and Control, p. 41.
22. MARLOW, M. J., and VINCENT, K., 'The Application of the Vardel Hydraulic Pump for use with the Roofmaster Hydraulic System and Possible Future Developments'. To be published.
23. 'Hydrostatic Transmission Units'. 1962 May Design and Components in Engineering, p. 8.
24. ZWETTLER, R. F., 'On some Problems in Missile Hydraulic Thrust Vector Control Systems'. Nat. Conf. Industrial Hydraulics, vol. 15, p. 176.
25. WILKINS, B. C., 'Protecting Hydraulic Systems against Pressure Transients'. 1962 August Control, vol. 5, p. 78.
26. FOGG, A., 'Fluid Film Lubrication of Parallel Thrust Surfaces'. 1946 Proc. Instn mech. Engrs, Lond., vol. 155, p. 49. also TOWLER, F. H., and WILKINS, B. C., 'Automatic Forging Controls'. 1961 Instn mech. Engrs, Proc. Conf. on Oil Hydraulic Power Transmission and Control, p. 170.
27. 'Collins Miner'. 1962, 23rd March The Engineer, vol. 213, p. 526.
28. 'Three-headed Mole guided by Nucleonics'. 1962, 28th May Product Engineering, vol. 11, p. 68.

Mr A. E. Bingham received his technical education at the South Western Polytechnic and Goldsmiths College, London. He was apprenticed to the United Aircraft Co. and served with the Rayal Flying Corps during the 1914—18 war. In 1926 he was appointed an Assistant Designer at Gloster Aircraft Co. and from 1928-1936 he was with International Combustion. He went to Cerac Ltd as Chief Engineer in 1936 and remained until 1939 when he joined Dowty Equipment. He was Chief Mechanical Test Engineer for Dowty Rotol Ltd and then Technical Information Officer. He is a Fellow of the Aeronautical Society. Retired 1965.

Automatic Control Theory

by C. R. Webb, PhD, CEng, FIMechE

Automatic control is not new but the recent enormous increase in its applications had to await the development of elementary theory as well as suitable engineering hardware. Various mathematical techniques have been adapted for dealing with the principal types of control systems. Since most practical systems are non-linear, only approximations can be looked for but these can indicate to the designer whether a proposed system will be stable or not and what accuracies can be expected.

There are different views among control engineers as to the definition of automatic control. At one extreme, the term is restricted to systems in which there is clear evidence of negative feedback. At the other extreme, the term is used to cover, in addition, all self-acting machinery as well as cybernetics and systems engineering[1, 2] including many aspects of computation and instrumentation.

It is clear that a modern community accepts systems which require the minimum of human assistance whenever they are reliable and improve the standard of living. At the same time, certain sections regard the continuing installation of automatic systems in factories as a threat to their livelihood. We still find many automatic plants being 'minded' by operators. Provided the installation of automatic control is an economic proposition, it is often possible to improve on the operator's performance and to avoid the consequences of his fatigue and boredom. However, in many cases it is desirable to use the human operator to control difficult systems, particularly where the plant reacts slowly.

Unlike their counterparts in North America and the USSR, many British students of mechanical engineering receive only brief instruction, if any, in automatic control. Before the 1939–45 war the subject was unknown in teaching establishments other than those of a military nature and consequently middle-aged mechanical engineers in the UK regard it as one fraught with mystery, best left to their electrical colleagues. In common with laymen, some have a qualitative appreciation of simple applications, *eg* the electric immersion heater, controlled by a thermostat, but are content to accept without understanding such systems as the directional control of space probes or the maintenance of uniform body temperature in animals.

The art of automatic control is not new. As with most techniques, however, the explosion had to await the development of an acceptable elementary theory and the wide dissemination between 1944 and 1948 of texts capable of being understood by interested parties without specialist knowledge. Most of these early popular texts discussed servomechanisms—a term the meaning of which is also still not agreed; it is often used to connote feedback but servo-techniques can also be used with open-loop control.

Because theory and know-how developed initially along separate (though similar) lines in different fields of control engineering, the BSI found it impossible to recommend a common terminology.[3] An entirely new Standard is nearing

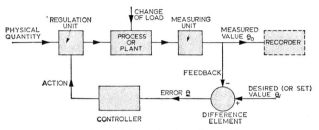

Fig. 1. The essential feedback loop takes data obtained from a process output to a unit which compares it with a set value and sends correcting signals, based on the deviation, to a regulating unit which acts on the process. Most types of controller (including the human one) incorporate both the difference unit and the correction computing unit

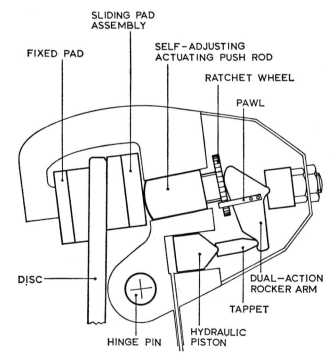

Fig. 2. Feedback need not be continuous: when the pad of this self-adjusting disc brake is worn, the clearance between rocker and plunger increases until it permits the spring pawl to engage the next notch on the ratchet: this unscrews it from the plunger, reducing the clearance

completion but this will not be as comprehensive as the current American standard[4] which has much to recommend it. In the short glossary printed here, rigour has been relaxed in favour of clarity.

Most introductory texts tend to restrict attention to particular branches of control technology but the author's volume[5] gives a treatment with wide applications.

What really is feedback?

In Fig. 1 a control system is represented by a block diagram. The controlled variable in some process may be a direction, a temperature, a potential difference, etc, and this is measured continuously by the measuring unit MU, to yield a signal called the measured value θ_0, in any convenient physical form. Sometimes this signal is fed to a recorder in order to provide a visual record. The primary reason for MU in automatic control is, however, to provide the feedback of θ_0 to the difference element where it is compared with the desired value θ_i of the controlled variable, fed in by the user. The difference, $\theta = \theta_i - \theta_0$, is clearly the error, and the control system is required to keep θ within certain limits in the face of a foreseen change in parameters.

The error signal is fed to the controller (which can take a number of different physical forms). In each case a regulating unit RU changes the flow of some physical quantity feeding the plant in such a way as to reduce the error.

Such arrangements in which θ_0 continuously influences the controller action are known as closed-loop systems or error-actuated systems. It may seem ironic that a system designed to minimise error cannot function unless error develops; however, for θ_0 to follow θ_i exactly at all times through any change of plant load or of other parameters is clearly unattainable and the best that can be expected is that θ never increases beyond an acceptable tolerance.

Glossary of Terms

Adaptive control. A system capable of adjusting its own control parameters to maintain the best performance.

Compensation. Improvement of a system by additional components to deal with potential instability.

Closed-loop (or feedback). The value of a controlled variable is used to provide a criterion for adjusting the system input. Thus, signals derived from a later stage in a process are fed back to control an earlier stage, not necessarily automatically.

Controller. The component which processes the information derived from a plant into signals to the mechanism which adjusts the input.

Damping ratio. The ratio of the actual viscous damping coefficient in the system to the value required to cause critical damping.

Describing function. The relationship between output and input of a non-linear process element as a function of the input amplitude.

Desired value. The value of the controlled variable which a controller is intended to maintain.

Deviation (or error). The amount by which the controlled variable actually differs from the desired value at any time.

Difference element. The component in a control loop which compares the desired value with the measured value to obtain the deviation.

Differential operator (D). The differential of a quantity with respect to time.

Error-actuated system. A closed-loop system in which the controller acts as a result of the deviation (or error).

Exponential lag. A first-order linear element having a transfer operator of the form $1/(1 + \tau D)$.

Feedback. (see closed loop).

Feed-forward. A signal depending on some early stage in a process used to control a later stage (to prepare it for changed conditions).

Finite delay. The delay of a signal by a constant amount of time; (sometimes called a 'pure' time lag or a distance-velocity lag).

Gain. The proportion of output to input of a controller.

Hold (or clamp). The component in a sampled-data system which attempts to reconstruct the signal received by the sampler.

Impulse function. A function of time representing a momentary pulse occurring (practically) instantaneously; the time derivative of a step function.

Laplace transform. A special integral of a time function (with respect to time) which converts it to function of s (the Laplace variable). Tables of standard conversions are available.

Laplace variable. Usually represented by s (or p) meaning a complex frequency parameter, $a + j\omega$, where a is the damping factor and $\omega/2\pi$ is the frequency.

Limit cycle. The steady state cycle which may be reached by a non-linear system.

Linear system. A system which can be represented by a linear differential equation with constant coefficients.

Load changes. External disturbances on a controlled system other than changes in the desired value.

Measured value. The value of the controlled variable as obtained from a sensing instrument (transducer).

Measuring unit (or sensing instrument or transducer). The component in a system which measures a variable and converts its value to a signal for the controller.

Natural frequency. The frequency at which a system oscillates following a disturbance.

On-off control system. A system containing a non-linear control element which has only two fixed output values.

Phase plane. A co-ordinate plane along the axes of which are plotted a variable and its rate of change with time; used in non-linear analysis.

Proportional band. The range of deviation acceptable by a proportional controller.

Proportional controller. A controller the output of which bears a straight line relation to the input.

Ramp function. A function which changes proportionately with the elapsed time.

Sampler. The component in a sampled-data system which samples a signal at regular discrete intervals.

Sampled-data control system. A system which uses data from the process at intervals rather than continuously.

Sampling period. The regular interval at which a sampler operates.

Signal. Any time variable conveying information from one part of a controlled system to another.

Stability. The absence of undesirable fluctuations in a controlled system It is judged by the rate at which transients subside, following a disturbance.

Steady-state error (or offset). A sustained departure of the controlled variable from its set value.

Step function. A function of time which changes instantaneously to a sustained value.

Time constant. A combination of the parameters of a linear system having the dimensions of time.

Transfer operator. A mathematical representation of an element in a linear system; which evaluates the output of the element as an operation done on the input.

Transfer function. Similar to a transfer operator but relating the Laplace transforms of output and input, given that the initial values of output and input and of all their time derivatives and integrals are zero.

Trajectory. The curve drawn on a phase plane to represent the changes undergone by a variable and its time derivative as time elapses. A family of trajectories with different initial conditions is called a phase portrait.

Transducer (see measuring unit).

Viscous damping. A condition in a system which gives rise to a resisting force, or torque, proportional to speed.

Z-transform. A further transformation of the Laplace transform, used in the mathematical analysis of sampled-data systems, based on $z = e^{sT}$, where T is the sampling period. z is a dimensionless complex variable. Tables of standard transforms are available.

In many public buildings, the continuous feedback principle is deliberately avoided in the control of heating. Such a long time-delay would occur between detection of a change of controlled temperature and its correction by a large heating system that the result would rarely be what was intended. Thus it is often better to use what is known as feed-forward from a changing non-controlled parameter to the process which will be affected by it: *eg* an outside thermostat detects a change of heat input and controls internal heating in good time to avoid large fluctuations of inside temperature. (Of course, if a fault develops, a substantial error could occur without the system taking any action because the controlled variable (inside temperature) is not itself measured.) The same principle may apply where it is not practicable to measure the actual controlled variable.

Feedback is found in many everyday applications which are taken for granted; some examples are the control of temperatures by thermostats, the re-filling of a water cistern employing a ball valve and the re-charging of the battery of a road vehicle. A recent very simple application is the automatic adjustment of vehicle brakes to compensate for wear of the linings or disc-pads. One form of this device[6] is shown, in Fig. 2. When a predetermined amount of wear of the disc-pads has occurred, the clearance between the rocking lever and the operating plunger, on releasing the brake, becomes sufficient to permit the spring pawl to engage the next notch in the ratchet wheel. At the following application of the brake, the ratchet wheel is unscrewed slightly from the plunger, thus reducing the clearance. The device is simple because it is required to function in one direction only; a similar principle was employed in Roberts' spinning mule[7] of 1830—so much for the novelty of automatic control!

Types of system

Many of the cheaper systems employ controllers the output of which is on/off, all or nothing. Others, *eg* the automatic domestic gas oven, employ continuous controllers which are capable of providing an output anywhere between zero and maximum. It so happens that elementary control theory applies only to the latter and it also assumes that the system dynamics always obey linear laws, *ie* that output is proportional to the input signal.

It would be wrong to infer from this that the on/off principle is nasty as well as cheap—it is preferred to the continuous principle in certain highly-sensitive systems, *eg* relay-operated position control. Many elements of control systems, besides the presence of an on/off controller, introduce non-linearities. Non-linear theory is therefore required for each piece of equipment that suffers from friction, backlash etc.

Fortunately, some straightforward methods of attack are available. One is the first-harmonic approximation[5] which uses the describing function, a relationship between system output and input which depends on the input amplitude (whereas the transfer function of linear systems depends on the frequency). Another is to use the phase plane technique due to Poincaré:[8] this method is restricted to second-order systems, however.

Most linear control elements are of the analogue variety, *ie* there is (ideally) a continuous relationship between input and output signals. But many of the newer systems work on the digital principle. Here, in effect, the controller is replaced by a digital computer and the output is in the form of a command to take up a definite position. There are three ways in which a digital computer may be used[9] to control plant.

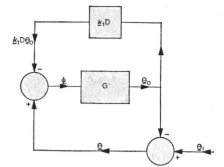

Fig. 3. Linear system with output velocity feedback: this stabilises the system

The first is off-line, where it collects and sifts data which are then supplied to the human (or automatic) controller for use in his own time. The second is on-line, where the computer outputs are used to reset the limits of analogue controllers.[10, 11] The newest method is direct digital control (ddc) in which the computer outputs act directly on the plant.[12, 13] For these (and for sampled-data systems which also involve discontinuous signals), difference equations (rather than differential equations) are needed and it becomes convenient to employ the Z-transformation.[14, 15] Like the Laplace transformation[16] used for linear continuous systems, it converts the system equations to algebraic form, making them simpler to handle. Standard lists of transform pairs are available for both transformations.[14, 16]

So far we have considered control systems in continuous contact with their sensing elements (even if the signals are transmitted in digital form). In certain applications, the measured value of the controlled variable is not available continuously but only at intervals: a well-known example is a radar scanner. Such sampled-data systems[17] are conveniently analysed by use of Z-transforms and have much in common with digital systems which also work in (smaller) discrete steps. Even where continuous data could be available, sampling methods may be employed because they economise in transmission channels and may be cheaper, or offer better performance: for example, a time lag is often troublesome in continuous systems because it easily produces instability but presents little difficulty to sampled-data systems[18] where it may be contained within the sampling cycle.

Much is now being written about self-optimising and adaptive systems.[19] Some knowledge of probability theory and of random processes[20] is a pre-requisite for their understanding. In recent years attention has been concentrated on the application of variational methods (calculus of variations, Pontryagin's maximum principle and Bellman's dynamic programming). Such techniques are sophisticated and, although a happy hunting ground of the mathematician, development of real systems is painfully slow. The relatively simple ideas behind computer control and its undoubted potentiality may herald a fresh approach to self-optimisation. The merits of direct digital control have already been demonstrated by Sutton and Tomlinson[21] in the adaptive control of a Ward-Leonard set subject to a stochastic load torque.

Complex plants have several variables which require controlling concurrently. For example, in an aircraft gas turbine, speed and intake pressure ought to be maintained at optimum values by adjustment of fuel supply rate and of intake geometry. Frequently the various control loops in a plant interact and it may be necessary to compensate for this interaction by cascades of controllers. Theoretically, the total number of such instruments required is the square of the number of controlled variables[22] but a digital computer can replace large numbers of them. One of the latest studies on interaction in distillation columns is reported by Rijnsdorp.[23] It seems fairly clear that such multi-variable systems are favourable targets for computer control.

Compensating a linear system

Suppose that we wish to to control the rotary displacement of a heavy gun of inertia $\mathcal{J}$. Assuming that friction is absent, the torque required at the gun for a displacement θ_0 is $\mathcal{J}D^2\theta_0$, where D is the differential operator with respect to time. This could be supplied by a simple proportional controller which provides torque equal to k times its input signal ψ. Thus:

$$k\psi = \mathcal{J}D^2\theta_0 \text{ or } \theta_0 = \frac{k}{\mathcal{J}D^2}\psi \qquad (1)$$

Here $k/\mathcal{J}D^2$ is the transfer operator of the box marked G in Fig. 3. θ_0 is compared with the desired displacement θ_i in the normal way, yielding the error given by:

$$\theta = \theta_i - \theta_0 \qquad (2)$$

θ is then fed to the input of G to form a closed loop.

Such a system would not be stable as it stands. We must measure the angular velocity of the gun $D\theta_0$ (by means of a tacho-generator, for example) and feed a portion of this signal $k_1D\theta_0$ negatively to the input of G. Hence:

$$\psi = \theta - k_1D\theta_0 \qquad (3)$$

Eliminating ψ and θ_0 from (1), (2) and (3) we obtain:

$$(\mathcal{J}D^2 + kk_1D + k)\theta = (\mathcal{J}D^2 + kk_1D)\theta_i \qquad (4)$$

One of the important requirements of such a position control system is that the gun should follow accurately an input θ_i which is changing at a constant rate, eg $D\theta_i = 1$. In the steady state (when θ has become constant), the left hand side of (4) reduces to $k\theta$ and the right hand side to $kk_1D\theta_i$. Hence:

$$\theta_{ss} = k_1$$

Thus the system of Fig. 3 follows with an error proportional to the speed. Clearly this would be unacceptable if we wished to fire at a moving target.

Now consider a further signal representing the velocity of the desired-value shaft $D\theta_i$ fed positively into the input of G, as shown in Fig. 4. Because of its sign, this new path is called feed-forward. Instead of (3), we have:

$$\psi = \theta - k_1D\theta_0 + k_1D\theta_i \qquad (5)$$

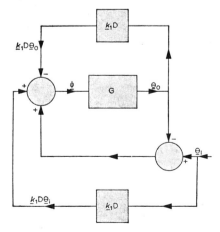

Fig. 4. A system which permits a steady error dependent on speed is clearly not good enough for tracking a moving target; by introducing feed-forward of the input signal rate, the target can be followed closely

Eliminating ψ and θ_0 from (1), (2) and (5), we obtain:

$$(\mathcal{J}D^2 + kk_1D + k)\theta = \mathcal{J}D^2\theta_i \qquad (6)$$

Now, for $D\theta_i = $ constant, we find that $\theta_{ss} = 0$.

This is an elementary application of the technique of compensation. In general, the requirements of stability and of accuracy are conflicting and some form of compensation is often necessary. An alternative method of eliminating a steady-state error, found frequently in process control, is to use a controller generating an additional output proportional to the time integral of the error.

Phase-plane method

A familiar non-linear system is the heating and cooling process in a domestic hot-water tank supplied by an electric heating element controlled by a thermostat. We can use the phase-plane method to predict how the system will behave in various conditions.

Suppose that a 3 kW-heater is used for a tank which suffers a uniform load of 1 kW. Thus, when the heater is switched on, the resultant rate at which energy is supplied to the process is 2 kW; when switched off, it is −1 kW. Suppose too that the heat capacitance of the process is 400 Btu/degF. Given that θ_0 is the actual temperature (°F) of the water, we may write:

$$\left\{\begin{matrix} +2 \\ \text{or} -1 \end{matrix}\right\} = \frac{400 \times 60}{3412}D\theta_0 \qquad (7)$$

depending on whether the heater is on or off (1 kW = 3412 Btu/h). Assuming that the desired temperature θ_i is kept constant, by (2) we may substitute for θ_0 in (7) to obtain:

$$\left\{\begin{matrix} +2 \\ \text{or} -1 \end{matrix}\right\} = -\frac{400 \times 60}{3412}D\theta \qquad (8)$$

So far we have neglected the time lags which inevitably occur. Assume a 5-minute exponential lag in the heat supply reaching or leaving the water; this is the common first order lag which here can be represented by the operator $1/(1 + 5D)$. The left hand side of (8) must be multiplied by this operator; hence:

$$\left\{\begin{matrix} +2 \\ \text{or} -1 \end{matrix}\right\} = -\frac{400 \times 60}{3412}(1 + 5D)D\theta$$

$$\text{or} \left\{\begin{matrix} -2 \\ \text{or} +1 \end{matrix}\right\}0\cdot142 = 5\ddot{\theta} + \dot{\theta} \qquad (9)$$

Equation (9) uses the Newtonian dot notation for time derivatives. In order to apply the phase-plane method to this non-linear application, we re-write the second-order equation (9) as a pair of first-order equations. The first of these is obtained by defining:

$$\phi = \dot{\theta} \qquad (10)$$

Then, since $\ddot{\theta} = \phi \, d\phi/d\theta$, we may write the second first-order equation as:

$$\frac{d\phi}{d\theta} = -\left\{\frac{\phi + 0\cdot284}{5\phi}\right\} \text{ when heater is on}$$

$$\text{and} \frac{d\phi}{d\theta} = \left\{\frac{\phi + 0\cdot142}{5\phi}\right\} \text{ when heater is off} \qquad (11)$$

The phase plane is a rectangular-coordinate plane with axes ϕ and θ. We draw a trajectory on the plane representing the continuous relationship between ϕ and θ with time, ie in this case the relationship between the error in temperature and its rate-of-change with time. The left hand side of (11) means the slope of the trajectory; it is evident that this depends on ϕ and that it changes as the heater is switched on and off. If we specify initial values for ϕ and θ, we can use a well-known construction for drawing the trajectory from (11).

Suppose that, initially, θ is zero and that $\dot{\theta} = \phi = 0.20$ degF/min, that the heater is on, and that there is a constant time delay of 1 min in the operation of the thermostat. This means that, when the temperature θ_0 rises past the set value of θ_i, a minute will elapse before the heater switches off; by (2), this corresponds to θ falling below zero. Thus the heater does not switch off immediately θ goes negative but after θ has overshot the zero value by an amount depending on ϕ, ie on how fast θ is changing. Similarly, the heater will not switch on, when θ is rising, until θ has again overshot the zero value.

The trajectory is shown constructed in Fig. 5, time increasing in the direction of the arrows. After several cycles, the transient converges to a fixed trajectory, so that, in the steady state, the system oscillates continuously according to a fixed curve on the phase plane. This fixed curve represents a limit cycle (in this case stable) and indicates the characteristic mode of operation of the system in the steady state.

A graphical construction, due to Diprose,[8] can be used to estimate elapsed time along the trajectory. For the present example, it is found that the limit cycle has a period of 16 min, one-third of which occurs with the heater on and two-thirds with the heater off; this is not surprising, since the power of the heater is given as three times the steady heat-loss rate. For steady-state operation, the phase plane portrait shows that the actual temperature varies from about 0·15 degF above to about 0·10 degF below the set temperature.

Obviously some simplifying assumptions have been made but, by-and-large, the result agrees with what is commonly observed. If a more costly controller were employed, the temperature could be held nearly constant but this application does not warrant closer control; cyclic behaviour is quite acceptable.

The limit cycle is a phenomenon quite foreign to linear theory, so that a control engineer having knowledge only of linear theory is inadequately equipped. Probably the commonest example of limit cycling is to be found in a clock or watch. If a watch were a linear system, we should have to provide a sinusoidally-varying torque from the energy source to cause the balance wheel to oscillate at a constant amplitude. This desirable behaviour is obtained in practice, however, by providing a constant torque from the mainspring and an escapement end stop.

Often a deliberately introduced non-linearity can improve the performance of an otherwise linear system; Towill[24]

describes such an arrangement which can be made self-optimising.

Sampled-data system

Although a sampled-data system works on discontinuous data, linear continuous theory can be adapted to it. Consider the motion of a body in a viscous medium, controlled by a simple proportional controller. The box marked G in Fig. 6 represents the forward path of the control system which will

have an operator given by: $G = \dfrac{\mu}{\tau D(1 + \tau D)}$ (12)

where τ is a time constant and μ is a dimensionless gain parameter. Suppose that the error is sampled by S to yield θ^*.

The process is illustrated in Fig. 7. It is convenient to regard the sampler as a device which generates a series of impulses (of unit magnitude) at a regular sampling period T. Thus the sampled error θ^* consists of a series of pulses of strength $a_0, a_1, a_2, \ldots$ at time $t = 0, T, 2T, \ldots$. This may be expressed as a linear difference equation:

$$\theta^* = a_0 u'(t) + a_1 u'(t - T) + a_2 u'(t - 2T) + \\ \ldots + a_n u'(t - nT) + \ldots \quad (13)$$

where $u'(t)$ defines a unit impulse occurring at $t = 0$ and hence $u'(t - nT)$ is a similar impulse at $t = nT$. We may

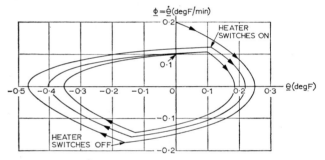

Fig. 5. The limit cycle of an electric water heater: a measure of the steady state fluctuation to be expected from the system

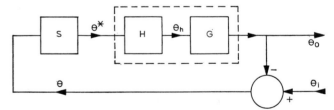

Fig. 6. Block diagram of a sampled-data system: S is the sampler which generates a series of impulses at a regular sampling period and H, the zero-order hold, reconstructs the sampled signal

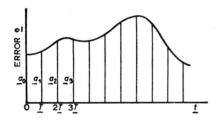
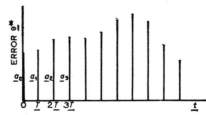
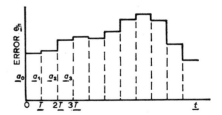

Fig. 7. Sampled data system: the L.H. curve shows the actual error signal sampled at intervals T to produce the instantaneous errors shown in the middle. On the right are the series of square output pulses, produced by the zero-order hold or clamp

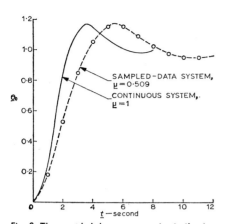

Fig. 8. The sampled-data response is similar in form to that of the equivalent continuous system

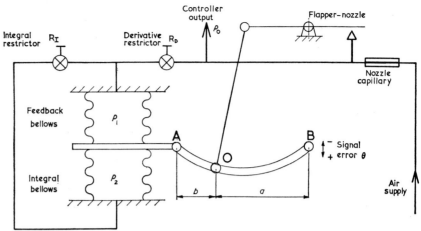

Fig. 9. A typical three-term pneumatic process controller, using bellows to convert both movement and position into air pressure signals

transform (13), term by term, using the Laplace transform to obtain:

$$\Theta^* = a_0 + a_1 e^{-sT} + a_2 e^{-2sT} + \ldots + a_n e^{-nsT} + \ldots \quad (14)$$

where s is the Laplace complex-variable.

For sampled-data systems, it is convenient to define a new variable $z = e^{sT}$; we may then regard the left hand side of (14) as the Z-transform of θ and write:

$$\Theta(z) = Z[\theta] = \Theta^*(s) = a_0 + a_1/z + a_2/z^2 + \ldots + a_n/z^n + \ldots \quad (15)$$

It is most important to realise that $\Theta(z)$ is not the Laplace transform of the continuous function θ but of θ after sampling, ie of θ^*. It describes the sampled error only at the sampling instants and is incapable of yielding information about what occurs between the sampling instants. Provided that Θ, the Laplace transform of θ itself, is a rational function, $\Theta(z)$ can always be written in a closed form.

We shall assume, for this application, that a hold or clamp, H, (of zero-order, for simplicity) follows the sampler (Fig. 6). The purpose of the hold is to reconstruct θ^* into a continuous signal which approximates to the original θ. Fig. 7 shows at the right the output θ_h from the hold which consists of a series of square pulses of width T and height equal to the sampled value at the beginning of each period. Such a zero-order hold has a transfer function given by $(1 - e^{-sT})/s$.

It is convenient to lump H and G together. Since the transfer function of G is obtained by substituting s for D in the operator given by (12), we obtain, putting $\tau = 1$ second:

$$HG(s) = \frac{1 - e^{-sT}}{s} \cdot \frac{\mu}{s(1 + s)}$$

Taking the Z transform and putting $T = \tau = 1$ second, we find that:

$$HG(z) = 0.368\mu \left[\frac{z + 0.717}{(z - 1)(z - 0.368)} \right] \quad (16)$$

Because we have taken the Z-transform of HG, we have automatically taken into account the action of the error sampler.

Whereas an analysis on the s complex-plane might be made for the equivalent continuous system (with the sampler

and hold absent), an exactly equivalent analysis for the sampled-data system may be carried out using the z complex-plane. This would show how the natural mode of the closed loop changes with the gain parameter μ. For example, taking $\mu = 0.509$, the natural frequency is 0.081 s^{-1} and the damping ratio is 0.5. If μ was increased to 2.40, the system would become unstable.

Again, by comparison with linear theory, the relationship between the Z-transforms of the measured and desired values, θ_0 and θ_i, may be written:

$$\Theta_0(z) = \frac{HG(z)}{1 + HG(z)} \Theta_i(z)$$

To find the transient response of this system to a unit step change in input, $\theta_i = 1$, we have to substitute the Z-transform of the unit step function, $z/(z - 1)$ for $\Theta_i(z)$ in the above. Then, substituting for $HG(z)$ from (16) and putting $\mu = 0.509$, we obtain:

$$\Theta_0(z) = \frac{0.181z^2 + 0.135z}{z^3 - 2.18z^2 + 1.68z - 0.503} \quad (17)$$

By long division of the right hand side of (17) we obtain a power series in z^{-1}, thus:

$$\Theta_0(z) = 0z^{-0} + 0.181z^{-1} + 0.530z^{-2} + 0.852z^{-3} + \ldots (18)$$

Each term on the right hand side of (18) may be converted to the time domain, using standard tables, to yield θ_0 values at $t = 0, T, 2T, \ldots$, the sampling instants; the values are the coefficients of the terms in (18). Fig. 8 shows the points plotted up to $t = 11$ second. The points are joined by a dotted line since no information has been yielded, by the above procedure, on the nature of the response between the sampling instants. However, it is evident that the frequency and damping of the system given earlier are confirmed by the shape of the curve.

For comparison, the unit step response of the equivalent continuous system, for the same damping ratio, is shown in Fig. 8 as a full line. For this, μ has to be set at unity and the frequency is $\sqrt{3}/4\pi = 0.138$ s^{-1}. For both systems, the output equals the input (unity) after the transient has faded.

Sampling methods can sometimes be used to overcome practical difficulties which continuous control cannot resolve. Davis[25] investigates a difficult telemetering problem in which phase measurement was made deliberately discrete and he shows the z-plane root locus of his system.

Hardware

One of the commonest conventional process controllers is the three-term instrument. A pneumatic type is shown in Fig. 9 but reliable electronic types are also available. The air pressure at the controller output varies over the standard range 3–15 lbf/in² gauge. The derivative and integral restrictors are conical valves capable of fine adjustment. Under dynamic conditions, air flow through the restrictors causes the pressure levels in the feedback and integral spring bellows, p_1 and p_2 respectively, both to be different from the output pressure p_0. End A of the curved link moves to balance the movement of end B caused by the process error θ. The flapper-nozzle is a high-gain amplifier which permits a pressure drop over the nozzle capillary in order to adjust p_0 over the whole of the standard range with negligible movement of the point O.

With R_I closed, linearising assumptions show that:

$$D p_1 = (p_0 - p_1)/\tau_d \qquad (19)$$

where τ_d is the derivative action time and depends on the setting of R_D. This leads to the controller operating equation:

$$p_0 - P = - k \frac{a}{b} (1 + \tau_d D)\theta \qquad (20)$$

where P is the value of p_0 when θ is zero and k is a constant depending on the compliance of the bellows assembly. In this mode, the controller generates proportional action, the sensitivity of which depends on the setting of a/b, together with derivative action.

With R_I partly open and R_D fully open, $p_1 = p_0$ and we obtain instead of (20):

$$p_0 - P = - k \frac{a}{b}\left(1 + \frac{1}{\tau_i D}\right)\theta \qquad (21)$$

The controller now generates proportional together with integral action, the latter being controlled by R_I which adjusts τ_i, the integral action time.

With both valves partly open, the controller generates three-term action but, due to interaction between the integral and derivative circuits, the time constants are not simply τ_d and τ_i as set on the restrictors. Integral action, though unstabilising, is often used in process control to avoid steady-state error and derivative control is useful whenever there are several time lags in the process.

In practice such pneumatic controllers behave linearly only for small perturbations; an insight into the large signal performance is given by Freeman and Bell.[26]

More precise investigation[27] of the same instrument shows that the compressibility of the air yields instead of (19):

$$D p_1 = (p_0{}^2 - p_1{}^2)/K \qquad (22)$$

where K is a constant depending on the ambient temperature; this equation applies only when p_0 and p_1 are absolute pressures. Comparison of (19) and (22) shows that:

$$\tau_d = \frac{K}{p_0 - p_1} \qquad (23)$$

Thus τ_d is not in fact constant but varies with the pressure level, being 68 per cent greater at the lower end of the standard range compared with the higher.

By complicating the mechanism it is possible to make a pneumatic controller more linear and to reduce the interaction of the terms. However, it is easier to avoid these troubles in an electronic instrument; the controller shown in Fig. 10 works between 0–15 mA for both input and output signals.

Fig. 10. A modern electronic process controller in which the interaction of the three terms is reduced

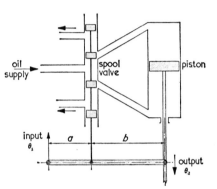

Fig. 11. A simple hydraulic servomotor, controlled through a spool valve: this is really a force amplifier but it also magnifies displacement with an exponential time lag depending on the system parameters

Analogue computers are used in simulation studies of control systems. These instruments, most of which are now electronic, comprise a collection of amplifiers, inverters, integrators, etc, with means for connecting them in circuits to simulate the system together with the usual peripheral devices. It is possible to make the model work faster or slower than the real system. This kind of computer functions in a continuous manner unlike a digital machine where operations are carried out by repeated addition and subtraction of digital data represented by discrete values of a physical quantity. The complexity of the system which can be simulated is restricted solely by the number of components available. Analogue computers are also used on-line, a typical application being the predictor device fitted between the radar and the gun in an anti-aircraft role. Stafford[28] describes the use of such computers to investigate hovercraft behaviour.

An hydraulic or pneumatic servomotor is used frequently as the regulator of a control system. A simple hydraulic servomotor is shown diagrammatically in Fig. 11. By making certain linearising assumptions, we may obtain:

$$\theta_2 = \frac{b}{a}\left(\frac{1}{1 + \tau D}\right)\theta_1 \qquad (24)$$

which relates the input displacement θ_1, requiring negligible force, to the displacement θ_2 of the piston where the force available depends on the oil supply pressure. Equation (24) indicates that the device is also displacement magnifying, according to the ratio a/b, but is subject to an exponential time lag given by τ which depends on the design and operation.

A critical assessment of such a servomotor is given by Lambert and Davies,[29] and the large signal response is covered by Nikiforuk and Westlund.[30] Another paper[31] shows that the synthesis of a pneumatic servomotor can be made with the aid of the root locus technique normally reserved for complete control systems.

Conclusion

This brief review has shown some of the analytical methods

Dr C. R. Webb *graduated with first class honours in Mechanical Engineering at London in 1941 and received two HND prizes from this Institution. He was commissioned in the REME and served in Italy and Yugoslavia on the maintenance of predictors and gunnery control gear. After a period with the Plessey Co, he joined the teaching staff of Queen Mary College, London, where he was awarded his PhD in 1952 for work on pneumatic controllers. Dr Webb was appointed Reader in 1960 and is now responsible for teaching and research in Applied Mechanics and Automatic Control in the Department of Mechanical Engineering. He is Vice-Chairman of the Automatic Control Group of the Institution. He has published many technical articles and is the author of a textbook on automatic control.*

available to control engineers. One of the primary concerns in synthesising a new system or in improving an existing one is the provision of an acceptable degree of stability. A system which oscillates continuously with uncontrolled magnitude whenever a disturbance occurs, is generally useless. The available criteria of stability are numerous; they are all inter-related and which one is used in an analysis depends upon the application.

Control systems are already part of our daily experience and increasingly wider use of control technology is inevitable. The kind of approach in which future general progress will be made is a matter for some speculation. At the third tri-annual Congress of the International Federation of Automatic Control, which will be held in London later this month, the expected large number of papers on self-adaptive analogue systems will be discussed. There is, however, a new emphasis on digital and sampled-data systems. It is already clear that the realisation in practice of self-adaptive analogue systems is difficult and there is an evident reluctance by control engineers to apply the difficult mathematical concepts involved.

The digital approach to self-adaptation is, however, much simpler to conceive and successful attempts to realise such systems are likely to increase in the near future.

In the conventional non-adaptive field, the digital approach is also very promising and the replacement of many analogue systems is not unlikely.

Should there be, in fact, a general move in this direction, much of the fundamental teaching of automatic control (based on transfer function) will become unnecessary and a new approach will have to be evolved.

REFERENCES

1. GOSLING, W. 'What is System Engineering'. June, 1965, *Trans. SIT*, vol. 17, no. 2.
2. MACFARLANE, A. G. J. Engineering Systems Analysis. 1964, Harrap.
3. BSS 1522, *Glossary of Terms used in Automatic Controlling and Regulating Systems*, 1960.
4. *ASA C85-1 Terminology for Automatic Control*, 1963.
5. WEBB, C. R., Automatic Control. 1964, McGraw-Hill, London.
6. 'Girling disc brake'. Nov., 1965, *Autocar*.
7. CATLING, H. *Kinematics and Control of the Winding Process in the Spinning Mule*. 1966, PhD Thesis, London University.
8. WEBB, C. R. Analysis of Non-Linear Control Systems, to be published in *Control*.
9. GRABBE, E. M. Review of Digital Computer Control Applications, June, 1962, *Trans SIT*, vol. 14, no. 2.
10. ROTH, J. F. On-line Computers for Process Control. June, 1962, *Trans SIT* vol. 14, no. 2.
11. HAMMOND, P. H. and BARBER, D. L. A. Use of an On-line Computer to evaluate the Dynamic Response of a Pilot Scale Distillation Column. Sept., 1965, *Trans SIT*, vol. 17, no. 3.
12. THOMPSON, A. Operating Experience with Direct Digital Control. 1964, *Proc. IFAC-IFIP Conf.*
13. *Direct Digital Control*. April 1965, Society of Instrument Technology Symposium.
14. JURY, E. I. *Theory and Application of the Z-transform Method*. 1964, John Wiley.
15. WILLIAMS, B. J., 'Advantages of Z-transforms in System Compensation'. Jan., 1966, *Control*, vol. 10, no. 91.
16. ZADEH, L. A. and DESOER, C. A. *Linear System Theory*. 1963, McGraw-Hill, New York.
17. LINDORFF, D. P. *Theory of Sampled-Data Systems*. 1965, John Wiley.
18. SOLIMAN, J. I. and AL-SHAIKH, A. 'Stability of Discrete Control Systems with Finite Delay. to be published in *Control*.

19. MISHKIN, E. and BRAUN, *Adaptive Control Systems*. 1961, McGraw-Hill, New York.
20. PARKS, P. C. and CLARKSON, B. L. 'Some Theory and Practice of Random Process Analysis'. 1964/65, Convention on *Advances in Automatic Control, Proc. I MechE*, vol. 179, part 3H.
21. SUTTON, R. W. and TOMLINSON, N. R. 'Adaptive Computer Control of a Simulated Second-Order System, 1964/65, Convention on *Advances in Automatic Control, Proc. IMechE*, vol. 179, part 3H.
22. MITCHELL, D. S. and WEBB, C. R. 'Study of Interaction in a Multi-Loop Control System'. 1960, *Proc. IFAC Conf.*
23. RIJNSDORP, J. E. 'Interaction in Two-variable Control Systems for Distillation Columns. 1965, *Automatica*, vol. 1.
24. TOWILL, D. R. 'Performance Advantages of Certain Non-Linear Servomechanisms. Dec., 1965 and Jan., 1966, *Control*, vol. 10, nos 90 and 91.
25. DAVIS, P. G. 'A Sampled-Data Control Example'. Nov., and Dec., 1965, *Control*, vol. 9, nos. 89 and 90.
26. FREEMAN, E. A. and BELL, D. 'Large Signal Responses of Pneumatic Controllers'. Sept., 1964, *Trans. SIT*, vol. 16, no. 3.
27. WEBB, C. R. 'Non-linearities in Pneumatic Controllers'. 1955, *Trans. SIT*, vol. 7, no. 1.
28. STAFFORD, J. 'Use of Analogue Computers as a Control Aid in the Development of Air Cushion Vehicles. June 1963, *Trans. SIT*, vol. 15, no. 1.
29. LAMBERT, T. H. and DAVIES, R. M. 'Investigation of the Response of an Hydraulic Servomechanism with Inertial Load'. Sept., 1963, *J Mech Engg. Sc.*
30. NIKIFORUK, P. N. and WESTLUND, D. R. 'Large Signal Response of a Loaded High-pressure Hydraulic Servomechanism'. 2nd March 1966, paper read, *ProcIMechE*.
31. BURROWS, C. R. and WEBB, C. R. 'Use of Root Locus in the Design of Pneumatic Servomotors'. To be published in *Control*.

The ever-increasing extent to which engineers are able to control plant and processes, often automatically, depends largely on the development of suitable instruments. Great progress has recently been made with digital instruments, suitable for direct input to computers, but the majority of instruments used by engineers are still of the analogue type.

Instrumentation for the Mechanical Engineer

by J. B. Chevallier, CEng, MIMechE

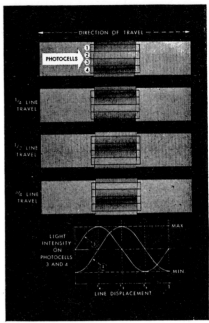

Fig. 1. In the moiré fringe technique two gratings are displaced at a slight angle; with relative movement, the light intensity varies sinusoidally, as shown below. The pairing of four photocells in the pick-up integrates the output

[Courtesy: Machine Tool Research and Devpt]

Instrumentation is a word of many meanings but in the present context it may be defined as the means whereby physical conditions are converted into states which can be discerned by our senses.

The performance of all instruments is assessed by considering five qualities: accuracy, sensitivity, stability, environmental compatability and adaptability. Extremes are reached of accuracy in the measurement of time; of sensitivity in the electron microscope; stability in gyroscopes for inertial navigation; and in compatability with the environment in nuclear reactor circuits. Adaptability is outstanding in the multi-function oscilloscope. In this case, adaptability means a loss of accuracy, sensitivity and stability, although this loss is minimised by the provision for plugging in different units for various purposes.

Recent progress in instrumentation has been directed into two main channels. First, to meet the requirements of measurement, results of which can generally be presented either directly or in recorded form. The requirements of accuracy and sensitivity on the one hand and environmental flexibility on the other, have stimulated much progress in this field. Secondly, instrumentation has been developed to meet the needs of system control. In general, progress in the latter field depends upon progress in the former, since the transducer is likely to be the same in both. It is in the subsequent processing of the information gleaned that the differences lie.

In what follows, the emphasis will be upon the first category since much of what is said will be common to both.

Length

Measurement of length has advanced because the accurate automatic positioning of the workpiece and cutter on a machine tool is fundamental to modern production methods.[1] Both digital (that is to say, counting) and analogue methods are used, the latter implying the conversion of length into a quantity which can be processed more conveniently, such as voltage or even gas pressure.

Early digital methods were based upon the accuracy of linear gratings ruled with, say, 1000 lines per inch and up to about 36 in. long. Counting was done by allowing light, usually from a collimated source behind the grating, to fall through a suitable slit on to a photocell fixed to the moving slide of the machine. This method suffered from two disadvantages: it was comparatively easy to lose the count, especially in the presence of vibration; and the accuracy of reading depended entirely upon the accuracy of each individual step.

More recent improvements of this are based upon moiré fringe techniques.[2] In a typical example, a small linear grating, usually of the same pitch as the main grating, is superimposed upon, but separated from, it by a few thousandths of an inch and is rotated in the parallel plane so that the two rulings make a small angle relative to each other. If the set of gratings is now illuminated as before, a number of light and dark bands will appear as a result of the mutual interference of light waves coming through the spaces between the rulings.

If the lines of the two gratings were parallel, then the passage of one over the other in a direction at right angles to the rulings would result in lines of light varying in width from zero to a maximum (when in phase).

On the other hand, if the lines of the two gratings are at right angles there will be a uniform chequerboard effect which alters only in position and not in size as one grating moves over the other. If an intermediate angle is chosen there will be a combination of both these effects, the chequerboard pattern taking a diamond shape, where the minor angle equals the inclination between the rulings. As one grating moves over the other, the diamonds move also, producing alternate light and dark bands travelling across the light source.

Fig. 1 shows the two gratings displaced at an angle so that there is just one fringe pattern across a significant area of the gratings which is scanned by four photocells. As the smaller grating is moved, the light intensity varies sinusoidally and output from each pair of photocells (1 and 3 and 2 and 4, respectively) are combined to form two sine waves. These signals represent an integration of the effect of many lines so that the omission of a few lines in either grating would only result in loss of contrast and not in entire loss of accuracy.

The signals are processed, usually by observing the four moments at which the curves cut the neutral axis. Two photocells would suffice for satisfactory functioning but by doubling this set, the order in which the cells are illuminated gives a direction-sensitive two phase output and an accuracy of about 0·0005 in overall.

For still finer subdivision it is possible to observe the relative phase of the combined signal with respect to the grating pitch. This finer subdivision is, however, normally only used in measuring instruments and not in the control of machine tools where the machine accuracy would not justify the additional complication. By thus measuring the phase angle, the accuracy can be made to depend only on the mean grating pitch variation over the area scanned and may be five times better than in the simpler method.

Recent methods of producing gratings have overcome the need for close juxtaposition of the two gratings and the limitation is now imposed by the angular aperture of the optical system. Linear gratings of 1000 lines per inch can now be made up to 14 ft long.

Rotating gratings are also in current use, notably for dividing machines and for some torque[3] and tachometric measurements.

For less accurate measurements, where length (or angular measurement) is required in digital form, encoders (or digitisers) may take one of two basic forms. Optical patterns rely upon marks and transparent spaces to interrupt or allow transmitted light to fall upon a series of photo cells, one for each concentric ring. The two states correspond to digits in the binary code and the pattern is arranged so that, for each angular position of the disc, a different binary number is produced.

These encoders may also be discs of insulating material (Fig. 2) coated locally with conducting material in a similar array of marks and spaces.[4] In this case, the marks may all be connected to a common lead and brushes of appropriate sizes 'read' each concentric ring, as did the photocells in the other type. Problems of contact resistance, brush and coating wear arise, but the conducting mark has the advantage of requiring

Fig. 2. To convert angular position into digital output, each circular track on this disc is read by a pick-up placed radially across it. Each angular position is thus automatically converted into a binary output, the value depending on whether the spot under the pick-up is insulating or conducting

[Courtesy: Baldwin Rotax]

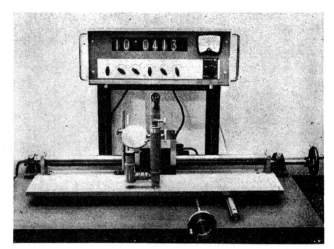

Fig. 3. This Sogenique differential capacitor length measurement rig checks components up to 20 in. long with a reading accuracy of one part in 100 000

little or no subsequent processing of the output before feeding an indicating or computing device. Typical angular converters have an accuracy of 30 min. of arc, but accuracies of ± 2 secs have been achieved.

Other varieties include cylindrical encoders, which are still angular converters but may have a geometrically more convenient form. Linear encoders, similar in principle but developed in parallel arrays of marks and spaces, are also used, especially where the accuracy required does not justify a fringe technique.

Analogue devices

For the standards room laboratory or inspection bay, the analogue device whose output will be directly related to the measured variable is likely to be preferred. Accuracy for accuracy, analogue devices are generally much cheaper and simpler. On the other hand it may be more difficult to use such devices for control and data processing purposes. But new and more useful transducers are marketed almost daily.

The paramount advantage of the analogue device is its ability to contend with wide changes of environment. No one would think of operating a digital encoding device within an atomic reactor vessel.

Accurate laboratory or inspection measurements can be made over lengths up to about 50 in. with a reading accuracy of 1 part in 10^5 and a repeatability of 1 part in 10^6.

Fig. 3 shows a rig incorporating one such instrument which is used to check (though not to calibrate) a precision clock gauge. A 3 Kc/s reference voltage is fed into an auto-transformer, provided with 22 taps. Of these, 20 are connected to individual annual segments of the linear measuring bar, the other two being connected to the end segments. The length of the latter is unimportant, but the length of the remaining segments is controlled so that the pitch, which includes the separating insulation, is 1 in ± 0·0005 in. An interpolating head, which surrounds, but does not contact, the linear bar, comprises two guard rings flanking a central detecting ring more than 1 in. long, so that it always straddles at least two segments. The guard rings serve to straighten the field and to improve accuracy.

The detecting ring of the interpolating head, acting as a

differential capacitor, assumes a voltage related to its position along the measuring bar. This voltage is compared with voltage-dividing transformers, each of which is provided with 10 taps corresponding to the integers 0–9 in the appropriate decade. A microammeter fed from the amplified output of the bridge circuit, indicates when balance is attained. Other examples of accurate measuring instruments for longer lengths use optical principles, eg, the Abbé length measuring instrument and the Zeiss Mirror Magic equipment.[5, 6]

Similar, if a little less accurate, machines have been constructed, based upon inductive and optical principles for obtaining the point of correspondence between the pick-off and the measuring standard. These are more suitable for control systems.

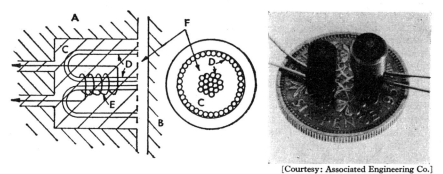

Fig. 4. For small distance measurements variable reluctance heads can be used, small enough to fit into the skirt of a piston or the web of a crankshaft

For small distance measurements, variable reluctance and capacitative transducers are available. The former can be made small enough to fit into the skirt of a piston or the web of a crankshaft, or even the eye of a needle. The principle is shown in Fig. 4, together with photographs of an actual head.

The core wires are bent round individually at the rear of the winding and returned to the front face to form a ring of individual wires, the whole assembly being encapsulated in epoxy resin or ceramic, leaving only the core and ring of wires at one surface. If a high frequency ac is fed to the winding disposed around the central core, an alternating magnetic flux will result between the core and the ring. Now let a metallic plate, or armature, be moved towards the plane surface containing the core and ring; the armature will affect the flux path and therefore alter the reluctance of the coil. Over distances, of say, 0·010 in., the response will be substantially proportional to the distance between the armature and core.

The capacitative transducer may take many forms;[7] one type was used to detect the amplitude of vibration of steam raising tubes, within the reactor circuit of an atomic power station. The great advantage of this type of transducer is that it can be made in virtually any combination of metallic electrode and non-metallic insulants. In the particular case the electrodes, which consisted of the pick-off element (a central disc) and guard ring were of stainless steel and the insulants of alumina, both of which are compatible with nuclear radiation. The purpose of the guard ring is to ensure that the electrostatic field between the central electrode and a mild steel plate, welded to the reactor tube, is linear and that the relationship between capacitance and distance is accurately maintained. The change in capacitance is compared with a standard in a bridge circuit, and the resulting current can be used to indicate distance directly. To indicate vibration amplitude, an integrating or smoothing network is included.

Lasers and holograms

Perhaps the most recent development in the measurement of distance[8, 9] is the laser. This device can be made to emit pulses of light energy over a very narrow band of wavelengths, much like a radar transmitter, so that a measure or range can be obtained over, say, 100 yards to about 1 foot. The disadvantages of this range finder are that it depends upon the maintenance of an unobstructed light path, free from turbulence, and also that direct viewing of the beam may result in serious damage to the eye.

Coherent, monochromatic light from lasers has several interesting properties, one of which at least, concerns the mechanical engineer. This is its suitability for holograms, such as the one shown in Fig. 5. A laser illuminates an object, a polished steel disc for example, and the light reflected falls upon a photographic plate, together with directly incident light. And an exposure is made.

The result is a hologram of the disc and, if the plate is replaced exactly in its former position, a three-dimensional image of the disc will be formed, coincident with the actual disc. If in the meantime, however, the dimensions of the disc have been changed by subjecting it to strain, for example, then the laser light reflected from the hologram towards the disc will travel a distance different from that when the hologram was made. The waves of coherent light will thus no longer be in phase so that interference fringes will result. These cover all illuminated surfaces of the disc and, by counting the fringes at any point, the amount of strain can be accurately ascertained.

This simple principle is now being applied to engineering inspection, measurement of creep, and studies of vibrations, for instance of loudspeaker cones.

In a variation of this technique the photographic plate is subjected to two or more successive exposures before development. In this case the fringes appear on the hologram itself, exactly as they would on the object being examined.

For accurate short-distance measurement pneumatic gauges have the advantage of requiring no mechanical contact. The principle is simple enough;[11, 12] a jet of air is made to impinge upon a surface. The rate of flow through the jet will depend upon the exit pressure at the jet orifice, which, in turn, depends upon the distance between the jet and the surface impinged upon. By fitting an orifice plate upstream of the jet, the pressure drop across it can be measured, hence the flow through the jet, and therefore distance from the measured object, can be determined.

An improvement on the basic principle makes use of a pneumatic bridge for comparing the pressure drop across the exit jet with that across a calibrated control nozzle. Critical pressure orifices may be used and then the flow is virtually independent of the inter-orifice pressure. A nearly linear relationship between rate of flow and impingement distance is achieved.

With the aid of pneumatic amplifiers, the change in pressure drop can be made to operate a chart recorder, with a sensitivity of the order of one micro-inch. Such a system is well fitted for use in high precision inspection equipment designed for production, where the range of distances to be measured will not greatly change: otherwise jets and calibrating scales must be changed for each new range.

Level

Measurement of level or of changes in level is of particular interest to the process engineer. Its basis has changed little in recent years but the difficulties of measuring in some environments, considered insuperable ten years ago, have been overcome. It is still not possible to measure, for example, the level of liquid sodium in a reactor vessel to the same order of accuracy as, say, oil in a tank. Nevertheless, it is now possible to obtain sufficient accuracy at least to indicate not

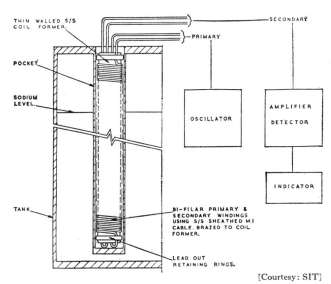

Fig. 6. To overcome the problems of measuring the level of very hot liquids, an HF current is passed through the primary of this instrument; the voltage induced in the secondary coil depends on the level of the fluid which effectively short-circuits the lower turns

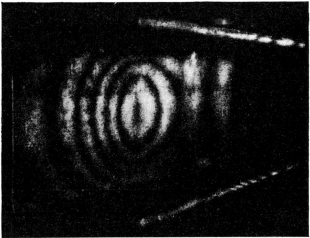

[Courtesy: NPL]

Fig. 5. Interference fringes indicating the differences between holograms of strained and unstrained sections of an RSJ channel clamped across its web

only when things have gone wrong but the trend which precedes that unhappy state. Fig. 6 shows a sketch arrangement of such a level transducer from which it can be seen that the mechanical engineer has much to contribute to the overall design. The difficulty of winding the mineral-insulated cable and brazing it to its former can well be imagined. An HF current passes through the primary and the voltage induced in the secondary depends upon the level of the fluid because the latter acts as a short-circuited turn, so altering the mutual inductance.

Other methods of measuring level include ultrasonics[13] and nuclear absorbtion: probably on economic grounds these have not met with wide application.

An entirely different type of level is used to replace the surveyor's level and, with suitable accessories, the clinometer. This consists of a balance beam, at the ends of which are fitted variable-reluctance transducers, similar to those described earlier. These measure movement of the beam with respect to the base and, by suitable amplification, can be made extremely sensitive. One particular advantage possessed by these instruments is that two can be connected to read differentially and it is thus a simple matter to read directly the relative inclinations of two planes randomly orientated with respect one to the other.

Roundness and surface

Roundness is generally expressed in the form of a diagram indicating departure from true roundness. It is of increasing significance, especially in very-high-speed spindles, hydrodynamic bearings, etc. The various instruments, of which the *Talyrond* is perhaps the best known, depend upon the use of a reference shaft and bearing whose departure from truth may be as little as 5μin. The object whose roundness is to be tested is located as nearly as possible coaxially with the master shaft and bearing.

A transducer is mounted on the reference shaft and a stylus bears upon the edge of the specimen: A very high magnification of stylus movement (and thus of departure from movement of the reference shaft) can be achieved, giving a resolution of a few μin. This is recorded on a circular chart the rotation of which corresponds to the position of the stylus on the test specimen. Where extreme accuracy is required, it may well be necessary to re-adjust the position of the test specimen in the light of the first results obtained.

A field in which measurement of length gives a quantitative description of quality is that of surface texture or surface finish. The need to provide surfaces of known texture for varying applications, such as valve lids and seats, and shafts running in seals, has become increasingly important. In some conditions too fine a texture may actually prove as harmful as too coarse.

Measurement is expressed in μin., or microns and records the variation in the microprofile of the surface due to grinding or turning marks. A complete record of the actual microprofile may be obtained[14] but generally only the departure of the surface from the mean height of the microprofile is required. In this way no account is taken either of the shape of the profile or of macro-errors, such as those in taper or roundness.

One instrument uses a piezo-electric cantilever, to the tip of which is brazed a diamond stylus of 0·0001 in. radius. Movement of the tip across the machining marks on a shaft causes it to bend and, due to piezo-electric effects, generate a voltage proportional to the rate at which it is moved. The

Fig. 7. Four photodetectors can be arranged to give outputs defining the coordinate position of a light source between them; the output of each will vary with the distance from the source. Here such an instrument is set up to measure shaft whirl in two dimensions

cantilever is located in a cell containing silicone fluid which, by damping, reduces the effects of the macro errors. The voltage is integrated to give a signal which corresponds to true profile; and integrated a second time, and further processed, to give a signal proportional to the deviation of the profile from its mean.

Other instruments use variable reluctance or capacitance in place of the piezo-electric effect.

It is usual to provide means for altering the degree of integration so that the cut-off (or highest value registered) may be adjusted to take into account the sensitivity of measurement required.

For very small or very flat surfaces a microscope is used.[15] This displays a magnified image of the interference fringes developed between reflection from a reference surface and from the surface under examination, of monochromatic light. The fringes are one wavelength of light apart and the finish of surfaces as narrow as 0·005 in. has been measured in this way.

Position

There are of course many ways in which position can be established but, for strain-free measurement, an optical method will be best. The simplest arrangement is to use a source of collimated (parallel) light to shine through a suitable lens system on to a mirror, the image from which is reflected to fall on a photocell if the position is correct. More accurate readings are obtained if the light image straddles the boundary of two photocells. A truly central position will give equal outputs from each cell. This arrangement can be made to compensate for variations in light intensity.

A more recent development[16] is a photocell, sensitive in two dimensions capable of detecting position over an area in co-ordinate terms. On the surface of a phototransistor are four contacts in pairs, at opposite corners of a square. If a spot of light falls on the centre of the square, the currents between the four contacts and the base will be equal; if the image is displaced from the centre, the balance will be upset, although the total current in the base connection will remain constant.

This device can readily be adapted for many purposes. In Fig. 7 it indicates the whirl of a shaft in its bearings when rotating at speed. Light from a point source attached to the shaft end forms an image on the photocell. Angular movements (as distinct from radial) of the principal axis of the shaft will give rise to a shift in image position which can be compared with similar movements under static conditions.

Weight and force determination

The mechanical engineer does not often have to measure mass as such but the measurement of weight (or force) is of paramount importance. Whilst weighing is of prehistoric origin significant improvements in the application of the balance principle have been made recently. One is illustrated in Fig. 8. Equilibrium is achieved by balancing the unknown against a constant tare. The advantage is that the same forces act on the points of the balance at zero load as at full load. Constant characteristics of deflection and friction are achieved throughout the weighing range, and these can be largely compensated for in the initial adjustments. This method lends itself more readily to single pan balancing since the taring is all done on the pan of the unknown weight is balanced against the constant load of the tare which can exist unseen. Such balances usually have a digital read-out and the finer fractions may be read from a projection on to a moving optical scale.

An improvement on the constant load balance is the force balance.[17, 18] This consists of a simple mechanical balance plus an electromagnet system which provides the remaining force to complete the balancing. The current in the coils can be varied until a nearly neutral balance is obtained; in some instances only a few milligrams of load remain to be supported by the pivots, thus reducing friction.

By this means it is possible to measure to one part in 10^6 and attempts are being made to increase the sensitivity to one part in 10^8. This sensitivity does not imply an absolute accuracy, for the weight of the tare will probably not be known to the required accuracy; but the weight of, for example, a chemical, added to a previously tared slide, can be ascertained to microgrammes or less.

These devices lend themselves—though at lower accuracy —to automatic weighing[19] and especially checkweighing. Samples may be rejected as high or low, or even sorted into weight categories. One such system is shown in Fig. 9,

[Courtesy: Oertling Ltd]

Fig. 8. By balancing the unknown load against a constant tare, the forces acting on the support are made constant and errors due to friction and deflection can therefore be compensated for in advance

Fig. 9. Fast and sensitive transducers are required for high-speed check-weighing: this equipment automatically rejects underweight packages of tobacco into a bin

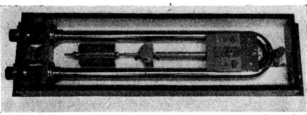

Fig. 10. If the density of the liquid in the tube changes, its weight will no longer be balanced and the deviation of the beam from the horizontal is detected optically. The additional current needed to restore the balance measures the change in density

At the other end of the scale, the load cell has become one of the essentials for measuring the heaviest loads. Many cranes have built-in cells to warn of overload, either static or dynamic, and load cells provide one method of achieving constant 'nip' between a roll and platen, again with automatic control. The transducers may be hydraulic pressure gauges but more recent types measure the distortion of a shape whose behaviour under load can be predicted, usually a ring, the distortion in which is measured by sensitive electric strain gauges.

Thus, in the measurement of weight and indeed of force in general, the parameter actually measured is generally an increment of linear strain, and the methods outlined in connection with length measurement can be used to meet particular requirements.

Density

A method of measuring mass, rather than weight, relies upon the use of a nucleonic source and detector. To a close approximation, certainly better than 1 per cent, it is possible to ascertain, from the absorbtion of beta particles, the mass of material interposed between source and detector. This mass may be considered as a mean density over the area exposed to the beta source.

If, on the other hand, the area is known but the density of thickness is variable, then the process may be inverted and this has been done, for instance, to gauge the amount of corrosion within tanks or the density of a web of paper or cloth.

The great advantage of the method appears to lie in the fact that no contact or stopping of the process is required so that it is readily associated with automatic control.

Measurement of density and especially the continuous recording of a fluid whose density may be changing has achieved importance in the control of many processes and various methods have been tried. One,[20] shown in Fig. 10, is to apply a force-balance to the measurement of a container—conveniently a 'U'-tube—through which the liquid is passed. The U tube is connected to the system through flexible bellows and allowed to pivot about an horizontal axis defined by the crossover of a crossed leaf spring (flexure pivot) type of suspension.

The beam is maintained horizontal by electromagnetic means. If the density of the liquid in the tube changes, the electromagnetic force will differ from the gravitational force and the departure from horizontal can be detected optically. Adjustment of the electromagnetic force can be made to restore balance and the change in current required will indicate the change in density.

If suitable taring is used, the sensitivity and accuracy of the instrument are likely to be governed only by the parameters of the read-out device. Accuracies of 1 part in 10^5 are claimed.

Pressure

In this field again the accent has not been upon new methods so much as on the perfection of existing means. In particular, the development of materials capable of changing their electrical properties under the influence of strain has resulted in small transducers with extremely fast response, which allow accurate measurements to be made of high-speed transient pressure rises, such as occur in explosive forming or hydraulic control systems.

At the other end of the scale emphasis has been placed upon the highest static accuracy, for example, the provision of

weighing a product familiar to all. As each packet passes over the weighbridge, it is photo-electrically detected and, at the appropriate positions, the packet and that part of the belt which is in contact with the bridge, is weighed by a force-balance which uses a photo-electric system to detect departure from a datum. Balance is restored by changing the current in the force balance coil and this current is also compared with preset limits. If the limits are exceeded, an electro-pneumatic system forcefully moves the defective packet into a reject chute, whilst allowing those within tolerance to proceed.

Checkweighing of more bulky and, particularly, more slow-moving items may be done by electric strain gauge and amplifier. Such an instrument will have neither the sensitivity nor the stability of the more complex force balance.

master gauges or aneroid capsules for use in the upper atmosphere.

In these cases particular care has to be taken in the selection of materials and the design of the instrument to obtain high sensitivity with both long-term stability and minimum hysteresis. Such a gauge will detect the air pressure difference between bench height and floor at ground level.

Regarding material stability, the main problem is to ensure full precipitation in the alloys used for diaphragms, either by prolonged or artificial ageing. The hysteresis is reduced by use of low friction plastics and special mechanisms of particularly low inertia.

A further development in this field is the servomanometer[21] which can measure changes in head of liquid, usually mercury, or water, to a resolution of 1 in 2000 in., typically over the range 0–60 in. Hg. The principle does not differ from that of the conventional manometer, but the method of measurement is advanced. A capacitance probe detects the level of the fluid in the tube and any change causes an unbalance in a servo circuit which drives the capacitance probe until it achieves a new zero position. At the same time the distance moved, and therefore the change in head, is measured. This method of reading is independent of the meniscus shape, provided that this is constant.

Humidity

Moisture control[22] is one of the applications of a new principle, nuclear magnetic resonance. Briefly, the sample is suspended within a very uniform and intense magnetic field produced by a coil surrounding the sample chamber. A signal of variable frequency is fed to the coil with, and without, the sample in place. The relationship between frequency and current shows marked depressions, indicating resonance, but at frequencies differing for varying substances. In particular it is comparatively simple to distinguish the presence of liquid-borne hydrogen atoms from solid-borne ones and for most practical purposes this means the presence of water. By calibration it is possible to indicate to, say, 1 per cent the quantity of water present in the sample, in a matter of a few seconds.

At the opposite end of the scale, the moisture content of air in ovens may be measured by observing its absorbtion of microwaves.

A simpler method involves the use of critical flow orifices and the measurement of the pressure drop associated with a known temperature drop between the inputs to two such orifices in series. If the humid atmosphere is pumped through the first orifice into a cooling chamber or condenser operating at constant temperature, then the exit pressure in it will be related to the moisture content of the incoming gas.

Measurement of time

Time is the basis of all measurement of rate, acceleration or higher derivatives. Recently significant improvements have been made in the accuracy of time measurement; first the advent of the quartz crystal vibrating in a closely controlled temperature environment, and subsequently the more powerful counterpart of atomic resonance. Quartz clocks are now available in a portable form, consuming 250 mA at 4V to give a service life between resetting, of 15 months. They are used as marine chronometers, *etc*, with an accuracy of 1 part in 10^6.

The *Accutron* watch is based upon the electric oscillation at 360 c/s of a small tuning fork. Being essentially mechanical,

Mr J. B. Chevallier *was educated at Sherborne School and at the Royal Naval Engineering College. He specialised in Advanced Ordnance Engineering at the Royal Naval College, Greenwich, and subsequently worked for the Director General of Weapons on the design of gunnery and missile fire control, and missile handling gear. He has also been Chief Engineer of a destroyer. In 1961, he joined Palmer Aero Products as special project engineer for test equipment. He has served as Head of the Engineering Department of Sira Institute, and now works in the Space Division of Hawker Siddeley Dynamics.*

this instrument suffers from positional inaccuracies but, since these are caused by the effect of gravity on the vibration frequencies of the fork, they can be predetermined. They are of the order of ± 5 s per day, according to the orientation of the watch. An accuracy of 1 part in 4×10^6 is guaranteed, which is likely to be ample for the requirements of most mechanical engineers.

Finally there is the atomic clock. From a suitable source caesium atoms are passed successively through a collimating magnetic field, before falling upon an electrode connected to an electrometer. If the frequency of the RF field is varied, a value can be found which corresponds to the transition frequency of the two atomic states of caesium. At this point a proportion of atoms undergo transition, and it can be arranged that they undergo magnetic deflection in the opposite sense from those which have not undergone transition. This results in a sharp drop in the beam current registered by an electrometer at the atomic frequency. An accuracy has been achieved of 1 part in 6×10^{10}, corresponding to an error of one second in 1800 years!

These clocks or simpler variants are not only of use in providing high-speed timing intervals of high repetition accuracy but can also be used to drive elapsed-time indicators or recorders which are becoming increasingly important for recording events, especially separations and sequences in time. MIDAS and similar aircraft event recording systems are examples.

Measurements of rate or acceleration[23] are derived by successive differentiation, often from data on high-speed photographs, showing changes in position of the object under investigation with respect to a reference, defined either by an object in the photograph[24] or, in some cases, by the frame.

Electron microscopes

The family of instruments which may be classified under this heading[25] is of rapidly increasing significance to engineers, although some 10 years ago only a few research laboratories had them. By now, however, not only have the basic microscopes become tools in development but the family has proliferated into such instruments as the electron probe micro-analyser,[26] the scanning micro-analyser[27] and the scanning electron microscope.[27]

The micro-analysing instruments are of special significance to the metallurgist for they proved to be a means for distinguishing the compositions of, for example, two adjacent crystals in steel.

A very fine electron beam is made to strike the surface under examination. This bombardment causes the emission of X-rays of wavelengths corresponding to the particular elements present (but not those below atomic number 6) which can be detected and analysed in an X-ray spectrometer. Quantitative examination is also possible if specimens of the pure elements present are mounted for comparison.

The usefulness of the micro-probe analyser can be extended if the beam passes through a scanning field similar to that used to move the spot on a television screen. The X-ray spectrometer can then be tuned to one frequency and a picture is built up of the distribution of that element over the scanned area. By retuning with a different frequency, successive pictures can be obtained to give the distribution in the scanned area of all detectable elements.

In the same fashion an electron microscope can be adapted to scan, the scintillation being detected by a counter. The output of this counter is used to modulate the beam intensity of a cathode ray tube, the beam being deflected to synchronise with the scanning raster. This mode of operation is particularly useful when solid specimens are to be observed directly instead of through replicas. Fig. 11 shows specimens which would be almost impossible to reproduce in the form of replicas but the scanning electron microscope can readily handle, at such a magnification that an optical microscope could not provide the depth of focus required.

Conclusion

The past decade has seen many advances in instrumentation and here I have tried to highlight some examples of particular interest to mechanical engineers. It is quite certain that the pace of advance will increase during the next decade. Scientists will demand greater accuracies and resolutions. The detection of new fundamental particles, perhaps with even shorter half lives, will certainly involve the extension of instrumentation beyond present capabilities.

The prospect of MHD generation of electricity will raise problems of environmental compatability which may dwarf those encountered in nuclear reactors. Electron microscopy will be required increasingly to determine the physical and mechanical structures of materials, and the mere handling problems associated with the presentation to the instrument of minute specimens will tax the ingenuity of the instrument engineer to the full.

In more general fields, it is likely that emphasis will be placed upon transducers with built-in digital conversion[28, 29] for direct access to computers. Instrumentation is now being developed with the help of solid state devices for the control of power by switching during the low voltage part of a cycle only. This will be applied to furnace control, welding processes and many feedback control systems.

Finally, the requirements for on-line quality control, whether of surface finish, density, shape or any other parameter essential to the product, will promote the development of new and more versatile, largely automatic systems of non-contact gauging.

Such progress is realised only by the closest scrutiny of all available methods, ranging over all disciplines and sciences. But the instrument engineer will still be left with the problem of ensuring that a solution is attained which is both technically sound and economically feasible.

REFERENCES

1. PUCKLE, O. S., and ARROWSMITH, J. R., *An Introduction to Numerical Control of Machine Tools*. 1964, Chapman and Hall.
2. GUILD, J., *Diffraction Gratings as Measuring Scales: a Practical Guide to the Metrological Use of Moiré Fringes*. 1960, Oxford University Press.
3. 'Designing Moiré Fringe Torque Transducers'. September 1966, *IBM Journal Res. Dev.*, vol. 10, pp. 412–415.
4. 'Contact Shaft Encoders'. July 1966, *Ind. Control Engg*, pp. 6–13.
5. KREIG, W., and WOSCHINI, H. G., 'Abbé Length Measuring Instrument'. April 1966, *DDR Feingeratae technik*, vol. 6, pp. 160–161.
6. 'Zeiss Mirror Magic Measuring and Precision Indicating Equipment'. 31st August 1966, *Machinery*, vol. 109, pp. 474–476.
7. LION, K., 'Capacitance Transducers'. June 1966, *Inst. Control Syst.* (USA), vol. 39, pp. 157–159.
8. ROULEY, W., and STANLEY, V., 'Lasers Applied to Automatic Scale Measurement'. December 1965, *Instr. Pract.*, vol. 19, pp. 1106–1107.
9. THOMPSON, J. B., 'Applications of Lasers'. July 1966, *Electron Equip. News*, vol. 8, pp. 70–77.
10. DICKINSON, A., and DYE, M. S., 'Principles and Practice of Holography'. February 1967, *Wireless World*, vol. 73, pp. 55–61.
11. 'The Dimensionair Differential System of Air Gauging'. 8th December 1965, *Machinery*, vol. 107, pp. 1263–1265.
12. MORGAN, I. G., and TOLMAN, F. R., 'Pneumatic Comparator for Measuring Reference Ring Gauges'. 22nd June 1966, *Machinery and Production Engineering*, vol. 108, pp. 1380–1389.
13. 'Steam Boiler Water Level Gauge'. May 1966, *Control*, vol. 10, p. 257.
14. OLSEN, K. V., 'The Standardisation of Surface Roughness Measurement'. 1961, *Bruel and Kjoer Technical Review*, no. 3.
15. DEHMEL, T. K., 'Optical Flatness Measurement'. June 1966, *Inst. Control Syst.* (USA), vol. 39, pp. 123–124.
16. WILLIAMS, T. L., 'Electro-optical Measurement and Detection of Small Displacements by a Light Spot'. March 1966, *Instr. Review*, vol. 13, pp. 85–90; also April 1966, pp. 150–152; May 1966, pp. 189–192.
17. BLAYDON, H. E., 'A Versatile Electro-magnetic Balance'. May 1966, *J. Scient. Instr.*, vol. 43, pp. 335–338.
18. CARSON, R. W., 'A Torsion Suspension Balance'. 10th October 1966, *Product Engg* (USA), vol. 37, pp 88–90.
19. ETIEME, R., 'New Answers to Continuous Weighing'. June 1966, *Mesures. Régulation Automatisme*, vol. 31, pp. 71–78.
20. WILLIS, J. E. d'A., 'Gravity Master in Petroleum Industry Experiments'. January 1966, *Ind. Electronics*, vol. 4, pp. 6–9.
21. MAIZEL, M. B., 'An Automatic Two Liquid Micromanometer'. Feb. 1965, *Instrument Construction*, pp. 8–10.
22. 'Humidity Measurement'. December 1966, *Inst. Control Engg*, pp. 32–42.
23. VON VICK, G., 'Accelerometers'. November 1965, *Inst. Control Systems* (USA), vol. 38, pp. 86–87.
24. HYZER, W. G., 'Introduction to High Speed Photographic Instrumentation'. January 1966, *Trans Instrmt Soc. Am.*, vol. 5, pp. 1–4.
25. *A Simplified Guide to the Principles and Practice of Industrial Electron Microscopy*. BSIRA, Chislehurst, Kent.
26. ADAMS, R. V., 'The Use of the Electron Probe X-ray Microanalyser for the Identification of Inhomogeneity in Glass'. June 1966, *Glass Technology*, vol. 7, pp. 98–105.
27. OATLEY, C. W., 'New Electron Probe Instruments'. August 1966, *Electronics Power*, vol. 12, pp. 282–285.
28. 'Digital Tachometer for Measuring Transient Speed'. 22nd April 1966, *The Engineer*, vol. 221, pp. 617–618.
29. SERRA, G. F., 'Torque Balance Principle Raises Accuracy of Digital Pressure Transducer'. August 1966, *Trans. Instrumt Soc. Am.*, vol. 13, pp. 51–54.

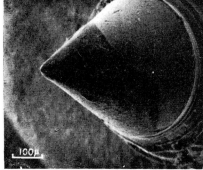

[Courtesy: Cambridge Instrument Co.]

Fig. 11. Stereoscan micrographs of a millivoltmeter pivot (× 580) and wool fibres (× 110) made with the aid of an electron microscope

Fig. 1. Digital machines have much greater storage capacities than analogue computers: the commonest type of storage is on high-speed magnetic tape reels which can be read at thousands of bits per second

Computer Application in Mechanical Engineering

by C. R. Edwards and B. P. Jeremiah

The value of electronic computers has been fully realised during the last decade; their very existence dates only from the 1940s. Today science, industry and commerce are using them as almost indispensable tools and mechanical engineering is no exception.

The fundamental difference[1] between analogue and digital computers is the way in which quantities are represented inside each machine. The analogue machine usually represents them as voltages which are measured while the digital machine represents quantities by a series of discrete steps—*eg* pulses which are counted. The two approaches are analogous to graphical and arithmetical solutions of mathematical problems respectively with broadly similar advantages and drawbacks.

Analogue versus digital

The digital machines have much greater storage capabilities, a good example of which is shown in Fig. 1, since it is easier to store coded numbers than voltages. Although both machines can carry out basic arithmetical operations, only the digital machine can be entirely controlled from within by storing its instructions just as it stores its other data. An analogue machine is more like a physical model in which parameter changes can be effected by the experimenter. Hence the digital machine is more accurate, more flexible— and usually more expensive. The analogue machine is used where great flexibility of parameters is not required and where its accuracy (usually some 5 per cent) is good enough.

Digital electronic machines are of various types, best distinguished by certain key characteristics such as the speed of basic operations; the size and type of storage; the type of input/output channels; and special facilities such as time sharing or interruption (*ie:* a priority can interrupt operations already in progress and many slower input/output devices can share one fast processing centre).

In addition there are important physical characteristics, notably size. Computers are now being built with such small components that the overall size, and therefore the lengths of circuits, are much reduced. Since operation times are of the order of microseconds, these shorter circuits actually increase the speed of the machine. Fig. 2 shows the central processor and core-storage of the Argus 400. It fits into a suitcase and is comparable with machines which would fill a room.

Nowadays building-block systems are being offered; the user may start his installation with a relatively small central arithmetic unit and increase its capability as required by adding more storage, input and output facilities—and he can also add additional interlinked arithmetic units, as required.

On a digital computer the solution of one problem may thus rapidly follow another; all that is needed is a new control programme with its attendant data. This contrasts with the analogue machine (Fig. 3) which has to be set up for each job by stringing together the appropriate electronic components. In practice the difference is the same as between a universal machine tool which, with its jigs, fixtures and control cams, may produce a wide range of products, and a series of more specialised machines which can make the same products. In both cases the economics will largely depend on the length of run of similar or identical operations. Digital machines have now been developed to such a stage that they can cope with much larger and more complex problems than their analogue equivalents. Peripheral equipment includes such input devices as card and tape readers and corresponding recording equipment as well as printers and auxiliary storage devices.

However, when a machine is used 'on line' (*ie:* permanently connected to other processes) the input may well consist of signals from a variety of devices, including special measuring transducers and the output signals may control pieces of plant.

It is not necessary or, in fact, always desirable, actually to purchase or rent a computer in order to use one. Computer manufacturers have their own data processing centres[2] where it is possible to buy time when required. This system is very attractive to companies who require large fast machines for a relatively short proportion of the available time.

Programming

The arithmetic unit requires instruction in very small steps, coded in binary numbers. However, the user who is not a trained programmer is now greatly assisted by the development of so-called automatic programming or 'computer

Fig. 2. Printed circuits and miniature semiconductors permit the construction of digital computers of suitcase size: not long ago this central processor and core storage might have filled a room ◀

Fig. 3. This 156 amplifier analogue computer has been designed for scientific and engineering calculations. ▶

[Courtesy: Ferranti]

[Courtesy: EAI]

languages' which enable him to use a limited number of ordinary words such as 'add' or 'stop'. These languages are of two types: interpreter or compiler. A programme written in an interpreter language is held in store virtually unchanged until it is required. Each part is then translated piecemeal into binary numbers by the computer itself. A compiler language, on the other hand, is completely translated into the basic computer language before it is required for running. This type of language is becoming more popular because it offers greater scope to the user and makes more efficient use of the machine.

Three examples of compilers:[3] *Fortran, Algol, and Cobol, Fortran* and *Algol* are scientifically slanted languages and include a large number of mathematical instructions. *Algol* is probably rather more efficient and offers a little more scope. However, few machines have the necessary equipment to accommodate both languages: the choice of language is usually made by choosing the machine manufacturer. *Cobol* is a language devised for commercial use. It suffers from trying to be too general and is never as efficient as the basic machine language. While this is true of most compilers, it is a particular shortcoming of those devised for commercial work.

Since commercial computer programmes are used many times, it usually pays to spend the extra time and effort needed to write them in the machine language, thus saving valuable machine time devoted to translation. Scientific programmes, on the other hand, tend to have rather shorter runs and therefore favour computer languages. There are also special-purpose 'languages', structured for the solution of a particular class of problem; *eg:* linear programming, critical path analysis or simulation.[4]

In many sections of mechanical engineering the demand on the designer for more efficient equipment, particularly prime movers, is resulting in design becoming much more an exact science than an empirical one. In developing plant towards the limit of performance, many more factors have to be taken into account and the theory has to be understood. It is no longer sufficient merely to make sure that each unit has the required capacity with due allowance for thermal losses, stresses, critical speeds, etc. The designer must aim at the optimum compromise between numerous requirements which can only be achieved by carrying out the calculations many times with different parameters on each occasion. For example, a tremendous amount of data is required for boiler calculations. Heat transfer, temperature gradients, flow characteristics and stresses in tubes, all must be fully analysed and related to each other for the optimum design. Taking the designer's problem a stage further, effective control systems for complex equipment demand an analytical approach.

On the other hand, the design of single components, such as crankshafts, gears etc, for which there are well established and accurate procedures, does not tax the engineer's ability as a creative designer but merely takes up a lot of his valuable time in repetitive and simple calculations.

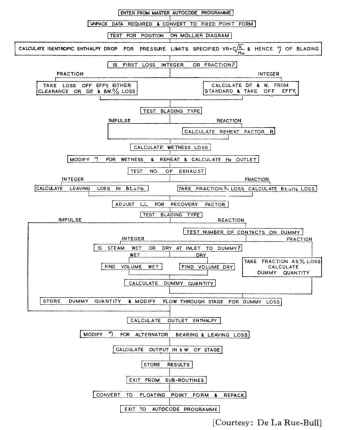

[Courtesy: De La Rue-Bull]

Fig. 4. Before a programme can be compiled, the procedure must be set out in a logical flow diagram: here is a turbine stage efficiency calculation

It is in these two categories of design that computers have been exploited most effectively because of their enormous speed. The designer cannot be replaced by a computer but, provided he knows exactly what it is he wants to do and can programme the computer accordingly, then, however long, tedious or complex the path to the solution, the machine can give an answer in minutes which would otherwise take days.

One of the main virtues of the analogue machine is its versatility as a system simulator. That is, by making an electronic model of a system, the effect of changed components on the system as a whole can be instantly determined.

In both cases, but particularly in that of the digital computer, the difficulty of programming must not be underestimated despite the introduction of simple 'languages' such as *Fortran* and *Algol*.[5] The designer is still required to understand the problem so that he, or a specialist programmer, can convert a logical procedure diagram, such as the one in Fig. 4, into a programme. This entails reducing the problem to the level at which the computer, with its limited number of functions, can handle it. However, once there is a programme for the general case, all similar cases can be solved by substituting new data for old.

There is a shortage of engineer-programmers which has prompted many companies to train their own engineering staff to programme problems. Ford (of Britain) for instance is training 200 of its engineers (minimum qualifications HNC) to make full use of a newly acquired digital computer.

System design

The design of commercial vehicle braking systems was formerly done on the basis of experience, testing and time consuming calculation. An American company[6] uses a digital computer to design its systems much faster while ensuring in each case the optimum compromise, taking into account all the variables. The programme flow chart, shown in Fig. 5, is designed to simulate the cumulative effect on the braking system of a large number of acceleration, cruising and braking cycles. For each cycle the computer calculates the variation in drum temperature, the self-energising effect of each brake, work done, and the wear on each lining, taking into account the effects of the lining coefficient of friction varying with temperature. These are the main parameters

which determine whether a vehicle will come to rest under controlled conditions during a normal or emergency stop, or will suffer from brake fade and differential lining wear.

The computer can carry out these calculations because formulae are available for each of the parameters so that, if the details of the vehicle, are put in the computer store together with a detailed list of instructions, then the values of the parameters for each cycle can be calculated.

The data for a four-axle vehicle (tractor and trailer), when analysed could, for example, indicate that the rear axle brakes become very hot or that the front axle brakes lock. Following the redesign of these brakes, it may be shown that one pair wear twice as fast as another or that the braking force on the trailer axles is higher than on the tractor.

In short, the designer can experiment with the mathematical model rather than the vehicle, until he reaches the best compromise.

Where the configuration of the system is relatively stable, analogue computers can be used with advantage. The problem may be to investigate the pressure variations in a gas pipe line resulting from the operation of a compressor situated part of the way along its length. The pipeline is some 193 km long and, therefore, a knowledge of the transient conditions along it, arising when variations in demand occur or when the compressor cuts in, is very necessary.

The method of approach is as follows:[7] a series of equations is drawn up which links together the variables applicable to fluid flow in pipes. By substitution, two non-linear differential equations are obtained, one relating pressure, length of pipe and volume flowing and the other relating pressure, time, volume and length. These equations describe the flow of the gas in the pipeline under steady or transient conditions. In order to use the analogue machine, the actual pipeline must be considered to be made up of several increments. The flows and pressures at each increment can then be considered and the computer can integrate these to give the overall conditions.

The conditions in each increment are found by solving a set of equations (one pair for each increment derived from the basic pair) by the use of 'transfer elements' which are electric circuits analogous to the equations, as shown diagrammatically in Fig. 6. To simulate the pipeline, a series of these transfer elements, including one for the compressor, are electrically connected, with switches to simulate the valves at the compressor and delivery ends of the line.

Initially, with the compressor and delivery valves closed, the line will be charged at bore hole pressure and the potential to the transfer elements is set accordingly. If the delivery valve is now opened, the line pressure will immediately start to vary along its length until a steady state is reached. The

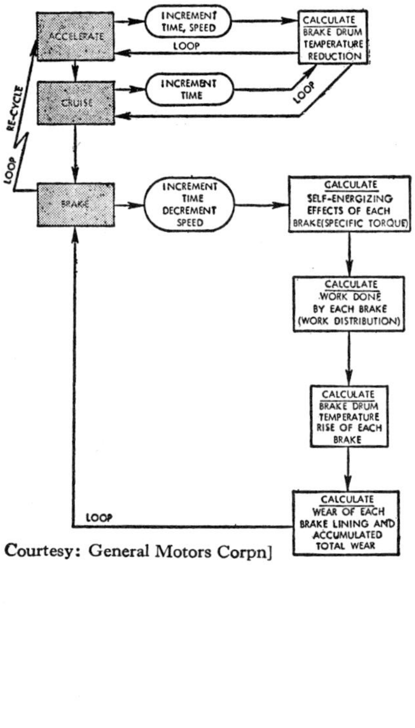

Fig. 5. Programmed in accordance with this flow sheet the computer evaluates vehicle braking systems

◀

Courtesy: General Motors Corpn]

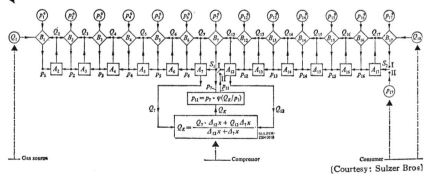

[Courtesy: Sulzer Bros]

Fig. 6. Each section of a pipeline is simulated with the aid of analogue elements, as are the valves and compressors: the effect of load fluctuations, etc, is then easily seen

computer will register the pressures and volumes at each increment of length and the transition curves can be plotted. The transition time, incidentally, is about 3·5 h.

By closing the switch which simulates cutting the compressor in, another transient period can be noted and curves of pressure variation plotted. Automatic curve plotters are available to work straight from the computer. To obtain these data at the design stage by conventional methods would be a very lengthy business and open to error.

In the design of automatic control systems computers are particularly valuable aids because they make it possible to simulate the projected system and test its stability under various conditions mathematically, rather than by building the actual hardware.[8, 9]

As another instance, the design of nuclear reactor concrete shielding requires a precise knowledge of the heat absorption by cooling pipes.[7] Computers are used to carry out the heat conduction calculations necessary for any configuration of cooling water pipes.

In the design of pumped storage schemes[10] a great deal of attention has to be given to calculating the effects that a power failure during the pumping cycle has on pressure distribution in the system. A badly designed system could cause severe water hammer problems when the main valves are closed, following a sudden reversal of water flow in the mains. The calculation of the pressure variations following a reversal of flow are complex and take into account many factors, such as the inertia of rotating parts, surge tank levels, junctions, valves, etc.

The programme for computer solution simulates the problem mathematically from a knowledge of pressure wave propagation and the hydraulic properties of all the component parts of the system, including the closing characteristics of the main valves. It then calculates values of pressure at key points in the system for small increments of time, until steady conditions are re-established. In this way several layouts and valve programmes can be evaluated to give optimum conditions very quickly.

Detail design

The design of a cam entails calculating the co-ordinates, velocities, accelerations and pressure angles at a number of points round its periphery. The work is tedious and, of course, repetitive. Because it is basically a series of similar calculations, it is readily done on a computer. At every angular increment round the cam the machine can compute the new values of the variables, because the programme will be written in such a way that the various formulae connecting the variables will have been reduced to a long-winded series of simple arithmetic. It is then only necessary to arrange for the computer to repeat this for each increment.

It is in this particular field of component design that computer manufacturers have been able to simplify data preparation to the extent that a company requiring a cam design merely fills in a form the basic information of radii, type of motion, rev/min, etc, sends it to the processing centre and receives back all the required information for any number of increments. Similar services[11] are available for calculating engine crankshaft loading, helical and hypoid gearing, prediction of critical speeds in shafts, pipe system stress analysis and many others in all branches of engineering.

These examples were chosen because they are relatively simple ones and show possible ways of utilising computers. Much bigger problems are of course being solved and reference to some published work is included in the bibliography. For obvious reasons a lot of information is not available in published form, but from correspondence with various companies it is apparent that computers are aiding designers in many diverse ways.

For example, computers are used in the stress analysis of steam and gas turbine wheels[12] and in the design of machine tool hydrostatic bearing systems.[13] Also the execution of rotor critical speed calculations[14] and complex gear design problems[15] are being made more straightforward with the aid of suitable computer programmes.

Boiler calculations, including the stress analysis of the tubes under non-uniform heating conditions, are now computerised. The main components of IC engines and compressors are being designed by computer, including the evaluation of stresses in the crankshaft resulting from torsional vibrations. One company which makes radial compressors uses a computer to optimise its designs for the widest possible duty range, maximum efficiency and minimum first cost for any particular specification.

In the car industry recent work[16] has proved the feasibility of using a digital computer for the stress analysis of an underbody pressing. This rather complex structure was considered to be made up from a series of interconnected beams and panels. By using a standard structure analysis programme, the stresses in each element can be calculated in terms of the deflections at the intersection of the elements. It involves the solving of several hundreds of simultaneous equations, the number depending on the number of elements into which the structure is divided. Comparison of computer results with measurements taken from an actual pressing, subjected to the same loading, was very favourable and justifies employing the technique as a design aid.

When one pauses to consider how much manual work is necessary for calculations of this sort one realises that the time and money involved in preparing a computer programme are well worthwhile, provided it is a recurring problem.

Computers can also help the designer in development work. They can be readily programmed to analyse test results, fed direct from a rig, and print out results, suitable for assessing a design. In this way parameters can be changed in the light of these results while the experiment is still set up, saving much valuable time.

Data collected from road vehicles[17] in analogue form can be later converted to digital form and be processed on a computer, thus yielding valuable information for improving or evaluating current models. Alternatively, data collected from an actual test run, sometimes by wireless transmission, can be used to control a test rig set to prove a new design. The speed with which comprehensive test results are made available permits a much more detailed and rigorous approach to development work.

Looking to the future, advances both in this country and the USA seem to foreshadow the automation of all but the creative aspects of design.[18] Equipment has been designed that automatically transfers a designer's work direct from drawing board to the computer where it can be processed into control tapes for numerically controlled machine tools.

One of the most impressive applications of this trend[19] is the manufacture of experimental turbine blades which previously took many weeks to produce, requiring detail drawings, templates, plaster work and then copy milling.

Machine tools

It is now possible for an experimental turbine blade to be produced by a numerically controlled milling machine, the control tape having been prepared from a minimum of design information.[19] Briefly, the designer specifies, say, nine sections through the blade in the form of easily defined arcs and straight lines. Suitably programmed, the computer then converts this information to co-ordinate data and can check these against known mechanical, aerodynamic and thermo-dynamic requirements. If the proposed sections are satis-factory the computer calculates intermediate sections by interpolation. Thus we get on punched tape a complete, three-dimensional numerical equivalent of the blade which can be fed to the numerically controlled milling machine.

Obviously the applications of this technique are numerous where the intermediate drawing work between a basic design and the finished article is extensive, as is the case with manufacturing motor car body dies, for example.

This example was chosen to indicate how the computer is used to assist the designer but it has also served to introduce computer service to production. Numerically controlled tools have a much wider application than the production of one-off components,[20] as was shown at the Machine Tool and Production Engineering Show at Chicago in 1966[21] where a large percentage of new machines were fitted with this facility. With the aid of a programme-controlled turret the machine can replace a whole transfer line in performing normal machining processes with some 60 tools. Numerical control is now applied also to the whole range of conventional machine tools as was seen in the first British exhibition of this kind earlier this year; and also to a numerically controlled measuring machine which will inspect components up to 5 ft × 3 ft × 2 ft and print out or display the dimensions.

It is possible to feed to a digital computer certain basic geometrical information for processing and conversion to a control tape that will operate the machine tool servos. Secondly, from the detail drawing of, say, an engine block, it is possible to draw up a machine programme and arrange for the computer to convert this to the instructions for con-tinuous motions of the machine tool in three axes, with due allowances for tool sizes. Here again, the user need not buy his own computer but can hire time in a service centre. PERA have produced a report on the economics of numerical control.[22]

Computers are not only used for design and control of machine tools but also for investigating certain of their basic characteristics. Birmingham University have been investigating chatter: this can be regarded as a three-phase exercise where data from actual cutting experiments is analysed, chatter hypotheses are developed, and theory is then tested by imposing certain forced vibrations back on to the machine tool, as shown in Fig. 7.

Experimental results are obtained by using a data scanner and periodically recording the values of the basic cutting parameters, which obviates having to wait for the so-called 'steady state' condition. In this way tests can be completed far more quickly and the results display the dynamic charac-teristics of the particular machine tool. They are also in a form ready for analysis by computer.

The results of such analysis suggest certain hypotheses which are developed in the form of a mathematical model. Further tests are then made, first to evaluate the model and then to note its behaviour under a wide variety of conditions.

This work is closely allied with work investigating the natural frequency expected from any particular machine tool.[23] Hence the vibration characteristics of a proposed design may be obtained, together with those of rigidity, stress and others, while it is still at the drawing board stage.[24]

Analogue computers are extensively used in applied research. PERA[25] has used one in an investigation of the dynamic behaviour of double-acting air cylinders. Informa-tion on this is increasingly required for air cylinders in-corporated in automatic systems. The analogue computer was used in parallel with a full physical test programme to facili-tate the theoretical work. The machine was set up to represent the equations connecting the different variables. Each run gave results for a whole family of cylinder/load conditions. These results served as a check on physical tests and quickly indicated areas for future investigations.

Linear programming and hill climbing

Operational research is now being applied more and more, both in business and industry. It covers problems such as deciding what plant is required, its layout, and how to sequence jobs for optimum results.[26] The purpose is to help management in deciding its policy and actions on a scientific basis and the computer has become the principal OR tool.

Most optimisation techniques take one specific objective and approach it by use of an appropriate mathematical model. Usually this takes the form of maximising profit, minimising cost or machine idle time, or some such tangible objective. One such optimisation technique is linear programming[27] of which a classic application is the transport problem. Consider three coal mines which must supply four power stations with coal. The output of each mine is known and the requirements of each of the power stations are also known. In addition, the different transport costs from each coal mine to all four power stations are known. The problem is to find the cheapest way to organise the transport. All constraints may be expressed as first order equations, hence the name 'linear programme'.

Briefly, the solution consists in starting from any plan by which all the power stations receive their requirements and no mine exceeds its capacity.[28] But this could give an ex-tremely high transport cost. Therefore one parameter at a time is varied incrementally until the transport cost decreases, this procedure is then repeated many times until the lowest cost is reached. It is the repetitive nature of this type of problem which makes it ideal for computer solution.

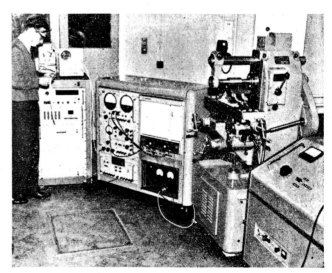

Fig. 7. At Birmingham University, chatter data from machine tools are first analysed on a computer, then evaluated and the resulting hypotheses tested by feeding imposed vibration back into the machine

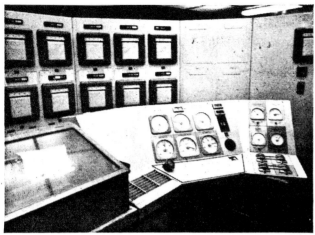

[Courtesy: Samuel Fox and Co.]

Fig. 8. This data logging installation monitors conditions in 18 soaking pits in a steelworks. Analogue chart recorders and set point controllers provide conventional instrumentation

All the major oil companies use this technique to minimise the cost of distributing their oil in all its stages but there are many more variables and the equations governing their relations are by no means always linear.[29]

Sometimes the problem is too complex for any rigorous optimisation procedure to be set up. 'Hill climbing' is a technique for introducing small random variations in parameters and following up those which tend to improve the desired performance until—using a 3-dimensional analogy—the top of the hill is reached. More complex problems are represented as 'n'-dimensional figures and, although hard to conceive, can be solved by the same iterative approach.

Stock control problems[30] are suitable for linear programming. In the simplest case, the object is to minimise the stock compatible with meeting expected demand. This entails balancing the cost of holding stock against the alternative penalties incurred by running out. The solution requires information of the pattern of demand, the lag time after re-ordering, and the cost of holding stocks, re-ordering, and running out. Full inventory control will note the changing patterns of demand and automatically indicate when new orders should be placed and for what quantity. Hence, a whole stores ledger may be held by a computer and controlled to much tighter cost levels.

Queueing theory for simple cases is well understood but the subject at present is highly theoretical and normal shop problems are too complex for solution by theory. However, queues may be investigated by means of simulation techniques,[31] *ie:* building an analogue model of the plant and of all the relevant constraints and interactions. Process times are sampled from appropriate frequency distributions.

The mathematical model may be used to try out the effect of changes which would prove costly if tried on the plant itself. Trials which can be programmed on the computer, could include: varying the plant layout, the number of pieces of plant, modifying a machine to achieve a different process time, or testing a variety of schedules or scheduling procedures to find the most suitable one.

This type of random sampling is an example of the Monte Carlo method[32] developed originally for the complex differential equations of gas diffusion. As a simple illustration of this approach, consider how to find the area of some complex figure. If the figure is enclosed in a square of known area and points within the square are chosen at random, the percentage of points which lie within the figure will approximate to the area of the figure expressed as a percentage of the area of square. The accuracy increases with the number of points. These techniques are ideally suited to the digital electronic computer; indeed, most would be useless without it.

Data logging and alarms

Computers of various kinds can continuously accept data from a series of scanning devices placed at strategic points in a process plant and can display relevant information for a controller to take action. Wylfa Nuclear Power Station,[33] for example, is monitored by analogue and digital scanners which can supply up to 2950 different signals from various parts of the plant for analysis by the computer.

This particular system[34] has two main duties; the on-demand display at a central control desk of the state of any given section of the plant (*ie:* when any one of a hundred push buttons is pressed); and, secondly, setting off alarms whenever a set value has been dangerously exceeded. The display also shows what corrective measures are to be taken. Set values are stored on the magnetic drum of the computer and, as each critical variable is scanned automatically, the computer checks it against the limit. Once an alarm condition is detected, a test procedure is automatically started to determine the corrective action to be taken. Scanning is continual and alarms always take priority over the controller's demand for information. Incorporated in the same control system is the facility to monitor the running up of the turbines.

A similar installation on board an oil tanker[35] monitors and records all the variables that affect the performance of the ship. This information is automatically punched on tape and is daily transmitted by radio to headquarters in London where it is analysed by a computer. Changes in performances are noted and the optimum time for taking the ship out of service for overhaul is calculated.

A smaller monitoring installation for the soaking pits in a steelworks, Fig. 8, is largely based on analogue controllers.

What is being widely applied in the process industries is, of course, also suitable for quantity manufacturing techniques which approach the continuous state. Thus, central monitoring and logging of machine shops or conveyor systems are now being introduced in many places.

Computer-controlled data logging and display systems have several advantages over conventional instrumentation. First, logged data in digital form is an ideal medium for future analysis by computer, *eg:* for evaluating the efficiency of a process or machine. Secondly, it is much more convenient to have hundreds of variables scanned every few seconds with the facility for automatic alarm and on-demand displays, than to have to use several hundred actual displays (one for each transducer) which require scanning by the man in charge. For this reason the computerised system may well be cheaper and more reliable.

Process control

As a calculator and optimiser the computer can often improve on the judgment of the human controller in situations where time is at a premium. In such cases the computer assumes part of the control function as against mere logging and alarm. For instance, in a steel company manufacturing billets, considerable waste occurred when long lengths had to be cut up to conform with various orders. With

Mr C. R. Edwards *was educated at Chester City Grammar School and Loughborough College of Technology where he gained an Honours Diploma in Mechanical Engineering in 1958. After a graduate apprenticeship with the United Steel Companies he wa appointed as an assistant engineer at their subsidiary Samuel Fox and Co. He spent four years on the installation and maintenance of heavy steelworks plant and was appointed Project Engineer for a £5.5 m. melting shop modernisation scheme in December 1964.*

Mr B. P. Jeremiah *was educated at Sevenoaks School and studied Production Engineering at Loughborough College of Technology. His first appointment was in the Production Control Department of Samuel Fox & Company and he was subsequently attached to United Steels Operational Research Department. In 1964 he studied computer applications in North America by winning one of United Steels Travelling Scholarships. Later he was involved with the department of an on-line order entry system. He is now the Computer Manager in the Special Steels Division of BSC*

a digital computer and suitable measuring equipment it is now possible to calculate the optimum length into which each billet is to be cut for minimum waste, compatible with orders in hand.[36]

The hot saw operator is able to feed into the computer the customer's length and range requirements from his console. The hot length is fed in automatically and, during the short time it takes for the billet to arrive at the saw, the cutting information is displayed and in some installations the saw is set automatically. Previously even experienced operators fell far short of the performance which this system permits. Here the computer is used 'on-line', ie: it is in effect part of the production line.

By building into the computer store the rules which an operator follows when faced with a given situation (or the rules which he ought to apply if he were capable and fast enough), we can automate the whole control function but we can still permit the human operator to interfere and overrule the machine. In such cases the computer is not only used on-line but completely closes the feedback control loop.

Open loop control very often precedes complete automation as a means of investigating and refining control techniques (by analysing the logged data for example) before the ultimate control system is installed.

Closed loop automatic control by digital computer of certain processes has received a lot of publicity in the last few years since the machine can be used for very sophisticated techniques, such as hill-climbing to optimise product quality. This should not be allowed to overshadow analogue systems which have long been in use for partial automatic control of plant or processes: they are usually much simpler and cheaper but, of course, less versatile. They usually employ analogue devices to measure the deviation from a set value and use this to control the input.[37]

For most purposes it is possible to design a control system based on either an analogue or digital approach. Both will need investigating by control theory to ascertain such things as frequency response and any tendency towards instability under prevailing conditions.

The digital computer has the advantage of flexibiliy in that it can take into account all sorts of parameters or it can even engage on iterative trial and error procedures if programmed accordingly.

The scanning of variables might be once every few seconds or on a sample basis. At present, analogue automatic controls are used far more extensively because they are simpler and usually cheaper but it is in overall digital computer control of processes that the most promising developments lie.[36] One of

these is 'feed-forward' as opposed to 'feed-back' control; here the computer is used to anticipate deviations from standard by sensing variations in the process parameters and allowing for them in its output signals. Clearly this provides smoother and more economical control. In an oil-fired furnace, for example, a drop in oil line pressure will ultimately affect temperature but this will be predicted before the temperature drops appreciably and corrective action taken in time.

Again, in a typical papermaking machine[38] there are ten variables in the process route between the hydropulper and the reeler which are all interdependent. If it is desired to increase the speed of reeling or correct for a variation in weight per unit area, then each of the variables must be controlled very precisely.

Therefore, it is clear that in designing a control system, one must first have a precise knowledge of the way in which each variable affects the others. Theoretically, at this stage, it is possible to design an analogue control system but the hardware would be enormous. It is far better to represent the system in the form of a mathematical model which includes, for example, the way in which the speed of the moving wire mesh must vary in relation to changes in the flow box feed rate for constant paper density. This model then forms the basis for control in as much as, when an error signal is received from the final product, the computer knows in what sequence, in what degree and with what time lags to apply corrective action to the process variables.

Similarly, after sampling the instruments, feed-forward control can be put into action.

In this installation the operator can set the machine up for any quality from his console, knowing that automatic control will ensure that the process is protected from unmanageable disturbances.

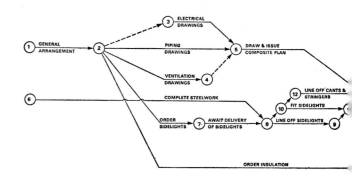

A steel plate rolling mill[39] in this country is completely controlled by computer. As well as determining the optimum mill setting for each slab, it also controls the slab reheating furnace, displays information at control points and also logs data.

The mill is set up for each pass to give maximum throughput consistent with quality and each set-up is calculated by the computer from rolling equations into which it substitutes data about the particular slab and also dynamic data received from mill instruments. From these it determines the slab pushing rate and the operator is also instructed on when to push each slab.

In the USA the Great Lakes Steel Corp. has satisfactorily automated a seven-stand hot strip mill. The computer sets up the stands for rolling a slab in six to eight seconds (manually it takes two minutes), scans the whole mill during the rolling process and immediately compensates for any variations from standard width or thickness.

An example of the successful combination of analogue and digital control is to be found in certain CEGB power stations. During normal running, conventional analogue controls maintain set conditions.[36] However, on start-up or shutdown, a digital computer controls the overall situation by continuously adjusting the set points of the analogue controllers to take into account the rapidly changing conditions.

Digital control of normal running would have to be extremely reliable because failure would affect all loops whereas, with local analogue feedback loops, a single failure will be confined to that section of the plant only. Still, experiments are continuing with full digital control because of the benefits of integration and flexibility.

Computers also play a big part in the design of analogue automatic control systems because of their simulation facilities.[8, 9, 40]

Planning

Computer-controlled alarm and data logging systems not only prevent disasters but further computer analysis can also assist maintenance in preventing or reducing outages.[41] These facilities make it possible to spot developing trouble early enough to take preventive action and, in extreme cases, shut down plant before complete failure occurs.

Planned preventive maintenance schemes can be based on schedules prepared by computer. Once maintenance in a complex plant is put on a rational basis, the control of maintenance stores by computer becomes a logical development.

The use of digital computers in connection with a relatively new planning technique, known as critical path analysis,[42] has been the subject of considerable discussion among engineers. Possibly this has centred on the merits or otherwise of the technique, rather than the application of the computer, which has been considered by some people to be the last word in gimmikry applied to an already gimmicky method of planning.

Critical path analysis is in principle a very simple and logical concept and the computer merely does all the very simple but tedious arithmetic which is necessary in using the technique. The project is divided into its constituent activities and their logical relationship represented graphically.

In Fig. 9 it will be seen that activity 31-32 cannot be started until activities 2-14 and 13-23 have been completed. Also, when 13-23 is complete, jobs 24-25 and 26-27 can be started, but 41-42 cannot start until these jobs and job 39-40 are complete. A network is drawn for the whole job and the activities numbered; each is given an estimated duration.

In each set of parallel activities converging eventually in one point, one will be the critical one or bottleneck, where maximum effort must be concentrated to reduce overall construction time. Computer analysis can thus easily discover a set of priorities for work-study, etc, as well as adding all critical paths together to estimate completion time. A list of all activities, together with full information regarding duration, starting times, degree of priority and trades required, can be printed out by a computer. Alternatively, the output could be listed in ascending starting times, so that a two-year project could be split into three month blocks and converted into more readily understood Ghant charts.

The computer is particularly useful for giving frequently revised outputs when it is required to know the effect that the over-running of certain activities is having on the whole project.

The method is being used in controlling a variety of schemes, ranging from steelworks modernisation, to the latest London underground extension and in some cases thousands of activities are interrelated so that only a computer can sort them out in a reasonable time.

Planning production in a factory where demand for various products can vary alarmingly is another good computer application, for example, a firm making diecast components for a variety of consumer durable assembly plants.[43] The most difficult problem is to assess in the long term what plant loading and labour requirements will arise due to probable changes in demand from assembly plants; also, rapidly to assess the effect that actual orders have on this plan and update it in the short term.

A computer can do this job more quickly and accurately than several clerks.

By virtue of the speed with which new information is made available, the management can plan at the earliest possible time to sub-contract work, install more plant or transfer men from one department to another.

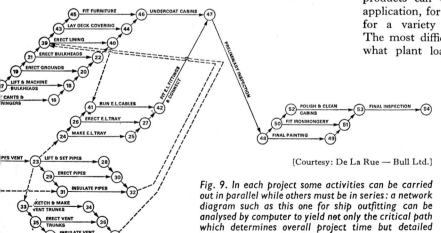

[Courtesy: De La Rue — Bull Ltd.]

Fig. 9. In each project some activities can be carried out in parallel while others must be in series: a network diagram such as this one for ship outfitting can be analysed by computer to yield not only the critical path which determines overall project time but detailed information on each activity

Conclusion

Computer use in mechanical engineering will develop along several fronts. Clearly, as the technology advances and computer languages become simpler, applications in design and research will increase very rapidly. Slowly, the conception of the computer as an 'electronic brain', to be used by boffins only, is being replaced by its acceptance as an advanced calculator which should be used as often as is economically justified.

Ordinary engineers and designers will use computers extensively in the same way that they use slide rules and desk calculators now. As programmes for a vast range of design problems become available in 'libraries', designers in all fields will have a correspondingly large store of experience to call on.

There will be considerable development in the linking of design offices with computers and numerically controlled machines, although, of course, there will never be automatic design in the widest sense because there is no substitute for the ability to conceive original ideas.

Computers will store and use, as necessary, what we call experience, but the designer will still be called upon to have new ideas.

The idea of creating a computer grid is a realistic one. Under such a system a user would enter the data for his particular problem at a console, together with processing instructions, and the resources of the grid would then be searched until the appropriate programme and any additional required data were located. This would allow organisations to plug into the grid for, say, the stress analysis of a turbine rotor. Such a system is in existence in the USA to connect several universities.

Advances in the use of computers for automatic control will undoubtedly be rapid, as human control, with its attendant lack of speed, accuracy and memory, becomes less and less suitable for process optimisation.

REFERENCES

1. HERSEE, E. H. W., *A Simple Approach to Electronic Computers.* Blacke and Son Ltd.
2. LYNCH, W. C., 'Description of a High Capacity, Fast Turn-around University Computing Centre'. Feb., 1966, *Communication of the ACM*, vol 9, no 2.
3. McKRACKEN, D. D., *A Guide to Fortran Programming; A Guide to Algol Programming; A Guide to Cobal Programming.* John Wylie and Sons Inc., New York and London.
4. TOCHER, K. D., 'Review of Simulation Lanuages': June, 1965, *Operational Research Quart*,. vol. 16, no. 2.
5. 'The Programming of Mechanical Design Problems'. Oct., 1963, *AEI Engg Supplement.*
6. 1964, *General Motors Engineering Journal.* 4th Quart.
7. 1964, *Sulzer Technical Review.* Nos. 1, 2, 3.
8. 'Analogue Computing—Hydraulic Servomotor Simulation'. April, 1964, *Measurement and Control.*
9. Process Optimisation Conference. April, 1964, *Measurement and Control.*
10. 'The Calculation of Water Hammer Problems by Means of the Digital Computer'. Nov., 1965, *ASME.*
11. *A Computing Service for Engineers.* ICT Computer Publications.
12. 'Analysis of Centrifugal Stresses in Turbine Wheels'. 1963, *JMES.* vol. 5, no. 1.
13. 'Hydrostatic Bearing System Design and its Application to Machine Tools'. *PERA Report.* no. 134.
14. 'Digital Computer Programme for Critical Speed Calculations of Non-uniform Shafts'. 26th May, 1961, *The Engineer.*
15. 'Solving Problems in Gear Design with Computers'. Sept., 1961, *The Tool and Manufacturing Engineer.*
16. 'The Analysis by Computer of a Motor Car Underbody Structure'. 1965-66, *Proc. Instn. mech. Engrs.* (AD), vol. 180, part 2A.
17. 'The Analysis of Transmission and Vehicle Field Tests Data Using a Digital Computer, 4th Quart., 1964, *General Motors Engineering Journal.*
18. 'Man-computer Design Team'. 19th Feb., 1965, *The Engineer.*
19. 'Computer-programmed Machining Speeds up Turbine Blade Development'. 16th Dec., 1964, *Metalworking Production.*
20. 'Numerical Control of Machine Tools'. July/Aug., 1965, *AEI Enginering.* Vol. 4.
21. 'The American Scene'. 3rd Dec., 1965, *The Engineer.*
22. 'Numerical Control—an Economic Survey'. *PERA Report.* No. 119.
23. TAYLOR, S. and TOBIAS, S. A., *Lumped Constant Method for the Prediction of the Vibration Characteristics of Machine Tool Structures.* Pergamon Press, London.
24. TAYLOR, S., *A Computer Analysis of an Openside Planning Machine.* Pergamon Press, London.
25. 'Dynamic Behaviour of Double-acting Air Cylinders. *PERA Report.* No. 135.
26. SCHUMAN, A., *Scientific Decision Making in Business.* Holt, Rinehart and Winston, Inc.
27. DANTZIG, G. B., *Linear Programming and Extensions.* Princeton University Press.
28. SCHROFF, G. L., *A Comprehensive Linear Programming System.* To be published by Banner, Moore and Associates Inc.
29. EILON, S., and DEZIEL, D. P., 'Siting a Distribution Centre, an Analogue Computer Application'. Feb., 1966, *Management Science,* vol. 12, no. 6.
30. DALLIMONTI, R., 'Automatic Warehousing and Inventory Control'. Feb., 1966, *Computers and Automation,* vol. 15, no. 2.
31. TOCHER, K. D., *The Art of Simulation.* English Universities Press Ltd.
32. ULAM, S., 'On the Monte Carlo Method'. *Proc. of 2nd Symposium on Large-scale Digital Calculating Machinery.*
33. 'Computer Controlled Monitoring for Wylfa Nuclear Power Station'. 1st May, 1964, *The Engineer.*
34. 'New Cathode Ray Tube Display System'. March, 1966, *Process Control and Automation,* vol. 13, no. 3.
35. 'Data Logger Monitors Ship Performance at Sea'. Dec., 1964, *Process Control and Automation.*
36. 'Digital Computer Applications to Process Control'. *Proc. of 1st Internat. Conf. of IFAC/IFIP.*
37. CONSIDINE, D. M., *Process Instruments and Controls Handbook.*
38. 'Method of Applying Supervisory Control in a Process Industry'. July, 1965, *Instrument Practice.*
39. 'Computer Control of Universal Plate Mill'. Dec., 1964, *Process Control and Automation.*
40. 'Analogue Computing for Control System Design'. April, 1964, *Measurement and Control.*
41. 'Computer Control of Power Plant Maintenance Programmes'. Aug., 1963, *The Steam Engineer.*
42. 'Critical Path Analysis'. 2nd April, 1965, *The Engineer.*
43. 'Geared to the Market'. April, 1963, *Data and Control.*

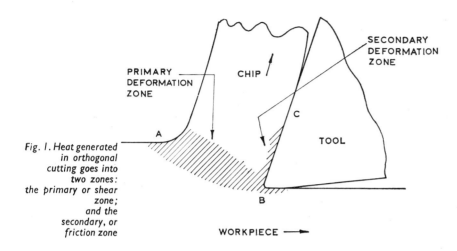

Fig. 1. Heat generated in orthogonal cutting goes into two zones: the primary or shear zone; and the secondary, or friction zone

[Courtesy: Mobil Oil Co.]

High-speed steel tool used on a capstan lathe to machine EN 5B at 50 ft/min

Metal Cutting Research*

by G. Boothroyd, PhD, BSc(Eng), MIMechE and E. M. Trent, MMet, PhD

In spite of much recent work on metal cutting, the fundamentals of the process are not clearly understood. Empirical investigations have provided some guidance to cutting forces and temperature distributions but the development of many new and difficult materials has made collaboration between engineers and metallurgists more necessary than ever, if tool life and machining costs are to become accurately predictable.

The development of high-speed steel tools and the rationalization of machining techniques which was recorded by Taylor[1] was the result of extensive experiments in the machine shops of a works where the developments occurred with a co-ordination of the work of metallurgist and engineer which has been a rare occurrence in the history of metal cutting research.

In the last 60 years metallurgists and engineers have dealt with different aspects of machining and, too frequently, the two functions have been kept apart. This has now got to the stage where it is retarding the development of metal cutting from an art to a science.

In this article the progress made towards understanding both the engineering and the metallurgical aspects of metal cutting is briefly reviewed and an attempt is made to indicate how greater co-ordination of effort between metallurgists and engineers will help to accelerate future progress.

Engineers are involved in the metal cutting problem in two ways: the manufacture of a machine tool for cutting metal; and the use of a machine tool in the workshop. The machine tool engineer is concerned with the cutting process primarily because it is at the tool edge that the forces arise which determine the structure of the machine. Problems of rigidity, vibration, etc., require knowledge of the forces acting upon the tool and much effort has been expended in developing dynamometers capable of measuring these forces

accurately.[2] The effects of many variables on the cutting forces have been determined—cutting speed, feed, tool angles, tool and work materials, cutting fluids, etc. This aspect of the cutting process has probably been more thoroughly investigated than any other and a most determined attempt has been made to understand the cutting process by studying its basic mechanics.

Cutting forces

Most investigations have been concerned with the process of removal of a continuous chip by a wedge-shaped tool under orthogonal cutting conditions. The term 'orthogonal cutting' is used to describe the special condition when the straight cutting edge of a wedge-shaped tool is perpendicular to the direction of relative motion of the tool and workpiece. Basically, the operation is one of shearing the work material to form the chip which then moves away over the rake face of the cutting tool.

The rate of energy consumption during cutting, W_c, is the product of the cutting speed, v_c, and the component in the direction of cutting, F_c, of the resultant force acting on the cutting tool. Thus

$$W_c = F_c v_c \qquad . \qquad . \qquad . \quad (1)$$

This energy is converted into heat in two main regions of plastic deformation as shown in Fig. 1; the region AB where the work material is sheared, known as the 'shear zone' or 'primary deformation zone', and the region BC, known as the 'secondary deformation zone', where large 'friction' forces cause severe deformation of the chip material.

*This lecture was given to the IMechE Industrial Administration and Production Group at the discussion on *Metallurgical and Engineering Problems in the Machining of Materials.*

Palmer and Oxley[3], using cine-photography to observe the cutting process, found that, when mild steel was machined at low speeds, the primary deformation zone had the form shown in Fig. 2. By photographing the cutting action, Nakayama[4] showed that this wide deformation zone had constant proportions for cutting speeds of up to 500 ft/min.

Nakayama,[4] using a technique for suddenly stopping the cutting action, produced quantitative evidence of a secondary zone of deformation within the chip, due to friction between the chip and the tool. Distortion of lines previously scribed on the copper specimen showed that the material adjacent to the rake face was being deformed as it moved up the tool rake face, and he was able to calculate the variation of strain rate within the chip. A typical result is shown in Fig. 3 where it can be seen that the secondary deformation zone is roughly triangular, with a maximum rate of strain at the tool point.

Investigations of this type, although leading to a better understanding of what is happening to the work material when metal is cut, are of little assistance when it is required to predict the forces acting on the tool during cutting. Before this can be done the geometry of the process must be determined and many attempts have been made to derive the basic laws governing the thickness of the chip produced in an orthogonal cutting operation. So far none of these analyses have proved successful and independent experimental work carried out at the National Engineering Laboratory and reported by Pugh[5] was compared with several theories and little agreement was found. It was shown, however, that for any tool-workpiece pair (Fig. 4) a relationship exists between certain important parameters: the shear angle ϕ; the mean 'angle of friction' at the chip–tool interface τ; the tool rake angle a. The empirical relationships obtained for several materials are shown in Fig. 4; in each case it can be seen that, for a given tool rake angle, a decrease in τ, which is a measure of the 'friction' between chip and tool, results in an increase in the shear angle with a corresponding decrease in the area of shear. Since the mean shear stress of the material in this region is constant for a wide variety of cutting conditions,[6] the force required to remove the chip will be reduced.

It is not yet possible, however, to predict the force required to remove a chip knowing only the physical properties of tool and work materials. It is puzzling why the theory of plasticity has not yet been applied successfully to the problem of metal cutting as it has to other problems in metal deformation, such as rolling, extrusion, etc. One possible reason is that in metal cutting the geometry of the process is not completely defined by the tool as it is in metal forming processes and the factors which control this geometry have not yet been determined.

The engineer is involved also in problems of production and, in this capacity, cutting forces and their relationship are of less direct interest to him. The production engineer is more concerned with increasing rates of metal removal, and improving accuracy and surface finish of machined components. His problems are centred on the life of the cutting tool and its effect on the economics of the process. These problems have been studied in engineering factories, laboratories and research establishments and the greater part of this work has been directed towards obtaining empirical relationships between tool life and cutting speed for different tool and work materials, using standardized testing conditions.[7-9]

Another more fundamental approach has been made by attempting to determine the conditions of stress, temperature and material flow at the wearing surfaces of the cutting tool

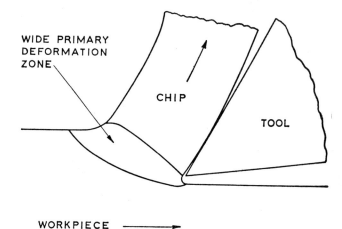

Fig. 2. With M.S. machined at low speeds, high speed photography showed this shape for the primary zone

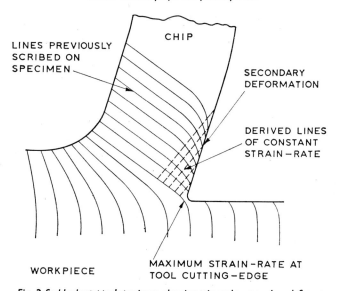

Fig. 3. Suddenly stopped specimen, showing triangular secondary deformation zone, with maximum strain rate at tool cutting edge

with the object of achieving a better understanding of those parameters which govern the life of a cutting tool in practice.

Friction

It has been suggested[10] that the idealized model of frictional behaviour between unlubricated surfaces, depicted in Fig. 5, applies to the frictional contact between chip and tool during metal machining.

With surfaces in sliding contact under relatively low normal pressures, the real area of contact between the surfaces, A_r, is less than the apparent area of contact, A_a. Under these conditions contact occurs only between the asperities of the two materials and when relative motion takes place the frictional force F is that required to shear the contacting asperities of the softer material. Thus $F = A_r S$ . . (2) where S = shear strength of asperities of softer material.

Under these circumstances the real contact area is determined by the normal force N and the bulk yield pressure of the softer material P: $A_r = N/P$ . . . . (3)

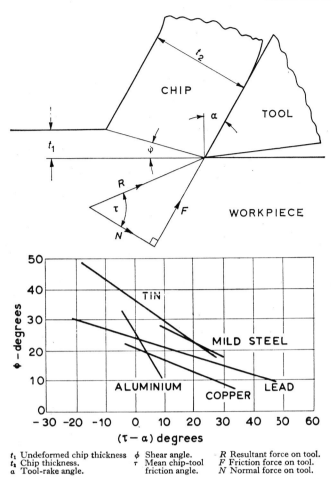

Fig. 4. Results of orthogonal cutting tests on various materials give these relationships, based on the shear plane model above

t_1 Undeformed chip thickness
t_2 Chip thickness.
a Tool-rake angle.

ϕ Shear angle.
τ Mean chip-tool friction angle.

R Resultant force on tool.
F Friction force on tool.
N Normal force on tool.

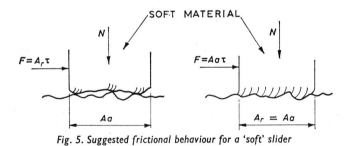

Fig. 5. Suggested frictional behaviour for a 'soft' slider

The coefficient of friction μ is given by the ratio $F/N = S/P$ and is constant for a wide range of conditions.

However, when the normal pressure between the two surfaces is high enough to cause the real contact area to approach, or become equal to, the apparent contact area (i.e. $A_a/A_r = 1$) then the real contact area has reached its maximum value and is constant. The frictional force F is still given by equation (2), but is now independent of the normal force N. Under these conditions the shearing action is no longer confined to asperities but takes place within the body of the softer material.

This description of frictional behaviour applies only when a soft slider is in frictional contact with a harder surface, because, with the alternative arrangement, the frictional force is increased when the asperities of the hard slider 'plough' the softer surface. In metal cutting, however, the slider (the chip) is always softer than the surface (the tool) and the ploughing term can be neglected.

It has been shown[10, 11] that, during machining, with a continuous chip produced and no built-up edge present, the normal pressures between chip and tool are sufficiently high to cause the real area of contact to approach the apparent area of contact and this has led to the model of orthogonal cutting shown in Fig. 6. Here the ratio A_r/A_a is equal to unity over the region of length S, adjacent to the tool cutting edge, termed the 'sticking region'.[10] In the length $(1 - S)$, extending from the end of the sticking region to the point where the chip loses contact with the tool, the ratio A_r/A_a is less than unity and, therefore, the coefficient of friction is constant; this region has been termed the 'sliding' region.

In previous work,[11] evidence of the 'sticking' mode of frictional contact was produced by examination of the undersurface of the chip on specimens where the cutting action had been suddenly stopped. It was observed that, in a region adjacent to the tool cutting edge, the grinding marks on the tool rake face were imprinted on the under-surface of the chip—indicating that no relative motion between chip and tool had occurred and that the real and apparent areas of contact were equal in this region.

Future fundamental work on tool wear should take account of these phenomena because the mechanisms and rate of wear under sliding and sticking conditions must be fundamentally different and governed by different laws.

The form of the stress distribution curves for the tool chip contact area shown in Fig. 6 was obtained by Andreev,[12] confirmed independently by Kattwinkel,[13] and later found to agree with the model of sticking and sliding friction described above.[10] This form of stress distribution is probably correct under the ideal cutting conditions used in these researches.

Quantitative information has been obtained on the frictional contact conditions occurring on the worn flank surface of a cutting tool. Okoshi and Sata[14] machined alloy steel (SAE 4140) with a carbide tool having an artificially controlled flank wear band with zero clearance angle. They observed that the frictional stress on the worn flank face of the tool was constant and independent of wear band width, undeformed chip thickness and normal pressure. The coefficient of friction on the wear band decreased with increasing wear band width and this led to the conclusion that the real and apparent areas of contact were equal in this region (i.e. 'sticking' friction occurred).

Kobayashi and Thomsen[15] machined free-cutting steel (SAE 1112) and alloy steel (SAE 4135) with tools having artificially controlled flank wear bands. It was observed, in general, that with gradually increasing flank wear band, the cutting forces were initially unaffected until a critical wear band depth was reached when the cutting forces increased roughly in proportion to the depth. This meant that the coefficient of friction was constant on the flank wear band and that 'sliding' friction occurred—a result contrary to that of Okoshi and Sata.

Temperature

Another important factor in wear of a cutting tool is the temperature distribution along the wearing surface.

It has been shown by Taylor and Quinney[16] that the

percentage of deformation energy remaining in a cold-worked metal as latent energy decreases as the deformation is increased. Thus, because in metal cutting the material undergoes severe deformation as the chip is formed, it may be assumed that all the cutting energy is converted into sensible heat. Thus

$$Q_s + Q_f = \frac{F_c v_c}{\mathcal{J}} \qquad . \qquad . \qquad . \qquad (4)$$

where Q_s is the heat generated in the primary deformation zone (shearing heat), Q_f the heat generated in secondary deformation zone (frictional heat) and $\mathcal{J}$ the mechanical equivalent.

This heat generation causes temperature rises in the workpiece, chip and tool and many attempts have been made in the past both to measure and to analyse theoretically the temperature distribution in the region of the tool cutting edge. In the experimental work, almost all the standard techniques have been applied to this problem but the experimental difficulties are severe, mainly due to the small and inaccessible areas over which the high temperatures occur and the extremely steep temperature gradients which exist.

In some very early work on cutting temperatures[17-19] it was realized that the tool and workpiece in metal cutting formed the two elements of a thermocouple, the thermo-electric e.m.f. generated at the cutting edge giving some indication of the temperatures existing in this region. By using this technique, the experimenters were able to show that an increase in cutting speed or undeformed chip thickness resulted in an increase in the measured temperature, the cutting speed having a greater influence than the undeformed chip thickness. As it was considered that the temperature in the cutting tool largely controlled the rate of tool wear, this experimental work gave an acceptable explanation of the effect of cutting speed.

Clearly, the work–tool thermocouple technique is limited in application since it can give no indication of the temperature distribution along the tool rake face. However, it has been used extensively in the past in obtaining purely empirical relationships between tool temperature and tool wear[20] and between tool temperature and the external variables.[21-23]

In an attempt to measure the temperature distribution, Reichenbach[24] buried thermocouples in a workpiece and arranged that, during the cutting test, they would be removed with the chip. In this way the temperature transients were recorded as the thermocouple passed through the deformation zones. From these temperature transients it was found possible to plot the isotherms in the workpiece and chip. However, doubt was expressed[25] whether the thermocouple junctions would be stable enough to give reliable data during transition through the shear zone and into the chip.

Infra-red radiation pyrometers,[24-28] temperature sensitive chemicals[29-31] and other thermocouple techniques[32-34] have been used in attempts to measure cutting temperatures.

Using an infra-red photographic technique, Boothroyd[35] was able to determine the complete temperature distribution in the workpiece, chip and tool during cutting and a typical result is presented in Fig. 7. It can be seen that as a point X in the material approaches the cutting tool, it is heated rapidly until it leaves the primary deformation zone and is carried away with the chip at constant temperature. Point Y passes through both deformation zones and heating is continued until it has left the region of secondary deformation; it is then cooled as the heat is conducted into the body of the chip and eventually the chip achieves a uniform temperature throughout. The point Z, which remains in the workpiece, is

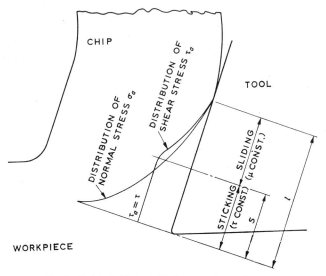

Fig. 6. Model of chip/tool friction in orthogonal cutting

heated as it passes below the tool cutting edge by conduction of heat, into the workpiece from the primary deformation zone. Some heat is removed from the secondary zone by conduction into the tool.

The maximum temperature occurs along the rake face of the cutting tool, some distance from the cutting edge, and this provides a further possible explanation of the characteristic crater wear of the rake face, i.e. the tool wears more rapidly where the temperatures are highest. Experimental evidence of the temperature distribution on the worn flank face of a tool has not yet been produced.

Yet another important factor affecting tool wear is thought to be the flow pattern of the work and chip materials in the region of the cutting tool surfaces;[36] very little is known about this and considerable effort will be required to obtain quantitative information on the subject.

Thus it can be said that the problems of adequate prediction and control of tool life have not yet been solved, because no valid general relationship has been established between tool forces, tool temperatures and tool life. Consequently it is not yet possible to predict rates of wear from knowledge of the properties of the tool and work material and the conditions of cutting. A second unsatisfactory feature is that even the 'standard' tests devised to measure tool life directly do not give an adequate guide to the performance of cutting tools on the shop floor.

A very large amount of excellent work in developing new methods of machining adapted to specific production problems is being carried out daily in factories throughout Britain. Some of this finds its way into the workshop magazines, much of the rest remains hidden as workshop tradition. Full advantage of this cannot be taken in the absence of general rules, guiding principles and theories integrating this workshop practice into a comprehensible whole. Consequently, workshop practice is very variable and will continue to be so until the process of metal cutting is understood and technicians can be trained accordingly.

Metallurgy of the tool

The metallurgist has been concerned with two aspects of the metal cutting problem—the tool and the work material.

Developments of metallurgical science and technology have resulted in new tool materials—cobalt based alloys of the Stellite type, cemented carbides and, more recently, sintered oxides—based on a search for materials harder and more like diamond. These have made possible large increases in cutting speeds, metal removal rates and productivity. Many potential tool materials have been tested and selection has resulted in the survival of the most suitable; the criterion being performance in the machine shop with the responsibility for the selection being placed on the shoulders of the practical machinist.

The work of the tool metallurgist has often been kept separate from that of the engineer using the tool. A great deal of metallurgical research and development work on tool materials has been carried out and some of this has been published. Much of this research has been directed to perfecting the processes of manufacture of tool materials or to investigating their structures, phase relationships, etc. Very useful advances have been made in this direction with both high-speed steels and with cemented carbides. Many studies have also been made of the physical and mechanical properties of tools and their relationship to composition, heat treatment and structure. The properties measured include hardness, hot hardness, transverse rupture, compressive and tensile strengths, coefficients of expansion, thermal and electrical conductivity, modulus of elasticity, Poisson's ratio, chemical and oxidization resistance, etc.[37] There is still scope for work in this direction, since these data are invaluable in fundamental metal cutting research.

This sphere of work has come to be regarded as the proper province for the tool metallurgist. The assessment of cutting performance of different tool materials has usually been carried out by the tool user, who is often little acquainted with the nature of the tool materials. The weak link in the chain is, therefore, the correlation between the structures and

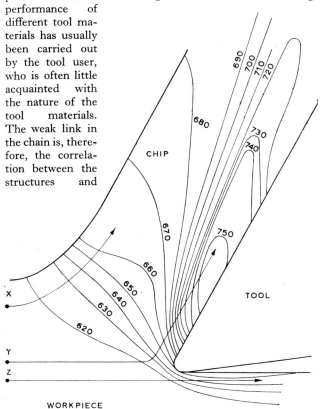

Fig. 7. Temperature distribution, mild steel; cutting speed, 75 ft/min; tool rake angle, 30°; undeformed chip thickness, 0·0238 in.; chip thickness, 0·036 in.; initial workpiece temperature, 611°C

properties of tool materials on the one hand and their cutting performance on the other. Because of the very large number of variables in metal cutting processes the influence of one factor such as tool composition is difficult to assess completely. Consequently, in the absence of definite knowledge of the mechanisms by which the edge loses it shape, explanations of correlations have often had to be based on conjecture.

Using various conjectures as working hypotheses, some experimental work has led to useful results. For example Dawihl[38] considered that much tool wear was due to welding of work material to the tool with subsequent tearing away of fragments of tool material. He therefore carried out measurements of the temperatures at which various work materials were welded to carbide tools by heating under load, and the strength of the welds thus formed. He was able to show that the presence of titanium carbide in the tool reduced the tendency of the carbide to weld to steel in this case.

More recently Prof. Opitz and his colleagues have paid considerable attention to the theory that tool wear is related to diffusion between the tool and work materials. They have heated contacting tool and work materials to study the changes taking place at the interface.[39] One very ingenious theory was put forward to relate the diffusion wear at the tool edge to the presence of large thermo-electric currents.[40]

Apart from experimental work based on such hypotheses, the metallurgist has been able to make a more direct contribution to knowledge of conditions at the tool edge. Making use of metallographic techniques, much useful information has been gained by studying tool wear as a problem of metallurgical failure. Studies of worn tools, used under workshop conditions, have shown that it is an oversimplification to ignore forms of deterioration other than flank and cratering wear. A large proportion of tools in service end their lives under the influence of other factors, including mechanical impact, deformation of the tool edge due to high temperatures and stresses, damage caused by break-away of the built-up edge, by thermal stresses, etc. Even flank wear on cutting tools may involve processes as different as diffusion, tearing away of small fragments and abrasion, each governed by its own laws.

These different forms of wear and the factors affecting them must be studied separately, and it must be known which ones are effective under any set of machining conditions before confident predictions of the life of tools can be made. A start has been made on this work with carbide tools[36] but there is much to be done. For example, many tools fail by chipping of the edge due to mechanical stresses. Practical measures are adopted to minimize this form of damage, such as chamfering or honing the tool edge or using a tougher tool material but theory lags so far behind practice that there is not even a satisfactory laboratory test for the measurement of toughness in this class of material. Some work is in progress to analyse the stresses and strains at tool edges and this could lead to the development of a toughness test and help to rationalize practices such as the honing of tool edges.[41]

The work material

The practical rate of metal removal, the tool life, the surface finish of the product, the disposal of the swarf, etc., depend to a large extent on the work material. These factors, taken singly or collectively, are often referred to as 'machinability' which, as a consequence, can have different meanings in different conditions. It is not surprising therefore, that improvements in machinability have been the result of empirical methods.

The deliberate addition of sulphur, lead or tellurium to non-ferrous metals and steels has made possible greatly increased production rates and improved surface finish; but how these additions function as they do is not completely understood. For most requirements machinability is not the critical consideration in an engineering material and metallurgical research to improve mechanical properties has had priority; consequently only half-hearted attempts have been made to provide a theoretical basis to explain the influence of structure and composition of the work material on machinability.

This lack of theory makes it difficult for the metallurgist to solve some important practical problems. For example, with increasing automation in machining processes, it becomes more important to maintain consistent tool life; but it is well known that very large differences in rates of tool wear occur in the cutting materials nominally of the same specification. These differences are often not related to the mechanical properties normally measured, nor to the changes in structure which are known to control these properties. Similarly new materials, such as those required for aircraft engines, are increasingly difficult to machine. The lack of machinability theory means that the metallurgist lacks guiding principles to help him to decide what changes in composition, heat treatment, cutting lubricant, etc., are most likely to improve cutting performance or consistency.

As in the case of tool materials, hypotheses have been put forward to explain observed correlations. For example, the action of sulphides has been described as one of lubrication and attempts have been made to explore this theory by experiments to measure the changes in coefficient of friction produced by the introduction of sulphides.[42] The results, which showed that the addition of manganese sulphide tended to increase the friction between sliding surfaces, were not conclusive, largely because the experimental conditions were not similar to those occurring at the tool face in practice.

More thorough attempts to correlate structures of the work materials with tool wear rates have been made recently at Aachen.[43] The changes in structure occurring in various steels as the chip was formed were investigated. Theories were advanced, suggesting that the rate of solution of carbide particles and the carbon content of the austenite formed on the under-surface of the steel chip, where it comes in contact with the tool, were factors responsible for changing rates of wear.

A number of workers[11, 36, 43, 44] have recently studied metallurgical changes taking place near the tool surface by metallographic examination of sections through the work and chip, when the tool was suddenly stopped in the cut, or of the metal left adhering to the tool. This metallographic evidence has confirmed the existence of conditions of sticking or seizure at the tool face near the cutting edge under a wide variety of conditions, as is shown in Fig. 8, and has also revealed the most extreme conditions of strain at and near the wearing surfaces of the tool. From this work it has become clear that tool wear and surface finish cannot be understood without study of the flow patterns of the material passing close to the tool, and the behaviour of different phases in the work material when subjected to these very extreme conditions of strain.

Suggestions for future work

Experimental investigations of tool life have, in the past, tended to over-simplify the problem by concentrating on measurement of the rate of wear and concerning themselves

Fig. 8. A section (with only the carbide etched), through the worn surface of a carbide tool with adhering cast iron, shows complete seizure between the two surfaces

mainly with flank wear and, to a lesser extent, with cratering. Much more information and better understanding of the laws governing the life of tools could be achieved if the quantitative measurements were accompanied by metallurgical investigation of the causes of wear on the tools used in tool life or force measurement tests.

To bridge the gap between laboratory tool tests and machine shop practice, more studies are needed of the processes and mechanisms of wear of tools used under machine shop conditions. Each wear process should be studied separately. Work such as that already in progress on the stresses and strains at the tool edge and on diffusion between tool and work material should be much developed.

The work done on surface finish as a function of tool geometry, speed, feed and work material should be supplemented by studies of the built-up edge and the pattern of flow of metal around the cutting edge of the tool. The mode of deformation of the work material should be studied under the conditions of sticking or seizure at the tool–work interface where extremely severe shear strains, high temperatures and rapid fluctuations of temperature give rise to unexpected metallurgical phenomena related to the conditions on the machined surface.

The action of cutting fluids as they affect surface finish can be explored by similar methods since it is clear that they operate through changes in the built-up edge and metal flow pattern. The extent of penetration of the fluids and gases at the interface and the interaction between the tool, the work, and the surrounding atmosphere should be investigated.

Machinability has a number of meanings and can only be studied scientifically if it is analysed so that swarf control and power consumption are investigated separately from tool life. This aspect of metal cutting is concerned with those differences in ease of machining which result from differences in the material being cut.

In the past, the guiding principle of work on machinability has too often been the search for some empirical test to give a 'machinability index'. This approach is of limited usefulness. In order to develop new materials with improved machinability it is not sufficient to discover a correlation between some factors in the composition and structure of the work material and the tool life and surface finish. It is necessary to study the processes by which changes in composition and structure exert their effect, for example, sulphide inclusions, or lead, or the changes from ferritic to austenitic structure, as in stainless steels. This type of investigation requires the closest co-ordination between the work of engineer and metallurgist.

Empirical methods have resulted in great advances in metal cutting. Still more rapid progress is possible if a thorough understanding of the process can be achieved. In the past, fundamental research into the cutting process has been carried out mainly as an engineering problem, together with a few isolated pieces of metallurgical research. However, this brief review of previous work indicates that many important problems remain, and progress in solving these will be more rapid if metallurgists and engineers co-operate in an integrated programme of research.

REFERENCES

1. TAYLOR, F. W. 'On the art of cutting metals', *Trans. A.S.M.E.* 1907 vol. 79, p. 31.
2. RAPIER, A. C. 'Cutting-force dynamometers', N.E.L. Plasticity Report no. 158, East Kilbride, Glasgow: National Engineering Laboratory, 1959.
3. PALMER, W. B. and OXLEY, P. L. B. 'Mechanics of orthogonal machining', *Proc. Instn mech. Engrs, Lond.,* 1959 vol. 173, p. 623.
4. NAKAYAMA, K. 'Studies on the mechanics of metal cutting', *Bull. Fac. Engng Yokohama Nat. Univ.* 1958 vol. 7, p. 1.
5. PUGH, H. LL.D. 'Mechanics of the cutting process', *Proc. of the Conference on Technology of Engineering Manufacture,* I.Mech.E. 1958, p. 237.
6. CHISHOLM, A. W. J. 'A review of some basic research on the machining of metals', *Proc. of the Conference on Technology of Engineering Manufacture,* I.Mech.E. 1958, p. 227.
7. GILBERT, W. W. 'Economics of machining', *Machining Theory and Practice,* A.S.M. 1950 p. 465.
8. BOSTON, O. W. 'Tool life testing', *Machining Theory and Practice*' A.S.M. 1950 p. 377.
9. *United States Air Force Machinability Report.* Curtiss Wright Corp., New Jersey.
10. WALLACE, P. W. 'An investigation on the friction between chip and tool in metal cutting', Fellowship Thesis, R.C.A.T., Salford, 1962.
11. ZOREV, N. N. 'Results of work in the field of the mechanics of the cutting process', *Proc. of the Conference on Technology of Engineering Manufacture,* I.Mech.E. 1958, p. 255.
12. ANDREEV, G. S. 'Photo-elastic study of stresses in a cutting tool using cinematography', *Vestnik Machinostroeniya,* 1958 vol. 38, no. 5, p. 54.
13. KATTWINKEL, W. 'Untersuchungen an Schneiden spanender Werkzeuge mit Hilfe der Spannungsoptik', *Industrie Anzeiger,* 1957, no. 36, p. 29.
14. OKOSHI, M. and SATA, T. 'Friction on relief face of cutting tool', *Scientific papers of the Inst. of Phys. and Chem. Research, Tokyo, Japan* 1958 vol. 52, no. 1493, p. 216.
15. KOBAYASHI, S. and THOMSEN, E. G. 'The role of friction in metal cutting', *Trans. A.S.M.E.,* Series B 1960 vol. 82, p. 324.
16. TAYLOR, G. I. and QUINNEY, H. 'The latent energy remaining in a metal after cold working', *Proc. Roy. Soc., London, A.* 1933. vol. 143, p. 307.
17. SHORE, H. 'Thermoelectric measurement of cutting tool temperatures', *Journal of Washington Academy of Sciences* 1925 vol. 15, p. 85.
18. HERBERT, E. G. 'The measurement of cutting temperatures', *Proc. Instn mech. Engrs, Lond.,* 1926 vol. 1, p. 289.
19. REICHEL, W. 'Das Temperaturfeld beim Zerspannen', *Masch.-Bau,* 1936 vol. 15, no. 17/18, p. 495.
20. SCHALLBROCH, H. and SCHAUMANN, H. 'Die Schnittemperatur beim Drehvorgang und ihre anwendung als Zerspanbark eitskenn-ziffer', *Zeitschrift V.D.I.* 1937 vol. 81, p. 325.
21. BOSTON, O. W. and GILBERT, W. W. 'Cutting temperatures developed by single point turning tools', *Trans. Amer. Soc. Metals* 1935 vol. 23, p. 703.
22. CHAO, B. T. and TRIGGER, K. J. 'Controlled contact cutting tools' *Trans. A.S.M.E.* Series B 1959 vol. 81, p. 139.
23. OLBERTS, D. R. 'Study of effects of tool flank wear on tool-chip temperatures', *Trans. A.S.M.E.* Series B 1959 vol. 81, p. 152.
24. REICHENBACH, G. S. 'Experimental measurement of metal cutting temperature distributions', *Trans A.S.M.E.* 1958 vol. 80, p. 525.
25. HOLLANDER, M. B. Discussion to reference 24, 1958.
26. SCHWERD, F. 'Uber die Bestimmung des Temperaturfeldes beim Spanablauf', *Zeitschrift V.D.I.* 1953 vol. 77, p. 211.
27. HOLLANDER, M. B. 'An infra-red-microradiation pyrometer technique investigation of the temperature distribution in the work piece during metal cutting', A.S.T.E. Research Report no. 21, 1959.
28. CHAO, B. T., LI, H. L. and TRIGGER, K. J. 'An experimental investigation of temperature distribution at tool-flank surface', A.S.M.E. paper no. 60-WA-87, 1961.
29. SCHALLBROCH, H. and LANG, M. 'Messung der Schnittemperatur mittels Temperaturanzeigender Farbanstriche', *Zeitschrift V.D.I.* 1943 vol. 87, p. 15.
30. PAHLITZCH, G. and HELMERDIG, H. 'Das Temperaturfeld am Dreh-meissel warmetechnisch tetrachtet', *Zeitschrift V.D.I.* 1943 vol. 87, p. 564.
31. BICKEL, E. and WIDMER, M. 'Die Temperaturen an der Werkzeug-schneide', *Industrielle organisation, Zurich* no. 8, 1951.
32. AXER, H. 'Temperaturfeld und Electrochemischer Verschliess am Drehmeissel', *Aachen Werkzeugmaschinenkolloquium* 1953 vol. 6, p. 23. Leiter H. Opitz Girardet, Essen, Germany.
33. KUSTERS, K. J. 'Das Temperaturfeld in Drehmeissel', *Aachen Werkzeugmaschinenkolloquium* 1954 vol. 7, p. 67 Leiter H. Opitz Girardet, Essen, Germany.
34. RALL, D. L. and GEIDT, W. H. 'Heat transfer and temperature distribution in a metal cutting tool', *Trans. A.S.M.E.* 1956 vol. 78, p. 1507.
35. BOOTHROYD, G. 'Photographic technique for the determination of metal cutting temperatures', *Brit. J. Appl. Phys.* 1961 vol. 12, p. 238.
36. TRENT, E. M. 'Tool wear and machinability', *Inst. Prod. Eng. Journal,* March 1959, p. 105.
37. SCHWARZKOPF, P. and KIEFFER, R. *Cemented carbides* 1960 (The Macmillan Co., New York).
38 DAWIHL, W. *A handbook of hard metals* 1955 (H.M. Stationery Office).
39. LOLADSE, T. N. 'Das Zusammenwirken von Werkzeug und Werk-stuckstoff beim Schneiden von Metallen', *Industrie Anzeiger,* Essen, no. 62, 1959, p. 991.
40. OPITZ, H. 'Wear on cutting tools', *Proc. of the Conference on Lubrica-tion and Wear,* I.Mech.E. 1958. p. 664.
41. BREWER, R. C. and PEARSON, J. H. 'Some considerations of toughness in hard sintered materials', Paper presented at the Second International Machine Tool Design and Research Conference, 1961, Manchester.
42. SHAW, M. C., USUI, E. and SMITH, P. A. 'Free machining steels: III, cutting forces; surface finish and chip formation', *Trans. A.S.M.E.* Series B 1961 vol. 83, p. 181.
43. OPITZ, H. and GAPPISCH, M. 'Some recent research on the wear behaviour of carbide cutting tools', *International Journal of Machine Tool Design and Research* 1962 vol. 2, no. 1, p. 43.
44. HEGINBOTHAM, W. B. and GOGIA, S. L. 'Metal cutting and the built-up nose', *Proc. Instn mech. Engrs, Lond.,* 1961, vol. 175, p. 892.

Dr G. Boothroyd *was educated at Stand Grammar School, Whitefield, and at the Royal Technical College, Salford, He graduated B.Sc.(Eng.), with first-class honours in 1957 and in March 1963 was awarded a Ph.D. by the University of London. From 1948 to 1957, he served an apprenticeship with Mather and Platt and then joined the Atomic Power Division of English Electric as a theoretical design engineer. From 1958 to 1967 he taught at the University of Salford and is now Professor of Mechanical Engineering at the University of Massachusetts.*

Dr E. M. Trent *received his early education in the United States and returned to England to study metallurgy at the University of Sheffield, where he was awarded the degree of B.Met. in 1933. He carried out research at the University for several years and received the degree of M.Met. for work on heterogeneity in steel ingots and a Ph.D. for research on the quench ageing of mild steel. From 1937 to 1942 he was employed by the Safety in Mines Research Board on investigations connected with wire ropes in mines. When this article was published he was Chief Metallurgist Research and Development with Hard Metal Tools Ltd (now Wickman-Wimet Ltd) where he had been engaged since 1942 in work on the methods of production and properties of powder metal products and of sintered carbides in particular. Much of this work was directed to studies of the behaviour of carbide tools in metal cutting. Since 1968 he has been a lecturer in the Department of Industrial Metallurgy at the University of Birmingham.*

Automatically Controlled Machine Tools

by C. A. Sparkes, MIMechE

Fig. 2. With the aid of automatic assembly machines, integrated into a production line, the whole manufacture of a part can be automated

Cutting machine tools still form the core of most metal-working processes and their automatic operation is therefore a condition of automation in many parts of the engineering industries. Automation in machine tools falls into three principal categories: the linkage of simple, single-operation equipment by automatic handling and interlocks; the copying of prototypes by mechanical and hydraulic devices; and, perhaps most revolutionary of all, the programming of servo-actuated machine tool movements by various devices, including punched tape that can be prepared on a computer.

Since the birth of engineering, efforts have been made to increase the accuracy of machine tools, while reducing the physical effort and mental fatigue of the operator. Pressure on the machine tool maker was increased by the need to manufacture interchangeable parts for naval and military components. These demands were often met by the use of simple single-purpose machine tools, in which the work was transferred from one machine to the next in trays moved along a simple roller conveyor.

Rapid industrial expansion before the turn of the century combined with an increasing demand for mass-produced bicycles, sewing machines and similar equipment led to the production of components in large quantities, with reasonable interchangeability and at the lowest possible cost. Accordingly, the machine tool industry produced capstan and turret lathes, automatics, and other types of production equipment which are nowadays taken for granted. In general, the movements on these machines were mechanically controlled by cams, turret stops, simple Geneva cross, and similar mechanisms.

Present trends indicate that machine tools can be classed into three broad groups.[1] The first is the single-purpose unit, often highly automated. In various combinations, it is now being used in increasing numbers for mass production. The second group covers the modern universal centre lathes, milling and planing machines, drillers and grinders. They are used for several types of operation but, to keep the cost down, automatic features are kept to a minimum.

Thirdly there is a range of machine tools, often high in capital cost, which incorporate features intended to reduce to

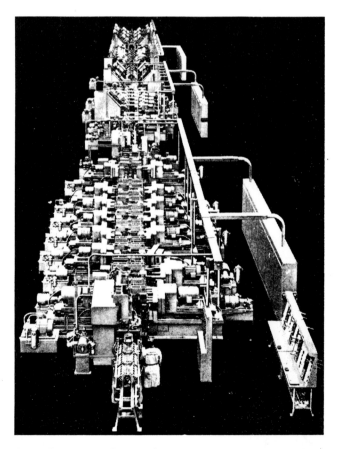

Fig. 1. This typical in-line transfer machine links a large number of unit heads, each carrying out simple interlocked operations, by means of a single shuttle bar transport device. Only car production and a few other types of quantity-produced consumer good justify this single-purpose type of investment

88

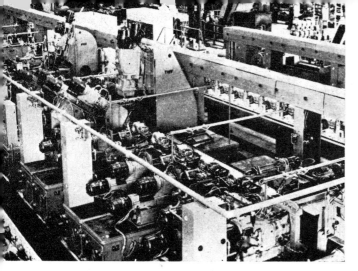

Fig. 3. With rigidly linked machines, handling is simplified but break-down of a single tool will stop a whole section. In this motor-car works, rigid linking of some machines is combined with intermediate buffer storage and a looser linkage between sections

a minimum the cost of the components they produce. Many of these machines are very versatile and need not necessarily be applied to quantity production.

Transfer machines

In the case of the first group, automatic control, so far as the machine itself is concerned, takes the form of simple cam or microswitch sequencing of operations, usually inter-locked with the clamping device holding the component. As might be expected, one of the largest users of this type of machine tool is the American motor car industry, now being closely followed in Russia and Europe.[2]

Fig. 1 shows a typical in-line transfer machine of this kind with associated control equipment. Each of the unit heads carries out a single operation, interlocked with all the others in the same section. The frontispiece of this issue also shows an example of this kind.

This type of machine cannot be fully utilized without efficient handling systems between one unit and the next, working in strict sequence control. For in-line transfer, in-dividual machining units are usually linked by a transfer system equipped with hydraulic or pneumatic fixtures and clamps, holding the pallets, on which workpieces of awkward shape are mounted.[3]

Although the output of the unit head machine can be ex-tremely high, it is, at present, often limited by the ability of the cutting tool to withstand long, continuous operation. Recent development work has, therefore, been directed to-wards producing better tools by means of special heat treat-ment or plating processes in order to reduce the down time on the transfer line. Considerable work has also been under-taken on the swarf removal problems, built-in inspection devices and the development of electrical and hydraulic equipment with an extremely high degree of reliability.

Meanwhile integration of the transfer machine with automatic assembly equipment is proceeding[4] and a typical example of this approach is shown in Fig. 2. Here a motor car shock-absorber is being automatically assembled. The machine is built round a rotary table fed from vibrating hopper magazines which, in turn, can be linked to transfer machinery by gravity chutes.

This combination of simple single-purpose machining units with work handling equipment and automatic assembly machines gives a completely automatic manufacturing line, reducing the need for progress control, transport and storage problems.

Two main systems of linkage are available: one is the loose type and the second a rigid linkage.[5] With loose linking, the separate machines in the line operate as independant units, each being automatic but capable of separate use. Between the machines any number of workpieces may be stored to provide a reservoir for periods of breakdown or servicing. When no longer required, the line can be broken up into individual machines which may be hand-operated.

With rigid linking, the machine tools in the line are con-nected by a common shuttle bar or similar transport device which simultaneously transfers all workpieces. Interlocks ensure that the transfer movement is made when all the machines in the line have completed their work cycle and tools are returned to their starting position. A centralized control system receives a signal from each machine as its cycle is completed and also initiates the feed movement at the appropriate moment, when transfer and clamping are com-plete.

Often a combination of the two systems is employed, as in Fig. 3, which shows part of the equipment installed in a motor car works, but this generally depends on the product and the time cycles.

Attempts to reduce the floor space occupied by transfer machines and to minimize handling problems led to the design of rotary transfer layouts. A typical four-station rotary machine[6] is shown in Fig. 4. This performs ten drilling and milling operations on an automobile camshaft at a rate of 288 per hour. The workpieces, which are machined in pairs, are held in hydraulic clamping and equalizing fixtures at the shaft end diameters and centre bearings.

The motor industry, which is at present one of the largest users of unit head and transfer line machines, is also mainly responsible for the development of the large automatic broaching machines for cylinder blocks used widely in the U.S.A. They are very expensive but, when employed on sufficiently large quantities, maintain very high rates of output which result in a low unit cost. By using long broaches, the metal removal per tooth can be kept to a minimum which results in low stresses being imposed on the component, combined with a close tolerance and good finish.

Research and development work is directed to developing built-in inspection systems and tool wear indication which is charted in a central control office.[7] Ultimately a completely automatic tool-changing sequence will be used for the produc-tion line. Changing of drills, etc., can already be carried out automatically as will be seen below.

Copying attachments

Foremost among our second group of machine tools is the centre lathe for which normal template type of mechanical automatic control has been available for many years. Generally it was used for simple contours such as bottle-shaped profiles. The maximum permissible angle between the contour and the centre line was limited to about 30 degrees.

Recently, however, copying attachments have been developed[8] which permit the machining of right-angled shoulders. This has been achieved by using a servo-motor, the control of which only requires a small proportion of the force which it exerts on the tool. These power-assisted servo attachments now make it possible to copy parts of practically every shape which would normally be encountered in centre lathe work.

The same general principle applies to a whole new range of copying attachments, and usually includes a tracing device rigidly mounted on the copying slide which carries the tool holder. The tracer is equipped with a stylus possessing either one or two degrees of freedom, which explores the template or model. Tracer, stylus and tool move initially in a direction parallel to the longitudinal axis of the machine as in the case of a normal centre lathe.

On reaching a curve in the template, a relative movement occurs between the stylus and tracer head. By the use of either hydraulic, pneumatic, optical or electrical devices, and sometimes a combination of these, the relative movement is used to displace the copying slide which carries the tool and tracer. The intention is always to reproduce the master as accurately as possible and, at the same time, obtain the highest possible ratio of cutting load to tracer contact pressure.

Although these copying attachments vary greatly in their design, they may be divided into two fundamental types. In the first, the tracing device only controls the movement of the transverse slide. The longitudinal feed is derived in the normal manner and is not changed during the copying process. In the second type, the copying tracer also influences the longitudinal feed which makes accurate three-dimensional reproduction much easier.[9]

In both cases the sensitivity of the servomechanism and its response rate determine the quality of the surface finish on the component.

Programme control by buttons

Automatic programme control[10] is now available on normal, vertical milling machines and on a number of other types. In this case, a bank of three hundred or more push buttons, ranged in rows, can be set to give vertical, longitudinal and transverse feeds or rapid traverse rates, and distances.

Fig. 5 shows a perspective view of a workpiece milled on this type of machine, with a diagrammatic layout of the programmed movements of the cutter relative to the workpiece.

Such machines are, in effect, numerically controlled but are very much less expensive than those taking instructions from punched cards or tape.

In addition to being applied to milling machines, programme control of this kind has recently been made available in package form for centre,[11] capstan and turret lathes.[9] The control board for a turret lathe is shown in Fig. 6. In this case[12] a plugboard is used to set up the necessary connections which determine the programme: this controls in sequence the pneumatic cylinders which replace the power and hand feed motions on the normal machine. The correct speed can also be selected for each.

On the plugboard, sockets and switches are grouped: the upper row being concerned with spindle speeds, feeds and movements of the cross slide, turret etc. The lower row of sockets are connected to the various microswitches on the machine and provide signals of certain operations having been completed so that the stepping switch wiper arm can be moved to the next contact.

Plugboard control is now also available for lathes, in which case the programme is capable of controlling the normal and rapid feeds[10] while the microswitches, actuated by trip dogs, limit the movement to any traverse.

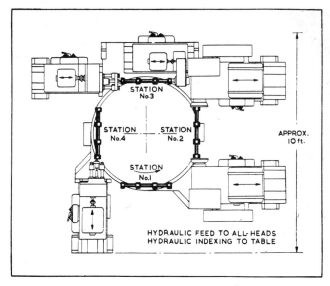

Fig. 4. The rotary transfer machine saves floor-space but usually requires manual loading and does not lend itself to straight line flow production. It is popular for smallish components made in large batches

Fig. 5. How a milling machine can be programmed to deal consecutively with a number of separate operations: only point-to-point straight line motion is required

Heat distortion

An interesting feature of modern automatically controlled machine tools has been distortion of the workpiece due to heat, which now has little time to dissipate at the high and almost continuous rate of metal removal made possible. This has proved particularly important in the case of precision gear-cutting and grinding machines.

In one case a machine of this type employs a master rack and gear which controls the rack cutter and component movement. Due to the cutting action, and friction of the cutter slides, a slight extension of the master rack takes place. In order to compensate for this, an automatic heating device warms the master rack, until the rack cutter and master rack are of the same length.

A thermostat mounted on the machine maintains the

Fig. 6. Like a telephone exchange, the plugboard can be used to set up connection which will programme the machine. Unlike an exchange, the board can sometimes be removed bodily and replaced with another, pre-programmed one to carry out a new set of operations

balance of temperature between the rack cutter and master rack. The thermostat is set from data obtained during the first series of gear-cutting operations and the automatic heating device is adjusted accordingly.

The Russian Machine Tool Research Centre is devoting considerable effort to the development of a programme controlled turret lathe[13] intended for machining small batches of components, which should possess the following properties:

1. Short setting time;
2. Simple tooling, permitting the use of standard tools and holders mounted on one detachable turret;
3. Simple programming, for any sequence of turret motions, the necessary traverse length for each and any feed and speed;
4. No need to change cams or any other component during operation.

On the latest Russian programme controlled automatic lathes a standard punched card stores the information for controlling the machine.

The setter records the co-ordinates of the points defining the positions of the machine units at the end of each movement. These are 'written' on parallel lines traced on a magnetic drum by means of recording heads travelling with the slide.

When the machine is operating automatically, each pass begins on the command from the punched card and is ended when the pick-up from the magnetic head feeds back to the control system a signal read on the drum. The combination of punched card programming with position feedback ensures simple programming of the machine and the accuracy of the component to 0·01 in.

Continuous numerical control

The third group of machine tools, which, as mentioned earlier, usually involve higher capital expenditure, incorporate features which are intended to reduce to an absolute minimum the cost per item produced. With workpieces of intricate shape, such as structural parts for aircraft, this means con-

tinuous control of all movement. The development of such machines,[14] even to their present stage, has been responsible for a complete revolution in the design of machine tools. The basic requirement of maintaining accurate co-ordination between the movement of the component and the cutting tool is an extremely difficult task for the transmission of the power to the cutting tools and the transmission of information instructing the various machine slides, involves very different principles.

In the manual system of control, the human operator provides the servo link and throughout the machining operation he is continually maintaining an accurate relationship between the workpiece and the shape he is expected to produce.

Similarly, a continuous supply of information must correct any error and it is not possible to arrange this with switches and simple interlocks, as in the programme-controlled point-to-point systems described above.

Generally, the movement of numerically controlled machine tool slides is now provided either by electrical or hydraulic servomechanisms. Information is stored on punched paper or magnetic tape which continuously gives the command signals for controlling the slide movements. By comparing this information with that obtained from a position-sensing system attached to the machine unit, a closed loop is formed which constantly provides the necessary correction.

The problem of an accurate and reliable position-indicating system was solved in Britain by the use of the Moiré fringe pattern which is obtained by superimposing two optical gratings one on top of the other. Fig. 7 shows a diagrammatic arrangement of this device.[15] When the short scale is moved, relative to the large one, the bands of light of the fringe pattern move in a direction at right angles to the slide movement. The distance traversed by the slides can be assessed by counting the number of times the light band passes a fixed reference position.

In the U.S.A. the *Inductosyn* system[16] of positional control has been developed and is now widely used in this country. Essentially it consists of a glass scale, about 10 in. long, with a single square wave metallic conducting pattern printed on it. The second, shorter element, known as the slider, has two printed metallic patterns, spaced at 90 electrical degrees or 0·025 in. apart.

A coarse system of a different type is usually employed for locating to 0·05 in. and final positioning by the *Inductosyn* is to 0·0001 in.

One of the first practical applications of the Moiré fringe system on a large machine tool was the joint effort of the Ferranti/Fairey organization early in 1950.[17] At that time aircraft designers were faced with the problem of considerably improving the aerodynamic shape of their structures. This led to the consideration of large integral structures instead of the then customary bolted assemblies.

Here was an ideal application for numerical control:[18, 19] manual machining would have been extremely expensive because of the large and complicated shape. On the other hand, numbers required were not large enough to justify copying templates, nor would the accuracy have been satisfactory.

The machine finally developed (Fig. 8) had a work table capacity 28 ft by 8 ft, longitudinal traverse 25 ft, vertical traverse 7 ft and maximum traverse rate 120 in. per minute. The use of traverse feed rates of over 100 in. per minute with a machining tolerance of $\pm$ 0·005 in. were set as target figures. This meant that, during the milling of a right-

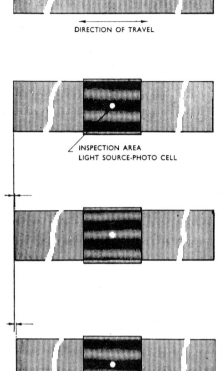

DIRECTION OF TRAVEL

INSPECTION AREA
LIGHT SOURCE-PHOTO CELL

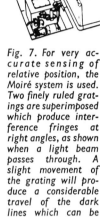

Fig. 7. For very accurate sensing of relative position, the Moiré system is used. Two finely ruled gratings are superimposed which produce interference fringes at right angles, as shown when a light beam passes through. A slight movement of the grating will produce a considerable travel of the dark lines which can be detected by photocells

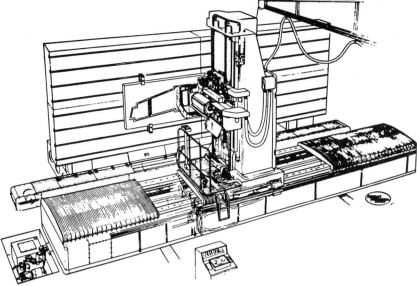

Fig. 8. This co-ordinate controlled machine for producing aircraft components has a 28 ft by 8 ft work-table

angled corner, the time available for the machine to receive its instructions and change direction, would be less than three milliseconds. Only continuous control of speed makes it possible to reduce it automatically, when a rapid change contour is approaching.

Mechanical factors

During the design study on this large machine, many basic mechanical problems were revealed which called for a radical departure from conventional machine tool practice. In any system where relatively large masses are subject to continuous positional control to a high order of accuracy, ultimate performance is limited by stiffness of the drive from the motor to the output motion. A springy drive introduces a time lag as well as setting up vibrations. The main sources of friction in a machine tool are the slideways. Normal frictional characteristics are non-linear at low speeds and there is a constant danger of slip-stick between the slides. A further problem is that of excessive backlash in the drive, in which case the motor rotation is not accurately followed at the output end.

There are also the questions of mass and structural rigidity to be considered and, equally important, the mechanical accuracy of the general structure of the machine.

Two of the major factors were successfully dealt with as follows. The friction of the slideways was reduced by a special hydrostatic support system. In addition to giving almost linear frictional conditions, this arrangement gives high damping and stiffness to the slides. The use of re-

circulating ball nuts and lead screws provides an efficient transmission which can also be pre-loaded to reduce the backlash. To reduce the backlash in the pinion gear box twin gear drives are loaded in opposite directions.

This machine proved capable of producing components up to 18 in. square to $\pm$ 0·002 in. with a repeatability of 0·001 in. On eliptical path milling, a 7 ft frame was machined to within $\pm$ 0·005 in. of the curve programmed on magnetic tape. These figures were obtained at the maximum feed rate of over 100 in. per minute in light alloy.

During these evaluation tests, the difficulty of operating the large main spindle bearings at high speeds and under continuous cutting operations, was shown.

A change in the military requirements from aircraft to guided missiles has recently reduced the interest in these large capacity machines. However, continuous control is finding new fields in the milling of wave guides for radar.[20] Another likely application would seem to be large motor-body dies but this has not materialized, mainly because their shapes are envisaged by artists rather than mathematicians. However, this is undoubtedly a field in which they will ultimately find favour as this type of control [21], [22] can give faster machining rates than those obtained with tracer methods from templates and models.

New developments often find their most useful application in circumstances for which they were not really intended. Very many production operations could be carried out by co-ordinate setting, particularly by the point-to-point drilling and boring devices now available.

As the result of close co-operation between users, machine tool manufacturers and electronics engineers, the world's first universal horizontal boring machine with numerical control was developed in the United Kingdom in 1953.[23] One of the latest models is shown in Fig. 9.

Considerable workshop experience has now been accumulated with these large machines.[24] They are used on shoe machinery, aircraft structures, diesel engines, cigarette making machinery, tile presses, general engineering and components for military equipment—to mention just a few of their applications. In these fields they have shown considerable saving in drawing office time, jigs and fixtures and

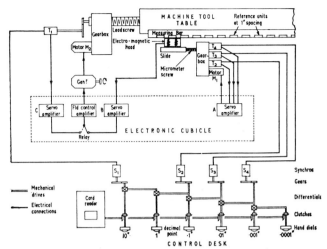

Fig. 9. The layout of the electronic co-ordinate controls for a large boring machine which can position the hole to an accuracy of 0·0002 in.

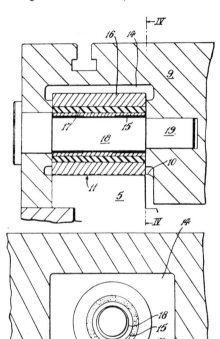

Fig. 10. Resiliently mounted pins turn on roller bearings and are adjusted by means of eccentric studs to take the full weight of the table. The weight of the load is then carried by the slide in the normal way

the development of prototype mechanisms.[25] They have proved capable of taking loads up to 10 tons and positioning the hole to an accuracy of less than 0·0002 in.

Accuracy

Essentially the co-ordinate setting system consists of a position controlled servomechanism operating the machine slides. The programme is set by means of dials on the control desk, or from standard punched cards, read by an automatic card reader; one card being needed for each co-ordinate.

In the largest of these machines[26] a combination of rolling and sliding elements is used to support the heavily loaded worktable, which reduces friction problems.

From the sketch in Fig. 10 it will be seen that resiliently mounted rollers move on needle roller bearings. These are adjusted by means of eccentric studs to carry the full weight of the worktable. The weight of the workpiece is carried by the two slideways under normal frictional conditions.

One important difference between continuously controlled machines such as are used for form milling and co-ordinate setting point-to-point systems is the oil film thickness permissible between the slideways. On the first type, films of 0·001 in. are easily acceptable but for really accurate co-ordinate setting, the oil film must be reduced to less than 0·0002 in.

The obvious way of automatically controlling machining operations is by motorizing the lead screws and simply measuring the rotation. When this was originally attempted it was soon found that normal lead screws are not accurate enough to do without position feedback from the worktable.

With improved manufacturing accuracies and various correcting devices, these basically simple systems have recently been re-introduced, with great price advantages but some loss of performance, compared with other systems.[27]

The *Jigmatic* punched tape controlled two-axis positioning table has been produced for drilling and boring gear boxes to accuracies of ± 0·001 in. between centres. The system operates by transmitting the angle of rotation of the machine tool lead screw.[28] Although the final accuracy depends on that of the screw, it is possible to use correction devices. Several systems of this type are now available in Britain.

Few existing machine tool structures are capable of utilizing to the full the high degree of accuracy which is now obtainable with electronic systems. This difficulty is particularly acute in the case of jig borers. The permissible inclination of a horizontal boring machine table in the traverse direction, is ± 0·0005 in/ft, corresponding to an error between centres of up to 0·004 in. at the extreme positions of a 2 ft traverse. If it is intended to achieve an accuracy between the top centres of ± 0·00025 in., the curvature of the saddle way, in its transverse direction must not exceed 0·00003 in. under load.[29, 30]

A research project sponsored by the National Research Development Corporation at the Manchester College of Science and Technology[31] is intended to study the fundamentals of machine tool construction with a view to improving performance under automatic control. Combined electronic and mechanical problems are being investigated which arise in building such machine tools. The aim is, for example, to be able to grind a straight edge from an outside precision reference on a machine tool. This means more accurate slides, spindle bearings which do not expand excessively due to heat, and no deflection due to the weight of the workpiece.

One method of control which has been proposed is that, in addition to providing input information on the mathematical shape to be machined, corrections would be superimposed to allow for the movement of the cutter spindle or machine tool slides from their nominal position.[32] In other words, a three-dimensional position feedback from the cutting tool itself, rather than from slides or lead screws.

New control systems

An interesting development is the Moog method of pneumatic co-ordinate setting.[33] Data from binary coded decimal punched tape is read pneumatically and passed by capillaries to an eight-level pneumatically operated binary selector valve, via a micrometer rotary actuator. This in turn operates a hydraulic servo valve in the table drive. A diagram of this system is shown in Fig. 11.

On comparatively small capacity machines,[34] the use of hydraulic servo motors, together with a cylinder and ram for the slide movement, are proving very satisfactory. They give high rigidity, minimum backlash, together with high transmission efficiency and a low total inertia.

In larger machines, recent developments are equally divided between electrical and hydraulic actuators, and re-circulating ball nuts and screws, or high efficiency worms and segmental nuts, are used to move the tables without backlash.

An important recent effort to evaluate particular types of numerical control has been the development of a test rig by the Production Engineering Research Association.[35] This enables a known degree of backlash, slip-stick and torsional stiffness to be introduced into the equipment, thus enabling the behaviour of the electronic control to be checked against a wide range of conditions.

Constant efforts are being made, both in Britain and abroad, to produce simple and inexpensive control equipment. A recent example is the capacitance pick-up developed by the Reilly Company. This employs a simple arrangement, as shown in Fig. 12. The static section consists of a number of cylinders or stators, insulated from each other and assembled on a tie-rod. By counting the pulses obtained when the sensing element passes over a cylinder, coarse positioning is obtained and the capacitance variation between the interpolater and the adjacent stator provides fine control.

At recent European Machine Tool Exhibitions many Continental makers were showing a marked tendency towards the use of optical transducers which, in view of their previous work in optics, is only to be expected. By employing the latest extremely compact photo-electric cell, Philips in Holland[36] have produced a simple numerical position indicator for machine tool slides. In principle, the small digital transducer is mounted either on the actual feed screw or on a separate drive, the number of revolutions of which are a measure of the linear displacement. By means of a photo-transistor the angular displacement of the drive is translated into electrical signals which are fed into a reversable counter, via a directional decoder. The position of the counter is shown by means of an illuminated display.

In the field of continuous control, the small and medium capacity milling machines form the principal application; the latest example is the Newall machine with A.E.I. equipment.[37]

Here, the electrical measuring element is normally a *Helyxin* arrangement which provides the direct measurement of linear movement. This device can be regarded as an electrical screw.

Like other systems now available, it allows milling cutter diameter correction, an essential feature for sound workshop practice.

The most recent development in the field of point-to-point numerical control is its application to machines with turrets which permits tool changing as part of the operation cycle. An example of this type of machine is illustrated in Fig. 13.

Conclusion

It now seems clear that machine tool makers can design and build fully automatic machine tools, which include tool changing and inspection, for single or batch production.[38, 39]

The recently formed Machine Tool Industry Research Association has plans for both long term and short term

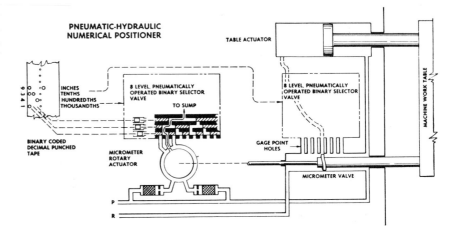

Fig. 11. In a pneumatic control system, data from tape is read by means of air-jets passing through the holes and actuating a binary selector which controls the hydraulic servo of the table drive

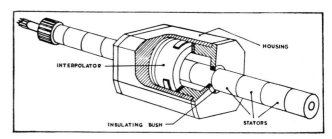

Fig. 12. New position transducers are constantly being developed: in this capacitance pick-up the stator consists of a number of insulated cylinders and as the slider moves, the joints cause pulses which are counted. The capacitance read at the end point is an indication of the distance along the last cylinder the pick-up has reached

Fig. 13. After numerical control of machining: control of tool choice. In this machine tape, instructions cause a transfer arm to pick the right tool from the magazine and insert it in the spindle which then drills the hole as with normal tape control

research projects on all types of machine tools.[40] It is destined to play an important role in the ultimate integration of the mechanical, electrical, hydraulic and optical systems into inexpensive, highly efficient and reliable, numerically controlled and other automatic machine tools.

Perhaps the need for these is most clearly demonstrated by a study of the cost of obtaining great accuracy, which increases exponentially after the ± 0.005 mark is passed. This tendency, combined with the ever increasing cost of labour, will accelerate the trend and there will be a growing appreciation of this in general purpose factories. Here, as in mass production, the installation of expensive capital equipment will require its continuous operation for maximum returns and the same careful planning will be needed.

A centralized control system may be the result with signals from each machine regarding break-down, tool changing, and material difficulties, automatically fed into a simple computer.

Mr C. A. Sparkes *received his early technical training at Newton Heath Technical School and the University of Manchester Institute of Science and Technology. He was awarded the Associateship of the College for Mechanical Engineering in 1926. In 1919 he joined H. W. Kearns & Co. Ltd, and after serving an apprenticeship in the works, he entered the Drawing Office, subsequently becoming the Company's Chief Designer in 1938. He has travelled extensively and given lectures in North America, the Continent, Australasia, Singapore and Russia. He was awarded the Joseph Whitworth prize of the Institution in 1952. In 1955 he was appointed to the Board of H. W. Kearns & Co. Ltd, as a Director responsible for design and development. He is a past president of the Manchester Association of Engineers and a founder member of the Machine Tool Industry Research Association. He was appointed Acting General Manager of Kearns-Richards Division of Staveley Machine Tools Ltd, and retired from this post to become Consultant to the Group in January 1970.*

REFERENCES

1. SPARKES, C. A., 'A Machine Tool Design Policy'. 1961 September 2nd International Machine Tool Design and Research Conference.
2. GALLOWAY, D. F., 'Recent Production Developments in the Soviet Union'. 1960 February *The Production Engineer*, vol. 39, p. 79.
3. Symposium on Recent Mechanical Engineering Developments in Automatic Control. 1960 January *I. Mech. E.*
4. GREEN, R. E., 'Automatic Assembly of Shock Absorber Piston Rods'. 1961, 20th December *Machinery*.
5. 'Applications of Work Handling Equipment to Linked Machine Tools—Guide to Work Handling Equipment for Machine Tools and Presses, Part 7'. P.E.R.A., Research Report no. 92.
6. FROST-SMITH, E. N., 'Applying Automation'. 1961, 25th August *Investors Chronicle*.
7. McRAINEY, J. H., and MILLER, L. D., 'Numerical Control'. 1960 October *Automation*.
8. SCHATZ, A., 'Automatic Machine Tool Control'. 1956 *Progressu.*
9. MOE, J. E., 'Magnetic Tape Control Turret Lathes'. 1961 November *Automation*.
10. BREWER, R. C., 'Programme Controlled Machine Tools'. 1960 June *The Engineers Digest.*
11. 'Programmed Controlled Machine Tools'. 1960 July/August *Machine Tool Review.*
12. 'Programme Control by Pneumatic Plugboard'. 1961, 8th November *Metalworking Production*, p. 72.
13. PADOGIN, A. A., 'Scientific and Research Work of ENIMS during 1960'. 1961 *Machines and Tooling*, vol. 32, no. 5. Production Engineering Research Association.
14. FROST-SMITH, E. N., 'Research on Machine Tools and Automation in an Industrial Research Laboratory'.
15. 'Ferranti Computer Control of Machine Tools'. List ES/A/1, Ferranti, Edinburgh.
16. 'The Inductosyn'. Farrand Optical Company, New York.
17. COLEMAN, R., GREGSON, J., and FLETT, A., 'The Fairy-Ferranti 3D Miller'. Fairy Aviation Ltd., England.
18. JOLLIS, G. S., 'The Advantages of Numerically Controlled Machine Tools'. General Electric Company, New York.
19. WILLIAMSON, D. T. N., 'Automatic Control of Machine Tools'. Conference on Technology of Engineering Manufacture. 1958 March *I. Mech. E.*
20. MORGAN, J. M., WILSON, J. W., and CARROLL, G. R., 'Cincinnati Numerical Control'. 1957 September Cincinnati Technical Activities Seminar.
21. ROBINSON, A. S., 'Machine Tool Control'. 1961 March *Electrotechnology*.

22. SIMON, W., On the Practical Design of Numerical Controlled Machine Tools. 1960 November *Werkstatt und Betrieb*, vol. 93, p. 255.
23. SPARKES, C. A., 'Electronically Controlled Machine Tools in Service, especially Co-ordinate Setting Horizontal Boring Machines'. H. W. Kearns and Co. Ltd., Manchester.
24. SPARKES, C. A., STOKES, J. A., and SHORT, M. O., 'Electronic Control of Machine Tools'. 1961 February *A.E.I. Engineering.*
25. JAEGER, J. J., 'How Numerical Control is Applied to a Jig Borer'. 1957 February *Machine Production*, Canada and U.S.A.
26. SPARKES, C. A., 'Machine Tools for Tomorrow's Engineering'. Paper delivered at East Midlands Branch Spring Meeting 1962.
27. 'Vero Automatic Drill Co-ordinate Setting Drilling Machine with Airmec Tape Control'. 1959, 2nd December *Machinery*.
28. 'Autoset Simplified Co-ordinate Table'. Leaflet no. 186B, Airmec Ltd, England.
29. SPARKES, C. A., 'Design and Development of Machine Tools'. 1959 *Proc. Instn mech. Engrs, Lond.*, vol. 173, p. 988.
30. HALSBURY, The Rt. Hon. the Earl of, 'Electro-mechanical Themes in Machine Tool Design'. 1958 November Lecture to the Manchester Association of Engineers.
31. KOENIGSBERGER, F., 'The Design of Automatic Machine Tools for Electronic Control'. 1958 May.
32. LEET, D. L., 'Automatic Compensation of Alignment Errors in Machine Tools'. 1960 September Conference on Machine Tool Design and Research, Manchester University.
33. 'Electro-hydraulic Servomechanisms in Industry'. Bulletin 110, Moog Servo Controls Inc., New York, U.S.A.
34. 'Hayes Ferranti Tape Master'. 1959 August *Metalworking Production*.
35. 'Numerically Controlled Machine Tools'. Report no. 75, Production Engineering Research Association.
36. 'Philips Numerical Indicator for Machine Tools'. Philips Eindhoven, Holland.
37. 'Numeritrol Continuous Numerical Control of Machine Tools'. Publication no. 4103/1, A.E.I. Ltd., Electronic Apparatus Division.
38. 'Automation in North America'. 1958 D.S.I.R. Overseas Technical Report no. 3, code no. 47, 195—3.
39. 'Automation'. 1957 D.S.I.R. code no. 47, 195.
40. DEBARR, A. E., 'Machine Tools in Transition'. 1961 November *Control.*

Modern manufacturing consists essentially of the forming or machining of materials, assembling the resulting components, surface-finishing either before or after assembly, and packaging. For a generation or so, forming and machining have been extensively mechanised, where economically justifiable. Only recently has much progress been made in applying mechanisation to the assembly process which now offers the maximum returns by way of cost reduction and quality improvement. The principal problems arise in orientating and placing the components: subsequent fastening is comparatively simple to mechanise.

Fig. 1. This machine automatically assembles needle rollers and washers on to a transmission gear. The needles are inserted by standard units and all parts are fed from hoppers

[Courtesy, Rehnberg Jacobson Mfg Co.]

Mechanised Assembly

by G. Wittenberg, AMIProdE, MIMechE

[Courtesy, Aylesbury Automation Ltd]

Fig. 2. With a disc rotating at 6·9 or 9·7 rev/min in a fixed container, this rotary feeder has a capacity of 0·7 ft³

Assembly consists in placing components into correct relative positions and fastening them together. The placing of the components entails handling operations like separating, orientating, transfer and positioning; the fastening together of the correctly placed components can involve riveting, welding, screwing, interlocking, bonding or similar operations. It is the difficulties experienced in the handling of components which for so long discouraged the engineer from mechanizing the assembling process. Even now, the rate of progress depends on the extent to which components of diverse shapes can be located and handled mechanically, as, for instance, in Fig. 1. By comparison, the problems of fastening are insignificant, amounting often to no more than the automatic actuation of well-known tools or a reduction of their cycle times to yield a balanced assembly cycle.

Apart from the essential operations of placing and fastening, certain others are sometimes incorporated in mechanical assembly, e.g., counting and inspecting. Indeed, the latter function may be regarded as a necessary part of automatic assembly. Some components, such as helical springs, are so awkward to feed and orientate, if introduced in their finished form, that it is often simpler to make them on the assembly machine. Thus, a few forming or machining operations may occasionally be found on what is primarily an assembly machine.

Where products are packed, either individually or in a pattern or batch, the packing operation may also be regarded as a stage in assembling.

Feeding

In machining or forming, the material is usually of a very simple shape, initially. In assembling, on the other hand, the process often starts with complex components and the shape of the growing assembly tends to increase in complexity. As engineering products become more specific in function, their shapes become more individual and varied. It follows that, even where a method of mechanical handling has been developed, it may have only a restricted application to another project. This makes for slow progress in the art of mechanized assembly.

Unless components are kept separate and correctly orientated at the end of the previous manufacturing process, (which is highly desirable but rare in practice) they must first of all be unscrambled and orientated. Hopper feeds are used for this purpose. About nine basic designs are known.[1] Of these, the rotary feeder (Fig. 2) is the most highly developed and has been fairly widely available for some time as a semi-standard unit, at least in the U.S.A.[2] Another, the centreboard hopper, shown in Fig. 3, is simple and minor modifications allow it to be used for quite a wide range of components; for instance, for glass rods used in pinch machines in the manufacture of electric lamps.[3] Both types, and some others, are now commercially available in Great Britain as semi-standard units.[4]

It remains to be seen whether their ready availability will now give these well-known designs a new lease of life; at present the vibratory bowl feeder, being simple, cheap, reliable and versatile, is the most widely used type of feed device.

A mechanism in the lower portion vibrates the bowl on top, which causes components near the edge to climb up the spiral flight attached to the inside of the bowl wall. After travelling along one or two turns, the components are discharged from the mouth of the bowl, correctly orientated by means of twisted rails and other devices which are fitted to the spiral flight.[5]

Feed rates of 40 to 60 small components per minute are common. At least one of the six or more companies who now supply vibratory bowl feeders in Great Britain offers the cast, segmented bowl construction in aluminium alloy as shown in Fig. 4, to provide increased adaptability;[6] another offers a quick-release bowl as small as 6 inches in diameter.[7]

The selection of vibratory bowl feeders, the design and location of orientating devices in and beyond the bowls, and the assessment, and possible modification, of components for suitability of handling in bowl feeders is rapidly becoming a subject of its own, of which the present state has recently been well described.[8]

In most assembly applications, vibratory bowl feeders are used only to separate one component from bulk, and to orientate it. In a development of this technique, however, components of more than one type are placed in a single bowl where they are separated from one another, selectively orientated and assembled, all inside the bowl. In one example, a $\frac{5}{16}$ in hexagon head screw is first assembled with a lock-washer and then with a heavy reinforcing washer at a maximum delivery rate of 1200 per hour. In another case, a steel rivet is assembled with a plastics roller at 4000 per hour.[9]

Instead of hopper feeds, magazines may be used to store and supply components. Magazines are usually simple, vertical or near-vertical devices, holding a column of components under gravity or spring pressure. They have to be loaded with correctly orientated components, either automatically at the conclusion of the preceding operation, or by hand.

Though the manual loading of magazines may entail a considerable amount of labour, it can nevertheless be more efficient than the direct manual feeding of an assembling machine as it allows the operator to feed and orientate

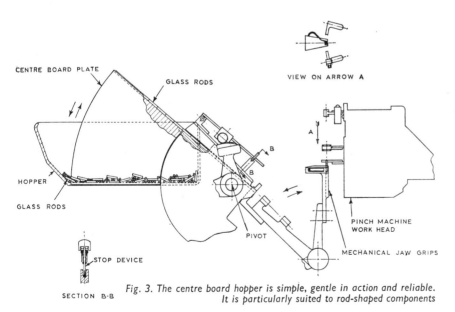

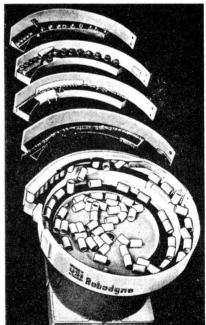

Fig. 3. The centre board hopper is simple, gentle in action and reliable. It is particularly suited to rod-shaped components

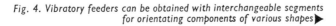

Fig. 4. Vibratory feeders can be obtained with interchangeable segments for orientating components of various shapes ▶

components at a high speed, independent of the time cycle of the machine.

Elevator feeders are similar to magazines in use and construction, and in their application to large components. Two pairs of elevator feeders are used to supply the lids and seats, respectively, in an assembling machine for toilet seats, one feeder of each pair being loaded with orientated components while the other is in operation. Each elevator feeder indexes the stack of components upwards by a system of counterweights and springs, and components are taken off sideways by simple, pneumatic transfer grabs.[10]

The elevator feeder should not be confused with the elevating hopper feeder, which is one of the nine basic designs of hopper feeds and is now available as a semi-standard unit.[4] The latter is essentially a belt conveyor, elevating relatively large components from an integral storage hopper and sorting them by means of slats attached to the belt and gating mechanisms mounted on the structure.

Transfer devices

Components form a queue at the hopper feed exit. From this point they have to be transferred, via a placing mechanism, into a work fixture or, more usually, into nests on a conveyor or indexing table which will present them to tools at successive stations. In almost all assembling machines gravity tracks or chutes connect the feed and locating mechanisms. The tracks may be twisted or looped, to turn the components over after they leave the hopper.

It is almost impossible to control the working cycle or the feed rate of the hopper feed itself accurately. Escapement mechanisms, such as the one shown in Fig. 5, are therefore used to synchronize the feeding of individual components with the subsequent operations and, incidentally, to separate from the queue the component about to be used. Standard escapements for certain limited applications are now commercially available in the U.S.A.[11]

As an alternative to gravity tracks, pneumatic conveying has recently been used for rivets, nuts, dowel pins, and similar small parts.[12] A pneumatically operated breech-block mechanism at the feeder exit serves the functions of separating one component and timing its delivery, thus acting as an escapement; compressed air then propels it through flexible plastic tubes. In pneumatic conveying the component travels at high speed, which reduces timing errors and promotes positive delivery (provided rebound is prevented). Unlike gravity tracks, the flexible tubes can run in any direction, thus allowing greater freedom in the positioning of the feed mechanisms relative to the tool stations.

Vacuum may be used in conjunction with compressed air, as in one case of transistor assembly, in which the emitter dot (a 0·009 in diameter indium sphere) is the first transistor component to be fed into the cavity of the assembly jig or 'boat'. Owing to the small size of the spheres and the softness of indium metal, they tend to adhere to most materials, making it difficult to use conventional escapement mechanisms for automatic handling. The emitter dot feeder developed for automatic boat loading incorporates a chamber in which a single sphere is isolated by a vacuum system. The sphere is then pneumatically propelled into a boat cavity through a glass tube.[13]

Where gravity tracks are used, and sometimes also with magazines, a considerable problem may arise in placing the part into the work fixture or conveyor nest. At this point, the part has usually to be moved both horizontally and vertically, and accurately timed. The difficulty lies in main-

Fig. 5. The hinged horizontal platform 'A' before the end stop is dropped by the solenoid and then replaced in the time it takes the next item to move into position

[Courtesy, Rhoden Partners Ltd]

taining positive control and orientation of the part under these conditions. Parts-placing mechanisms usually employ one of three basic methods: gravity fall, pick-and-place or push-and-guide. Besides placing, they may also provide the escapement function.

Because such devices are needed frequently, yet are difficult to develop without an accumulation of experience, attempts have lately been made to produce standard placing mechanisms. The *TransfeRobot* unit shown in Fig. 6 will grasp, lift, move in a straight line, lower, and release a part in a widely variable combination of movements. It is, for instance, being used to grasp simultaneously two spring contacts and transfer them into position over a previously placed plastic base, in the assembling of butt-end fluorescent light sockets.[14]

The Herbert mechanical arm (Fig. 7) has no horizontal straight-line motion but swings horizontally through 90 or 180 degrees, in addition to lifting and lowering from $1\frac{1}{2}$ to 4 inches at any swing position, opening and closing fingers for inside or outside grip, and twisting through 90 or 180 degrees. It consists of three units: the arm itself, a hydraulic power pack, and a console. Only the grip has to be specially produced to meet a given requirement. The arm is very substantial and may be used for transfer machining as well as for assembling.[15]

The Swanson-Erie Corporation has developed an alternative approach to maintaining positive control at this difficult transfer point. They use a chassis with a vertically reciprocating overhead tool plate, for mounting and actuating escapements, as well as orientating devices, probes, unloading mechanisms and other tooling. The plate is powered by the main drive and synchronized with the work transfer index.[16]

A centre-mounted turret on the Camatic assembling machine[17] contains a system of cams and linkages. The cams provide basic horizontal and vertical motions, which are transmitted to the area above the work circle of an indexing table.

Horizontal placing motion is brought about by a grooved face cam and transmitted by an oscillating disc. Vertical motion for placing parts into the nests is obtained from a cylindrical cam. The cam drive shaft is mechanically synchronized with the indexing; both are driven off the same power supply.

A third, optional motion, also cam-actuated, facilitates the clamping of parts. The same motion can be utilized for external or internal collets or for actuating jaws. The parts can also be held by vacuum, supplied through flexible tubes, or by permanent magnets.

Conveyors

In the simplest assembly machines some of the components —usually the fasteners—are supplied automatically by mechanisms of the type described above, but the parts to be fastened together are manually placed in a single work station directly below the assembling tool[18] or mounted to slide into and out of that operating position. In more sophisticated machines, a number of fixtures or nests are linked to form, or are mounted on, a conveyor which takes them past tool stations at which successive assembling operations are carried out. The term conveyor is here used in the widest sense, to include not only vertical and horizontal in-line conveyors, but also dial feeds, rotary indexing tables, and other circular devices.

The distinction between dial feeds and rotary indexing tables is not always clear; it is perhaps largely historical in that dial feeds were introduced more than 60 years ago, to facilitate the safe and rapid handling of components into and out of power presses, while rotary indexing tables are a later development, generally independently driven by an electric motor, through a Geneva movement or other intermittent-motion device, or by fluid power. Some of the earliest dial feeds used for assembling served single tool stations and had two dial positions only. Later dial feeds had an increasing number of positions, for hand loading by several operators. In an early assembly machine for toy locomotives which was equipped with a dial feed, five operators loaded and transferred components and sub-assemblies to and from fixtures for operations such as staking and crimping in a press. An unusual feature of this machine was that four revolutions of the dial were required to complete one assembly.

In modern applications of dial feeds and indexing tables, the associated use of hopper feeds is, of course, much more in evidence. For example, five hoppers are used to produce 48 conveyor roller bearings per minute; they deal with a housing,

[Courtesy, U.S. Industries Inc.]

Fig. 6. The standard positioning arm mounted at the back can be programmed to grasp (in the clamps), lift and move the pressings as required with accurate timing of the operations

[Courtesy, Alfred Herbert Ltd]

Fig. 7. Another transfer arm for larger loads not only moves them about but can also up-end them as shown

a lower race, a centre core, 11 balls, and an upper race. The dial feed carries 16 fixtures, each of which passes two inspection units where limit switches check the number of balls and the correctness of assembly. There are two press stations, at which the edge of the housing is first crimped and then flattened.[19]

How to select a standard dial feed or indexing table, or how to design a special one, is an engineering problem only second in importance to the correct choice of feeding and placing devices. For example, the type and sequence of operations has to be considered: heavy vertical loads may have to be absorbed in press work, great accuracy or high speed of indexing may be required, the number of indexes per revolution may have to be changed quickly, or rotation in either sense may be necessary. In the U.S.A., and to a lesser extent in Great Britain, a wide choice of fluid-powered and of mechanically operated, standard rotary indexing tables is now available. A typical one of these is shown in Fig. 8.[20] It is probably true that none of these is completely satisfactory in respect of all desirable features, including those just described. In general, design faults or weaknesses to beware of are lack of accuracy or a tendency to overshoot indexing positions, inadequate bearing or impact-absorbing arrangements or complexity of design, leading to premature wear or maintenance difficulties; and unsatisfactory methods of control, resulting in stoppages, jamming, or incorrect sequencing.

Intermittent motion is an essential feature of the rotary units so far described. It allows most, if not all, operations to occur while the work fixtures are stationary, with obvious advantages. Nevertheless, on general engineering grounds, intermittent motion has serious drawbacks for this type of work, because of the acceleration forces arising, the additional mechanisms required, and because part of the cycle time is generally not utilized.

It follows that improvements can often be obtained by using continuous, in place of intermittent motion, as in the case of a pinch-making machine for electric lamps, which recently underwent trials at an output rate of about 3500 per hour. Earlier, intermittent-motion machines for making pinch-seals operated at 1500 to 1800 units per hour and could not conveniently be run faster, not only because of the forces involved in starting and stopping but, in this particular case, also because of shortage of time at given indexing positions to complete certain heating and cooling operations.[21] To handle components at the high speed required by this continuously rotating pinch-making machine new mechanical feeding and placing mechanisms had to be developed.[3] One of these is shown in Fig. 9.

The rotary construction of assembly machines has certain useful features. The bearing arrangements can be simple, the frame structure and its alignment present few problems, and it is not too difficult to maintain the working surface accurately level as the machine rotates. On the other hand the working spaces, consisting as they do of sectors of a circle, cannot conveniently be fully utilized, the area near the centre being generally wasted. It would appear at first sight that the central area of a rotary machine could be well utilized by accommodating hopper feeds, services, and other equipment; but these would then be difficult to reach. On the other hand, if hopper feeds are ranged round the periphery of the machine, access to the work stations is poor.

In fact, the difficulty of arranging both the work stations and the feeds and other supplies so that there is adequate access to all is one of the inherent weaknesses of the rotary layout, particularly where operators too must be accommodated. A further disadvantage, possibly more serious than any so far described, is that, with a given table diameter, there is no scope for increasing the number of operating stations once the machine is built (except by decreasing the index angle, which is usually impracticable). This makes it difficult to add further operations, such as inspection.

In-line assembly machines do not suffer from these drawbacks. The simplest design is based on a straight chain conveyor; more elaborate ones follow those used in transfer machining and may have horizontal and vertical paths. Standard in-line assembly machine beds incorporating conveyors[22] are now offered in Great Britain (Fig. 10), and a rather wider variety of standard conveyor indexing chassis[23] and in-line transfer tables[11] is available in the U.S.A.

Very generally, assembly machines with few, small tool stations and work fixtures tend to be rotary, while large machines tend to be of in-line construction, but there are many exceptions. Rotary electric lamp assembling machines have as many as 60 work fixtures.[3] A rotary assembly machine for shock absorber pistons and rods has 30 fixtures and an annular platform of 6 ft diameter.[24] In many cases, of course, the choice between rotary and in-line construction is governed by some specific process requirement which may override the above design considerations, as in the case of a machine for miniature lipsticks, in which a Swanson-Erie standard in-line transfer machine chassis was used to accommodate a refrigerated tunnel extending over several stations.[25]

Tooling

Riveting, staking, screwing, welding, soldering, bonding, and doweling are typical operations used to fasten the components of an assembly automatically. They and similar operations have been listed and grouped to show their relative importance and suitability for automatic assembling.[26] These being conventional operations, as used in manual assembling, the tools used are also largely conventional, except for the methods of actuation and certain aspects of inspection.

Interlocking is the neatest assembling operation, as it requires no separate fasteners and is quick, reliable, and involves no heating, drying, or similar process times. To be successful, the interlocking movements must be short and simple and this, in turn, depends on the design of the components to be assembled.

An interesting approximation to interlocking is used in assembling cast iron inserts for inlet and exhaust valve seats

Fig. 8. A double-acting air or oil cylinder operates racks for providing up to 24 indexing motions per revolution of the table. The smaller cylinder effects the change-over from the forward to the reverse rack. Hydraulic control also permits work between stations

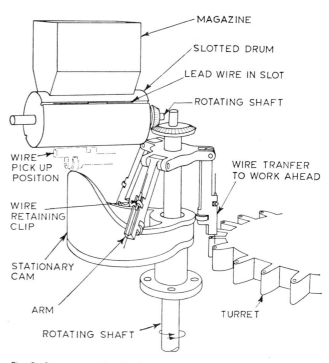

Fig. 9. Arrangement for the feeding and placing of lead wires in lamp manufacture. The pivoted arms on the shaft retain the wires fed from a magazine until they can be dropped over the turret

to the aluminium cylinder heads of Buick Special V-8 engines by shrink-fitting without positive fastening. The inserts are press-fitted into the heads at 350 to 375 degrees F, ensuring a tight shrink fit after cooling. The assembly set-up not only includes heating facilities for the cylinder heads but also means for rapid placing of the inserts before cooling can take place. Two vibratory bowl feeders are used, as well as air hammers to drive home inserts at the rate of 120 or more cylinder heads per hour.[27]

Welding, particularly projection and other forms of resistance welding, involves no separate fasteners and often lends itself very well to incorporation in automatic assembly sequences. It is particularly suitable for fastening thin or small components in electrical assemblies. Soldering, gluing, and other forms of bonding approximate to welding in their convenience and reliability as fastening operations, having regard to the strength limitation in each case. The use of self-adhesive surfaces is also gaining ground.

Fasteners

Of the methods which involve separate fasteners, those using linear thrust are preferable to those requiring torsion. This leads to the extensive use of rivets and eyelets, including the types which can be placed and secured from one side of an assembly without access to the rear. *Prestincerts*[28] is a family of fasteners which include threaded bushes, rivets, and pins, all of which pierce their own holes into sheet material, including steel, thus obviating the need for a separately punched hole. Apart from saving a press operation, this has the advantage, in mechanized assembly, of eliminating a lining-up process in those cases in which highly accurate positioning of the fastener is not essential.

Screwing fasteners into threaded holes, and nut-running, are relatively elaborate operations because of the difficulties of starting and stopping the travel and avoiding crossed threads. Chamfered threads, pilot ends, and free-running threads on the one hand, and slipping or automatically reversing torque-limiting tools with flexible or resilient heads on the other, are of course used but, in practice, are often still so difficult to set or use as to make them unacceptable, or at least unpopular.

From the user's viewpoint, if not from the manufacturer's, screwed fasteners have the advantage that they can be removed and replaced—for inspection and repair of the assembly—while riveted ones are generally destroyed in dismantling. Certain specialized types, notably Dzus fasteners,[29] combine the advantages of non-destructive removal with fairly easy assembly, by twisting instead of screwing or by spring-retained axial loading, but there appears to be scope for additional types of simple, rivet-like fasteners, capable of being applied and secured by simple and auto-

matic tooling, the fasteners to exert a good retaining force on the assembly but to be re-useable a number of times without appreciable deterioration.

The feeding and orientating of fasteners is, of course, only a special case of handling small components, to which much of the early part of this article has been devoted, but because fasteners are highly standardized, both for shape and dimensions, equipment for handling them automatically is more highly developed. Commercially available rivet hopper feeds and associated gravity tracks have been in use for a long time,[30] though the remainder of the assembling operations may have been entirely manual.

In equipment which can be added to a standard electric welding machine, a *Weldnut* is supplied from a vibratory bowl feeder and held just below, and in the path of travel of, the upper electrode by a pair of jaws, which release it just after it has been entered by a pilot pin protruding through a location hole in the component.[31]

Riveting, particularly with semi-tubular rivets, also lends itself well to automatic control, especially when using compressed air to time the rivet delivery and operate the riveting press or other tool. An incidental advantage of multiple-riveting[12] is the elimination of distortion of the assembly, which is sometimes introduced when closing the various rivets one at a time. Again, the same automatic method which is used in riveting to control the supply of the fasteners and to actuate the tooling can readily be adapted to dowel pins[18] or to the newer, split, hollow *Rollpins*,[32] which act as resilient dowel pins.

Despite the difficulties described concerning screwed assemblies, there is American semi-standard equipment for feeding and driving screws and nuts,[33] and British equipment for simultaneously feeding and driving single screws, is shown in Fig. 11.[34] The British equipment uses a mechanical vibrating tray as a hopper feed or, in a more recent version, a reciprocating hopper feed, permitting up to 6 screws to be fed and driven simultaneously in 3 seconds.

A German electronic-pneumatic screwdriving machine has jaws fitted with a device which, if it senses a minor fault such as crossed threads, will order a new screw and restart the machine cycle. If the fault is serious, the machine will stop.[35]

In the machine for assembling shock absorber pistons and rods already mentioned,[24] the hexagonal piston nut is automatically screwed on to the rod, using a sliding carriage as the placing mechanism and a nut runner which engages the top face of the nut with four spikes instead of the normal socket.

Among more specialized forms of automatically controlled tooling, various forms of dispensers, for lubricants as well as for adhesives, are gaining importance. Lubricants frequently have to be introduced into a product part-way through the assembly process, as in ball bearing manufacture,[36] while adhesives are required in the bonding methods already referred to. Adhesive dispensers may have to mix and apply two substances simultaneously and at the same point, when accelerators or catalysts are specified.

Machining combined with assembling

There is a growing tendency to carry out some forming or machining operations during assembly, where this eliminates an awkward orientating or alignment problem. In the manufacture of Kodak *Brownie 20* cameras, the red plastic button for film positioning in the camera is fabricated from tape and assembled and heat-sealed to the camera back during the camera assembly process. Similarly, the aluminium clips that

[Courtesy, P. C. Payne Ltd]

Fig. 10. A standard in-line conveyor base for various assembly operations can take as many tool stations as required

hold the camera halves together are partially blanked in a hydraulic press but retained in orientated pairs attached to a distance piece and fed in strip form to a magazine loading station. These two examples[37] superseded vibratory bowl feeder methods which, though rapid, sometimes gave trouble that delayed production. A reduction in the cost of handling, avoidance of damage in transit, and the elimination of chips and malformed parts are advantages of the new method. Production control is also improved.

In the automatic production of stators for alternators for the electrical systems of cars, Chrysler Corporation use a combined forming and assembly line in which a winding machine form-winds phase coils from copper wire. The coils are placed in the slots of a stator, the stator is then transferred to a wedging unit where strip material is inserted in the slots to retain the coils and, in a final work station, the windings are press-formed to size and the assemblies ejected automatically.[38]

A punched paper-tape-controlled assembly machine for small batches of printed circuits has a station at which holes are drilled in panels up to 10 inches square, prior to the automatic insertion and clinching of the components.[39]

Inspection and packaging

The purpose of inspection in connection with automatic assembly is threefold: to ensure satisfactory quality; to prevent damage to the machinery; and to prevent unscheduled stoppages in the assembly process.

Inspection for quality has first to establish that the components supplied to the assembly machine are made to specification and will fit together; and secondly that the assembly operations are carried out correctly. Prevention of damage to the machine is often even more important as it is likely to be much more costly than the loss of one, or even a batch of assemblies. Inspection, therefore, has to ascertain that components do not obstruct the tools or other machine parts, by failing to take up their correct positions, both relative to each other and to the machine. For fully automatic operation, automatic tool inspection is also required, to ensure that broken tools will not cause scrap. Broken or blunt tools can also cause damage to tooling further along the line.

Comprehensive automatic inspection of this kind is complex and expensive for any but the simplest assembly. So far, it has been restricted to the largest and fastest lines, for instance, electric lamp machinery and the Russian automatic ball-bearing production line.[36] A further example is a 16-station machine, used by IBM for assembling wire contact relays, which monitors the presence of components and their proper location. If a tool breakage is detected, part-assemblies by-pass subsequent stations and are discharged.[40]

Viewed as part of a comprehensive manufacturing process, wrapping, cartoning and some other forms of packaging are often closely related to the preceding assembling operations. Moreover, the engineering problems arising in mechanized packaging are frequently identical to those encountered in mechanized assembling. Indeed, it is not unreasonable to regard packaging as one assembly operation, particularly where the product consists of units which have to be handled, orientated, and generally arranged with some accuracy in relation to the package.

As packaging becomes more sophisticated, the resemblance between packaging and assembling becomes more striking. When drawing pins are mechanically packed into small cartons holding 36 or so, it may be questioned whether the automatic erection of the carton, the filling by pouring and weighing, and the automatic closing and labelling of the

Mr G. Wittenberg received his technical education at the Northern and Northampton Polytechnics in London. After an engineering traineeship with Selman and Son Ltd, he carried out research and development work in a number of small engineering companies, on vacuum pumps, food machinery, sheet metal products, and packaging equipment. Since 1953 he has been a director fo Rhoden Partners Ltd, taking a particular interest in development projects involving production mechanization, including assembling and packaging. He undertakes advisory work, writes, and lectures. He became an Associate Member in 1953.

carton constitute an assembly process, but when the same pins are automatically orientated in a regular pattern of 9×4 rows, automatically pressed half-way into a carefully aligned, thick card which takes the place of the conventional carton, and automatically over-wrapped in transparent plastic film (as in a display pack popularized by American self-service stores and now produced in Great Britain[41]) mechanized assembly techniques are certainly involved.

Pressure-packs (aerosols) may be the best example of the integration of assembly proper and packaging, since the pack comprising container, valve, and cap has to be assembled in close co-ordination with the filling (packing) process. A production line comprising 14 major machines, for feeding, cleaning, filling, pressurizing, testing, labelling, and case loading containers of Johnson's Wax products operates at a rate of more than 200 containers per minute.[42]

An example of more conventional automatic packing in conjunction with automatic assembly is to be found in the manufacture of electric lamps. A 70 ft line for assembling 3000 incandescent lamps per hour at Westinghouse comprises, apart from four major assembly machines (stem-and-mount machine, sealing-in machine, basing machine, and silica coating machine) and a monogramming unit, two packaging machines (for sleeving, and for packing into cartons on an automatic case loading and sealing machine), integrated by conveyors, parts feeders, loading and unloading devices.[43]

The integration of assembly is of particular value in the manufacture of pharmaceutical and of surgical products, as it helps to maintain cleanliness. Sterilizing, e.g., by irradiation, just before or after sealing the package, often forms part of the automatic process.[44]

Conclusion

In the ever-continuing search for more efficient manufacturing techniques the mechanizing of assembly processes now deserves, and to an extent receives, the greatest attention. Completely or nearly automatic assembly, as in the manufacture of ball point pens, dry batteries, some car accessories, bearings, and electric lamps, is still rare, mainly because of difficulties in automatic inspection.

In mechanized, as distinct from automatic, assembly, which implies some manual operations, the main difficulties centre on the handling and orientating of components, rather than the fastening and other tooling operations. Some semi-standard hopper feeds are now commercially available, notably vibratory (bowl) feeders, but there is need for many more types of standard equipment. Standardization, by reducing the development risk in new projects, accelerates progress in this field.

Despite the increasing versatility of standard hopper feeds, escapements, conveyors, and similar units, mechanized assembly equipment still tends to be inflexible; that is, a given machine cannot readily be designed to assemble more than a narrow range of shapes or sizes. While this lack of flexibility prevails, the best applications for highly-mechanized assembly will be found in long runs, say over one million per batch or per year continuously, of products comprising few components, simple and accurate in shape, with few fastening operations, preferably interlocking, welding, or riveting.

To meet the development of standard assembly machinery half-way, the product components will have to be designed to be suitable for such techniques. This consideration need not clash with functional requirements. As a simple example, components which are slightly asymmetrical are obviously more difficult to orientate than symmetrical components, but they are also more difficult to orientate than components of pronounced asymmetry; they should be avoided. Again, sorting by shape or weight is relatively easy; sorting by colour or other surface characteristics is much more difficult and the need for it should therefore also be avoided. Attempts have been made to classify component shapes by complexity of features,[45] and while these are too primitive to be of great practical value so far, they indicate a useful approach towards making, sorting and orientating a predictable subject.

More types of standard fasteners will probably be developed specifically for mechanized assembly. Their success will partly depend on the simultaneous development of hopper feeds for them, perhaps by the fastener supplier rather than by the user. Other new fasteners may be manufactured in linked form or adhering to feed tapes, to obviate the need for subsequent orientation.

Packaging should in many cases be regarded as an extension of the assembly process, using the same engineering techniques. In mass production, and in industries in which cleanliness is vital, this view is already becoming accepted.

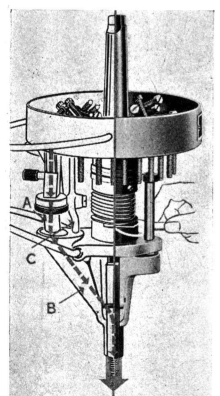

Fig. 11. The screws fed to this automatic driver descend to where tube (A) enter delivery member (B). When the preceding screw has been driven home, trip level (C) drops the next one into the driving nozzle

[Courtesy, Russel Auto-feed Screwdrivers]

REFERENCES

1. 'Hopper Feeds as an Aid to Automation'. Machinery's Yellow Back series, no. 39. Machinery Publishing Co. Ltd.
2. Precision Detroit Co. Catalog 60. Detroit 34, Michigan, U.S.A.
3. GARDINER, H. W. B., HEYWOOD, E. A., and WORSDELL, W. J., 'Machines for Electric Lamp Manufacture'. Proc. Instn mech. Engrs, Lond., 1963 vol. 177 separate no. 24.
4. 'Small Component Feeder Units'. 1963 February Light Production Engineering, p. 27.
5. MCKINSEY, W., and MORAN, J. P. 'Application of Vibratory Hopper Feeders'. 1957 May The Tool Engineer, vol. 38.
6. 'The Segmented Vibratory Bowl Feeder' leaflet. U.S. Industries Inc. Engineering Ltd, Welwyn Garden City, Herts.
7. 'Syntron Small Part Feeders'. Publication R 598. Riley (I C) Products Ltd, London, W.C.1.
8. SMITH, F. E., 'Applying Vibratory Bowl-Type Feeders'. 1962 November pp. 97–102 and December pp. 79–85 and 1963 January pp. 78–85 Automation.
9. BURGESS, W. C., 'Handling with Vibration'. 1962 July Automation, pp. 76–82.
10. AIDLIN, S. S., and AIDLIN, S. H., 'Conveyor-fixture links 8 Stations in Producing Hinged Assemblies'. 1961 February Automation, pp. 77–78.
11. 'Dixon Parts Escapements''. Bulletin ES-B5, and 'Dixon Standard Components'. Dixon Automatic Tools, Inc., Rockford, Ill., U.S.A.
12. 'Machine for Inserting and Closing Multiple Rivets'. 1961, 20th April Machinery, pp. 882–83.
13. 'Assembling Transistor Parts for Furnace Alloying'. 1961 March Automation, pp. 87–89.
14. 'Robot Lubrication and Assembly'. 1961 September Control, p. 122.
15. 'Mechanical Handling'. 1962 March–April Machine-Tool Review, pp. 26–28.
16. SWANSON, DOUGLAS, 'Precision Feeding and Handling Operations'. 1962 August Automation, pp. 77–81.
17. SIMONE, R. J. 'Standardized Parts Placing Mechanisms increase assembly machine adaptability'. 1963 February Automation, pp. 61–64.
18. 'Machine for Feeding and Inserting Dowel Pins'. 1961, 1st March Machinery, p. 502.
19. SCHUG, W. W., 'Applying Dial Feeds for Presswork and Assembly'. 1962 February Automation, pp. 86–89.
20. RADNOR, GLEN, 'Rotary Transfer for Increased Productivity'. 1963 April Mass Production, vol. 39 no. 4 p. 41.
21. 'Electric Bulb Machine Assemblies on a Turntable'. 1960 January Automation Progress, p. 25.
22. 'Semblamatic Automatic Assembly Units' leaflet. P. C. Payne and Co. (Keynsham) Ltd, Willsbridge, Somerset.
23. 'Conveyor Indexing Chassis'. Leaflet 3/61-Y-DR. Sylvania Lighting Products, Salem, Mass., U.S.A.
24. 'Automatic Assembly Machine'. 1962 Feb. Automobile Engineer, p. 78.
25. 'Indexing Machine Moulds and Assembles Lipsticks'. 1962 April Automation, pp. 78–79.
26. AMBER, G. H., and AMBER, P. S. Anatomy of Automation, p. 70 1962. Prentice-Hall Inc, Englewood Cliff, N.J.
27. CHASE, H. 'How Valve Seat Inserts are Assembled to Aluminium Heads' 1962, 15th May Automotive Industries, pp. 85 and 126.
28. Textile Weekly 1961, 14th July.
29. HERRIDGE, F. W., 'Production of Dzus Fastener Parts'. Machinery 1962, 20th June, p. 1409.
30. 'Phoenix' leaflet. Hunton Ltd, Phoenix Works, London N.W.1.
31. 'Automatic Feeding Equipment for Nut Welding Machine'. 1957, 17th May The Engineer, p. 767.
32. 'Fasteners and Fastenings—Part 2'. 1963 February Light Production Engineering, pp. 4–8.
33. 'Dixon Auto-Torque Driver'. Bulletin SD-B1. Dixon Automatic Tools Inc., Rockford, Ill., U.S.A.
34. 'Russell Screw or Pin Driving Machines'. Leaflet ANH/KP. Russell Auto-Feed Screwdrivers, Studley, Warwickshire.
35. 'An Electronic-pneumatic Screwdriver'. 1963 February Design and Components in Engineering, p. 68.
36. 'Automation for Ball and Roller Bearings'. 1960 October Automation Progress, pp. 350–351.
37. 'Assembling and Testing Camera Parts'. 1962 May Automation, pp. 57–60.
38. BIDDISON, J. M. 'Automatic Assembly System produces Wound Stators'. 1961 March Automation, p. 74.
39. 'Robot Builds Small Batches of Circuits'. 1962 22nd November New Scientist (no. 314) pp. 440–441.
40. LARUE, A. J. 'Stress Quality for Successful Automatic Assembly'. 1960 June Automation, pp. 73–78.
41. 'Drawing Pins in Colour'. 1963, 15th February Office Equipment News, p. 4.
42. BARRETT, J. P., 'Packaging pressurized products'. 1962 April Automation, pp. 61–65.
43. WILBURN, J. E. 'Making Incandescent Lamps'. 1961 November Automation, pp. 64–70.
44. 'Automated Handling of Packaged Syringes in Gamma Ray Sterilisation Plant'. 1963 January Industrial Handling, pp. 13–17.
45. NOY, P. C. 'Check-Chart for Single-Part Complexity'. 1959 Product Engineering Design Manual. McGraw-Hill Book Co., pp. 150–151.

Mechanical Developments in Nuclear Power

The nuclear power industry's main contributions to mechanical engineering techniques in general have been in the field of heat transfer, control mechanisms and the practical application of new or previously unfamiliar materials. Remote handling techniques and the effects of radiation on materials have presented new challenges to mechanical engineers.

by H. Kronberger, OBE, BSc, PhD

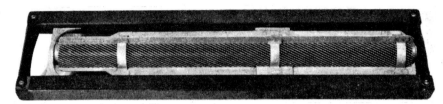

Fig. 1. Convection is increased by the polyzonal design, here shown in a Berkeley fuel element with its helical finning

The nuclear energy industry is a very young one, being less than thirty years old. The early period of its existence was governed by military requirements, when the prime consideration was to get a plant or reactor into operation by a certain date, and it was necessary to accept the economic penalties of this philosophy. The design of plant in those days was characterized by the 'belt and braces' approach to guarantee satisfactory operation, and this was reflected in the mechanical engineering aspects.

The nuclear energy industry has now become a competitive one, competing not only against conventional sources of power but within itself; new types of reactor or processing plants will only be developed if they show a clear economic advantage. The skill of the engineer is taxed to get the most out of the materials and techniques available, while observing the safety criteria which are still considerably more stringent than in normal industrial practice. There have been a few radically new mechanical developments, but in the main development has been concerned with the extension of existing techniques.

Much of the development reviewed here is not unique but two aspects are peculiar to atomic energy applications and have had their repercussion on design. One is the change of physical properties of materials under irradiation, which had to be allowed for in design, very often on the basis of extrapolations from limited data. The second aspect is that certain plant components became inaccessible in operation or only accessible at high cost. This has led to the requirement of extremely high reliability, and to the development of remote handling techniques.

Prediction of the performance to be anticipated from a nuclear power station is necessarily more complicated than that for a conventional station. There are a greater number of alternatives to examine, including even possible variations in basic data, so that a larger number of evaluations will be required in support of a given design. It is just this sort of situation that the electronic digital computer is ideally suited to meet, and today it is doing so on a scale that is consider-

able, and likely to increase. Once performance predictions are made in this way, a natural step forward leads to the 'parametric survey' technique in which a 'cut-and-try' process first gives a near-minimum cost specification. After refinements are added, a large number of variations of this specification can finally be displayed. In this way, the designer can select one which not only combines several desirable features but also keeps close to the lowest generating cost.

Heat transfer

Atomic energy is now well established as one of the main forcing points in the field of heat transfer. Increases in heat rating and coolant temperatures constitute important advances in the development of power reactors. The margins between the design temperatures of the fuel elements and the permissible metallurgical limits are necessarily small so that the fuel element temperature distributions must be established in great detail, with considerable accuracy. The conditions which give rise to disastrous failures of the heat transfer process, such as instabilities and burn-out phenomena in boiling heat transfer, must be very thoroughly established.

It will be found that heat transfer work in the atomic energy field contains a higher ratio of basic experiments to *ad hoc* tests, than is usual in other fields.

Perhaps the most interesting thermal conduction problem has been that of heat transfer across the interface between two solid surfaces. Its investigation is important in minimizing temperature differences between the fuel and its cladding. Excessive fuel temperatures could lead to undesirable phase changes in the case of uranium metal and to high internal fission gas pressures in the case of uranium dioxide. The work of Cetinkale and Fishenden[1] has been of particular importance in clearly distinguishing the two parallel heat transfer paths across an interface: solid conduction across the contact points and fluid conduction elsewhere (radiation is usually unimportant); and in relating the conductances of the two paths to the properties of the materials (including their surfaces) and the intermediate fluid. Application of

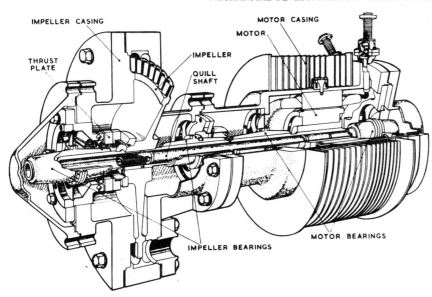

IMPELLER CASING

MOTOR CASING

MOTOR

IMPELLER

THRUST PLATE

QUILL SHAFT

MOTOR BEARINGS

IMPELLER BEARINGS

Fig. 2. The totally enclosed motor and impeller of this carbon dioxide circulating compressor are both mounted between two gas bearings

these methods led to remarkably good predictions of the main features of uranium-magnox interface heat transfer: the effect of contact pressure, of temperature (through its effect on the hardness of magnox), and of the dominance of solid conduction, as against conduction through the interface gap, at the higher temperatures and contact pressures.[2]

Interfaces between steel and uranium dioxide present a very different picture. Solid conduction is ineffective, owing to the hardness of the materials and the poor thermal conductivity of uranium dioxide, and therefore, gaseous conduction becomes very important. A correlation has been developed which appears to account quite well for the influence of gas pressure in a wide range of gases and surface finishes.[3]

The great bulk of the work on convection has been concerned with forced convection from finned surfaces.[4] Although much of it has been empirical in nature, it has required a nice appreciation of thermodynamic similarity in situations where fin efficiency enters as an additional complication. Ingenious heat transfer correlations have been developed which not only cover a given finned can under various operating conditions, but also embrace a very considerable range of can geometries.[4]

The fuel cladding developed by British industry for the magnox power reactors represents a substantial improvement in heat transfer performance as compared with the transverse-finned cans used at Calder Hall. This has been achieved by augmenting the random, turbulent heat transfer by a systematic secondary flow process in which hot gas is continuously removed from the can surface and replaced by cool fluid from remote parts of the flow passage. The polyzonal arrangement, of elements, shown in Fig. 1, which embodies this idea, has flowered into many shapes—helical, straight and herringbone pattern fins and, as Ritz[4] has made clear, its possibilities are by no means exhausted. The complex polyzonal flow pattern has led to problems, notably in undesirable point-to-point variations of heat transfer coefficient, but the progress being made is encouraging.[5]

Investigations of heat transfer from systematically roughened surfaces have been made by the A.E.A. at Windscale and elsewhere over the past five years.[6] They were started in the belief that secondary surfaces would seriously decline in value, owing to diminishing fin efficiency, with the trend to more highly rated reactors. Roughened surfaces are indeed finding application in advanced gas-cooled reactors.

The development of reactor steam raising units has been supported by extensive heat transfer measurements on various arrangements of finned and studded tubes in cross-flow, and empirical performance correlations have been presented.[4]

Burn-out is probably the most important aspect of boiling heat transfer, since it imposes a limit on the quantity of heat which can be safely removed from the reactor. It occurs if the heat flow to a particular coolant channel is too high and results in the formation of an insulating layer of vapour on the heated walls. The wall temperature rises, usually above the melting point of the materials which can be used. Knowledge of the phenomenon is empirical and so far confined to particular geometries and pressures.[7, 8] Other interacting phenomena, such as bubble nucleation, two-phase flow patterns, oscillations, etc., have been reviewed.[9, 10]

Fluid dynamics and gas circulation

The development of reactor systems has brought to greater prominence an aspect of fluid mechanics which has not previously received much attention. This concerns the movement of bodies under more or less parallel flow.

Two conditions can be considered. The body, if unrestrained mechanically, may rattle around in its constraining channel as a rigid body under the action of the gas forces. This freedom of movement can be an important aspect of reactor design where clearances exist to allow expansion of fuel or where fuel elements require changing under load conditions and charge and shut-off mechanisms need to operate freely. With these problems it is generally necessary to assess each case on its merits, and *ad hoc* testing is usual.

Alternatively, the body may be constrained in some manner, but the gas flow past it may cause elastic vibration. This problem has been examined for simple rods by Burgreen *et al*[11] who have deduced non-dimensional groups of parameters and shown their applicability to vibration.

Great strides have been made by the U.K.A.E.A. in the development of hydrodynamic gas bearings.[12] Compressors utilizing gas bearings have been operated successfully in the Capenhurst diffusion plant for a number of years. A circulator has been developed for circulating carbon dioxide in irradiation loop experiments. The totally enclosed electric driving motor and the driven impeller are each mounted between two gas bearings, the drive being taken through a quill shaft connecting the two, as shown in Fig. 2. The type of impeller used is one which incorporates peripheral blades passing through a peripheral channel. This circulator can operate with an inlet temperature of 300°C; a new design is being developed to raise the temperature of operation to 500°C. The advantages of gas bearing compressors, in addition to the high permissible temperature of operation, include elimination of contamination from non-gaseous lubricants and reduction in maintenance requirements. Gas bearing mounted compressors are now manufactured in Britain, France and Switzerland.

The problem of circulating gases which need to be sealed from the atmosphere is not always solved by the use of totally enclosed compressors and gas-lubricated bearings. In some cases it is more appropriate to use compressors running in conventional bearings, in which only the impeller is immersed in the plant gas.

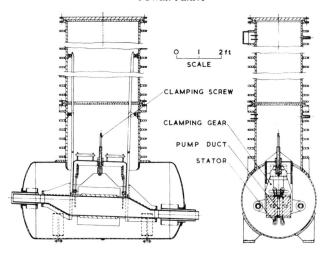

Fig 3. One of the 48 electromagnetic pumps for the sodium-potassium coolant, installed in the Dounreay fast reactor

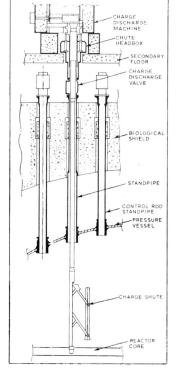

Fig. 4. The charging chute at the Bradwell nuclear power station can operate through one of a limited number of access openings

The rest of the machine, and the atmosphere, are separated from the gas by a shaft seal. The stringent limits of leakage rate, coupled with the reliable long-term operation often required in atomic energy applications, have led to the development of new types of shaft seal.

Whitley and Williams[13] discuss four types of gas-lubricated seal, in each of which the sealing action depends on the limitation of gas leakage between parallel surfaces, separated by distances of the order of 2×10^{-4} in. The parallel surfaces can be on discs or concentric cylinders. In the case of the parallel discs, the small gap is maintained by means of a gas-lubricated thrust-bearing which may be either hydrodynamic or hydrostatic. The concentric cylinders rely on similar journal bearings. The rate of leakage across such seals can be reduced to a few cm³/min at S.T.P. in some cases. Whitley and Williams conclude that, in general, the disc-type seals are more useful than the sleeve seals, because it is possible to obtain lower operating gaps.

The seal developed for the Calder circulators[14] is not of the gas-lubricated type, although the principle is basically the same. It is a hydrodynamic disc seal, with an oil-lubricated thrust-bearing portion.

Pumps—canned and electromagnetic

Pumps for water-cooled power reactors have recently been reviewed.[15] The most commonly used type is the 'canned-rotor' one in which rotor and impeller are contained in a single, leakproof unit; both rotor and stator are sealed in corrosion-resistant non-magnetic alloy liners. An auxiliary impeller circulates a little water between the stator and rotor which cools the motor and lubricates the bearings.

Recent developments have been concerned with improvements in the *Graphitar* bearings, thickness and material of cans, and in hydraulic performance. In Britain, a totally submersible pump has been developed in which the motor windings are in direct contact with the water being pumped; polythene has been used for wire insulation and joints. Removal of the stator-can makes possible higher motor efficiencies but the performance of the windings at high temperatures, in the presence of radiation or radiolytic products, presents formidable problems.

High heat transfer coefficients and relatively high boiling points make liquid metals attractive reactor coolants, particularly for highly rated fast reactors. Electromagnetic pumps have been used to circulate them. They have the advantages over conventional mechanical pumps of having no seals or moving parts.[16-18]

So far the larger pumps have been of the linear induction type, analogous to the squirrel cage induction motor. The two forms developed are the flat form in which the pump tube is flat and there is a wound stator on each side; the alternative is the annular type. The prime use of electromagnetic pumps, within the U.K.A.E.A., has been in the Dounreay fast reactor, with sodium and sodium-potassium alloy as coolant. Forty-eight flat linear induction pumps, such as the one shown in Fig. 3, each with a flow capacity of 400 gal/min. and a head of 30 lb/in², were installed in the reactor and have operated without significant difficulty since mid-1959.

The main aim of development has been to increase the size of pumps, and to raise the operating temperature of the stator insulation materials whilst retaining a long reliable life. A design study for a DC conduction pump with a capacity of 10 000 gal/min was made for the U.S. EBR.II reactor.[19] However, the lower efficiencies of electromagnetic pumps have biased the present trend in liquid-metal-cooled reactors towards the mechanical pump.

Refuelling

A large uranium-magnox power reactor needs more than 30 000 fuel elements and each one has to be placed in its precise position in the core, and subsequently removed, with a high degree of reliability. The nearest point of access to the core is the outside of the biological shield, usually at least forty feet away. The activity of a newly discharged fuel element is measured in hundreds of kilo-curies.

This handling problem would be easier if access could be provided individually to each fuel channel. But the provision of several thousand such access openings would be costly and bring with it severe mechanical difficulties in the design and construction of the pressure vessel and shield. The reactor designers have, therefore, almost invariably chosen to provide access through a reduced number of openings, each serving a group of between twelve and sixty-four channels.[20] This has necessitated the development of the charge chute shown in Fig. 4, which can be inserted through the opening when required and manipulated to connect its inner end to the particular channel to be refuelled.[21] A later development of this principle[22] makes use of a cluster of fixed tubes, permanently installed above the fuel channels. This arrangement cuts out the handling of the chute and reduces the moving parts to one simple rotating component.

In most designs the fuel elements are lifted in and out of the core through the chute by a grab suspended from a rope or cable. Mechanical, electrical and pneumatic actuation of the grab have been developed to differing degrees.[20]

Control and shut-down devices

The present uranium-magnox reactors are controlled by means of vertical rods which absorb neutrons. The rod positions are varied to regulate the reactor power, and are arranged so that they are automatically and rapidly inserted into the reactor to shut it down, should an emergency arise. Any mechanism used must be able to withdraw the rod slowly, and allow it to fall quickly when the actuator mechanism is de-energised, preferably without causing permenent damage to the rod.[23, 24] The mechanism developed includes a small amount of gearing through which a very slow synchronous motor drives a rope winding drum. During an emergency control rod insertion a braking force is provided by an eddy current brake which permits a rapid initial insertion, followed by a slow approach to the support at the foot of the channel.

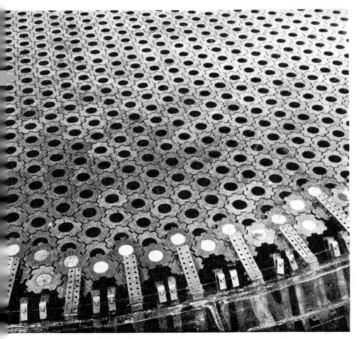

Fig. 5. The interlocking graphite brick arrangement of the Tokai Mura reactor permits some play while maintaining constant lattice pitch

In a reactor environment it is not generally possible to use a conventional hydraulically-damped impact absorber. The most promising development consists of a thin-walled steel tube of which the upper end has been rolled back on itself about a radius. The rolled over part is welded to the base of the control rod and the bottom end of the collapsing tube is stiffened by a support ring. In operation, the energy is absorbed by the tube turning inside out like a sock as it reduces in overall length. The actual work done is in bending and subsequently restraightening the tube as it passes around the radius, and in stretching the tube to a larger diameter.

Prompt primary protection devices have been developed which will rapidly actuate the normal shut-off mechanism under emergency conditions, e.g., loss of coolant. One such device consists of a reservoir built into the top of the control rod, into which is placed a bellows unit, complete with orifice plate.[25] Normally the gas pressure within the reservoir will be the same as the main gas circuit pressure. If the reactor pressure falls, the pressure within the reservoir will fall at a slower rate, dependent upon the size of orifice and reservoir volume. If the rate of pressure change in the reactor is large enough, sufficient differential pressure will result between the inside and outside of the bellows unit to compress it. This bellows can then operate a release mechanism to disengage the rod completely from the hoisting gear and let it fall freely into the core.

The control and shut-down circuits of nuclear reactors have been designed for highly reliable operation. However, the possibility exists that under some extreme event all the primary shut-down devices may fail to operate. This has led to the design and development of a second device, independent of the primary, to reduce the risk of simultaneous failure to negligible proportions. In one form a large number of small-diameter steel balls containing boron, held in a hopper above the charge pan, are released into the reactor when abnormal conditions are detected.[26] The feasibility of installing a pressure-sensitive device in the reactor core has also been examined.[27]

Attempts have also been made to develop automatic devices responding to neutron flux and coolant temperature, as well as to coolant flow.[28, 29] None of these have yet reached the stage of incorporation into power reactors.

Graphite structures

Recent developments in graphite structures of uranium-magnox reactors have been aimed at conferring high stability on the fuel channels formed by the core graphite and at achieving a constant overall size of core with a uniform lattice pitch throughout the lifetime of the reactor, while still permitting the individual components of the core structure to undergo dimensional changes induced by irradiation.

These requirements have been met, to a great extent, by the development of the core-keying system, in which columns of bricks are interlocked in such a way that each can contract or expand while maintaining constant lattice pitch. A variety of such designs have been developed, all depending on the principle of providing an adequate gap between neighbouring columns of graphite whilst contact is made only through mating keys and keyways which are in sliding contact. Such a design[30] is illustrated in Fig. 5. The introduction of fully keyed graphite structures has been accompanied by improvements to the core support and restraint systems.

One of the outstanding features of the Windscale advanced gas-cooled reactor has been the development of the re-entrant core design. A proportion of the gas at inlet temperature is allowed to flow through the Wigner gaps and through the annular space between the fuel element sleeves and the graphite bricks in the opposite direction to the main coolant flow. The effect is to smooth out axial temperature variation in the graphite, minimizing the differential growth and shrinkage that occurs at different levels in the core; furthermore, since the temperature of the graphite is kept essentially constant at that of the gas inlet, the nuclear control is greatly eased.

Pressure vessels and containment

All British gas-cooled reactor pressure vessels and containment buildings to date have been fabricated from mild steel. The Calder reactor vessels represented an extension of existing experience with the use of 2 in. thick plates in a 37 ft diameter vessel. In the subsequent civil magnox stations, plate thicknesses of over 4 inches have been used.

The principal developments have been associated with safeguarding against possible failure[31] since the integrity of the pressure circuits, or of secondary containment vessels, is of paramount importance for the safety of the plant. All pressure vessels have been designed to B.S. 1500, and considerable effort has been devoted to examining conditions where this standard would not provide adequate safeguards against failure.

Although creep deformation is unlikely to lead to catastrophic failure, it could result in excessive distortion of standpipes in the top domes of vessels and hence limit their useful operational life. Establishment of creep data on thick plate gave limiting stresses less than the permissible stresses specified in B.S.1500.[32, 33]

In addition to laboratory testing under biaxial or complex stresses, model pressure vessels have been tested with special emphasis on the effect of relaxation of stress concentrations due to nozzle welds, etc., on the rupture behaviour.

Considerable effort has been devoted to investigating the effect of using thick plate in pressure vessel construction and providing a rational basis for a design safe against brittle failure. It was concluded[34] that absolute safety could only be guaranteed by maintaining the vessel above the highest temperature at which an initiated brittle crack could propagate.

This temperature is assessed from crack-arrest tests, based on the technique devised by Robertson[35] in plate cut out from pressure vessel duct openings, etc. The 4000-ton capacity apparatus,[36] shown in Fig. 6, has recently come into operation and will allow full-size plates (up to 96 in. length × 5 in. thick) to be tested and so eliminate possible errors due to size effects. The effect of irradiation in raising the brittle/ductile transition temperature has been evaluated, using Charpy V notch specimens, and more pertinent values are being obtained by irradiating crack test specimens.

Since the construction of the Calder vessels, plate thicknesses have increased to over 4 in. and are approaching the practical limits of adequate inspection techniques. However, greater load-carrying capacity can also be obtained by using steels of higher tensile strength and a development programme has been carried out with the aim of combining higher U.T.S. with improved creep strength, and to give greater notch toughness and so allow a lower, more economical, minimum operating temperature. A number of low-alloy steels which have been made in plate thicknesses up to 5 in.[37-39] approach these objectives.

Pressure vessels for water-cooled reactors are, in general,

Fig. 6. This 4000 ton machine for crack-arrest testing of full-sized plates eliminates errors due to size effect in plates

much smaller than for gas-cooled reactors, but operating pressures are very much higher to obtain reasonable steam conditions.[40] The first large pressurized water reactor at Shippingport in the U.S.A. is 8 ft in diameter and has a wall thickness of $8\frac{3}{8}$ in. for a design pressure of 2500 lb/in²; subsequent development in water reactors has been towards larger diameter vessels and lower operating pressures. An important feature of pressure vessels for water reactors is the use of carbon steel which is internally clad with austenitic steel.[41] For heavy-water moderated reactors, pressure tube designs are favoured.[42]

Steel containment vessels differ fundamentally from reactor pressure vessels in operating at or near atmospheric pressure and ambient temperature, in the absence of irradiation. Design problems are very different[43] but many manufacturing problems are common to both types of pressure vessels.

The merits of reinforced or pre-stressed concrete for reactor containment buildings have been obvious for a number of years;[44] more recently the possible advantages of using pre-stressed concrete for the pressure vessels in gas-cooled nuclear power reactors have become apparent. Design has commenced of a pre-stressed concrete pressure vessel for the Oldbury station.

It will be subject to far greater duty than has ever been achieved before with steel vessels. This advance in the recognition of the pre-stressed concrete vessel in the U.K. has been founded on theoretical design studies[45] and on research work with large models.[46] However, vessels of this type are already in operation in France.[47]

Remote handling

A comprehensive review of manipulators has been published.[48] The early models of the master slave manipulators, developed for use in hot cells, were manufactured as complete units, consisting of master arm, slave arm and through-the-wall tube. The unit was installed through a horizontal liner cast in the wall of the cell, and was capable of seven independent motions.

A technique for covering or 'booting' the slave arm has now been evolved. This has two important advantages: improved sealing against leakage from the cell into the operating area; and spreading of contamination on the slave arm is prevented, thereby easing the maintenance problem. To achieve this booting, the jaw of the slave end is supplied with a detachable wrist seal, to which is attached a P.V.C. sleeve, which covers the arm and the through-the-wall tube, and is secured to a flange on the outside wall at the master end.

With the increasing use of manipulators in hot cells, where fuel elements are examined and fuel is broken down prior to chemical reprocessing, there was a natural demand for a more rugged version. The Mark 9 heavy duty manipulator in Fig. 7 was designed to fulfil this need. The jaw squeeze motion is performed by a chain and all pulley wheels are fitted with ball or roller bearings. At the master end there is now a pistol grip. This manipulator permits additional rotating motion.

To meet the needs of hot cells, where closely controlled inert atmospheres are necessary because the irradiated materials are self-inflammable or atmosphere-sensitive, a sealed master-slave manipulator has been introduced. An extended reach manipulator incorporates a 'dual-telescope' slave end, which gives twice the normal expansion, thus allowing the manipulator to cover the cell floor area while the

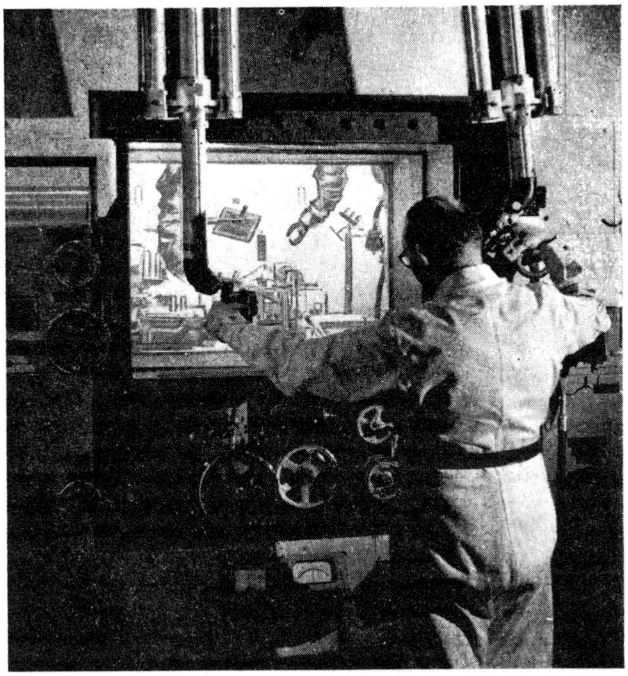

Fig. 7. The latest master–slave manipulator is of more rugged design, without impairing accurate control over the working jaws: jaw-squeeze is transmitted by a chain

operator works at normal height. An intriguing development is a robot which can move around freely in the hot cell.

Fabrication of new materials

Atomic energy has provided an incentive for the development of fabrication methods for a considerable number of the newer metals, e.g., niobium, zirconium, beryllium; the principal reasons for the choice of these metals for reactor applications are their low neutron absorption cross-sections and their corrosion resistance and irradiation stability in the chosen environments.

The development of zirconium alloys, stimulated by the requirements of water reactors, is one of the outstanding features of nuclear metallurgy.[49] Zircaloy II is now widely used for fuel cladding and pressure tubes in light and heavy water reactors; other alloys (e.g. with niobium) show considerable promise.

The developments in vacuum engineering applicable to the large-scale production of reactor materials are demonstrated very clearly in the case of niobium which was chosen as a canning material for the Dounreay fast reactor. Purification requirements for the production of the ductile metal as plate include both high vacuum and a high rate of pumping:[50] typical conditions used on a production scale are a temperature of 2250°C, pumping speed of 50 000 l/hr, and final vacuum better than 3×10^{-6} mm of mercury.

Conventional dies of the size required for processing would have been extremely expensive, both to make and maintain. As an alternative, picture frame techniques have been developed in which hollow frames of aluminium are filled with niobium and pressed under an 8000 ton load. The aluminium frame is deformed during pressing and is subsequently removed mechanically before sintering.

Beryllium has been extensively studied as a fuel canning material. Fabrication difficulties which were overcome included the development of improved lubrication techniques for the extrusion of the metal, and the development of direct extrusion of small-bore tubing to bore accuracies of ±0.001 in.[51] The low ductility of the metal at room temperature has been extensively investigated.[52]

Major problems of machining in the atomic energy field have not in the main been concerned with the mechanics of cutting but with associated problems such as irradiation hazard, toxic dust, the possibility of fire, or nuclear explosion[53]. The latter hazard involves an organizational problem in providing with absolute certainty that critical masses of fissile material are not created during processing.

Welding

Welding developments have been of special importance in the atomic energy field, particularly for fuel element cans, reactor pressure vessels, and chemical plants.[54] Fuel elements canned in magnox have been produced in quantity for several years for the Calder and civil reactor programmes.[55] Development has continued to improve the integrity of the fuel can welds. For this purpose the research machine, shown in Fig. 8 which uses the inert gas tungsten arc technique, has been developed at Springfields. Variables on it can be set with great flexibility and its response rate is high.[56] With its aid the settings of production machines can be specified in absolute units of time, speed and current.

Since metals such as beryllium, uranium and zirconium are deleteriously affected when heated in air, welding has to be carried out in an inert atmosphere. A chamber, capable of evacuation to approximately 10^{-3} mm mercury, and then filled with argon or helium, is preferred for general work. This means that the glove boxes have to be designed as vacuum vessels which introduces problems in the design of the welding equipment inside the vessel. These problems and their solutions have been discussed previously.[57] The six welding processes currently used for fabricating fuel element bundles for the U.S. plutonium recycle programme have recently been described.[58]

The trend in pressure vessel construction has been to increase the use of automatic welding, not only in manufacturers' own works, but also in workshops erected on site and, indeed, during the field erection of the vessels. This was due to the greater thickness of material, and the larger size of vessels which made control of manual techniques very difficult and uneconomical.

In the Calder Hall pressure vessels all the welding was done with a rutile type of electrode.[59] It has now become the practice to use low-hydrogen types to ensure the production of radiographically sound, crack-free deposits. Weld metal deposited in automatic processes must be capable of providing the same guarantee.

Apart from submerged-arc welding, which has been practiced for many years on Class 1 pressure vessels, two other automatic techniques have recently come into prominence. These are 3 o'clock welding[60] and electro-slag welding.[61] The former involves making multi-run welds comprised of small individual runs, with the plates to be welded lying in the vertical plane, suitably supported. Slag welding is virtually submerged-arc welding, carried out in the vertical position with the aid of moving, water-cooled dies.

In the U.K.A.E.A., 18 per cent Cr/13 per cent Ni/Nb stainless steel was used for construction of the first chemical separation plant at Windscale some ten years ago. Its use has been continued in later chemical plants,[62] but the welding methods have been improved. One joint design employed makes successful use of a copper backing strip as a means of controlling the shape of the penetration bead.[63]

Valves and ancillary equipment

The criteria determining the design of a valve for normal engineering apply in nuclear power applications but sometimes with more stringent requirements.[64] For example, in extreme cases valves would be required to remain in place during the life of a reactor, say 20 years.

It is sometimes necessary to have a valve which combines flow control with positive shut off and this is a particularly difficult problem in large diameter duct valves. Much development work has been done to achieve low leakage under these conditions. The use of flexible metallic seats has been extensively used. For smaller applications, the non-rotating spindle has found favour. For inaccessible positions, where servicing is difficult, a hard plug, mating on a soft seat with applied loads sufficient to cause seat deformation, is used if the number of closures required is limited.

Conventional packed spindle glands can be used under certain circumstances but the trend is towards bellows sealing and, to a lesser extent, diaphragm sealing. In some applications leaks must be detected and interspace drains, thermocouples, etc., have been introduced.

Valves are sometimes required to handle fluids up to at least 600°C and to function correctly after thermal cycling. The main problem here is that the plug and seat should continue to form a seal. The use of bellows as a gland seal assists the provision of a flexible plug and spindle combination. Extensive use is made of extended spindle length and

finning of bonnets to reduce temperature at the gland or handwheel end. Fluids such as liquid sodium tend to reduce the surface oxide film and promote welding together of plug and seat. Hard facing alloys such as the *stellites* have proved successful.

Naturally, most of the development on special valves has been done by the manufacturers and is not generally published; valves used for gas-cooled reactors[65] and water-cooled power reactors[66] have, however, been recently reviewed. In the U.K. the A.E.A. have developed, amongst other valves, a stop valve with water at 325°C and 2500 lb/in² which is illustrated in Fig. 9.

In the atomic energy field there is a frequent requirement for leak-tight systems containing process fluids; new techniques for reducing leakage from gas-cooled reactors have recently been described.[67] Where such systems have to incorporate joints which must be broken and re-made, it is sometimes possible to use edge-welded flanged joints, but it is generally more convenient to use flanged joints incorporating a gasket. Such joints have been developed over the past few years[68] and feature thin annular gaskets, clamped between machined faces by bolts and backing flanges. The gasket material used depends upon the working temperature and the fluid to be contained. PTFE is satisfactory up to 250°C, aluminium up to 400°C and silver up to 600°C. A difficult problem is the joint between mild steel and *Zircaloy* (because

Fig. 8. This research welding machine for fuel element cans is used to determine optimum settings for production machines

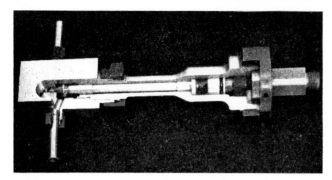

Fig. 9. A stop valve for high pressure water systems, operating at 325°C is one of the ancillary components developed by the U.K.A.E.A.

of differing thermal expansion), which has been developed for use in pressure tube water reactors. In this application the resistance of the joint to temperature cycling is of paramount importance and joints up to 6 in. diameter, using silver gaskets are satisfactory at temperatures up to 350°C and indefinite amounts of temperature cycling. A radically different solution is a rolled joint between the two materials.[69]

In earlier gas-cooled nuclear power stations, relative movements and expansions between fixed portions of the coolant circuit were accommodated by the use of bellows capable of angular deflection. Various types of such bellows have been developed, some of which include a flexible tongue to withstand the axial load, and multiple inclined tension rods. The use of ring reinforcements to the bellows convolutions has also been introduced and results in more flexible convolutions. In the Windscale A.G.R. the co-planar support of co-axial ducting and heat exchangers eliminates vertical expansion effects and therefore the requirement for bellows.[70]

Since, very frequently, mechanisms in nuclear reactors cannot be lubricated, much work has been done[71, 72] to study the friction and wear of materials at high temperatures in dry carbon dioxide and other gases. The use of gas bearing compressors for intermittent (stop-start) duty for reactor and in-pile loop applications, has highlighted the problem of the choice of bearing materials to meet the very stringent requirements.

There is, in this field of sliding friction and wear, growing evidence that encounters between surface asperities are essentially elastic with little plastic flow occurring, and that an important mechanism in the formation of wear particles is akin to a surface fatigue rather than adhesive welding.[73] Similar considerations apply to friction and wear work in liquid metal environments.[74, 75]

Vacuum techniques used on a number of atomic energy projects were recently discussed by the author.[76] The biggest of these projects was the diffusion plant at Capenhurst, and vacuum techniques were well suited to prove the tightness and cleanliness of this huge plant. In recent years, however, the emphasis in the nuclear power programme has been on the testing of components destined for operation at high pressures (and temperatures) where vacuum techniques can be used to prove the tightness. The main developments have therefore been the increasing application of 'sniffer' techniques,[77] and the more widespread use of high-sensitivity mass spectrometers as gas detectors.

The methods of measuring and producing ultra high

vacua (10^{-9} torr) with reference to thermonuclear devices have been reviewed by Munday.[78] In this field of plasma research the evaluation and removal of residual gases is of primary importance.[79]

Inspection techniques

Inspection techniques developed up to 1957 have been already described;[80] only subsequent development is considered here.

For detection of laminar defects in thick plate, ultrasonics is now generally accepted as satisfactory. In the inspection of plate for reactor pressure vessels,[45, 81] pulsed ultrasonic waves are injected at one side of the plate, laminations being detected by their reflective capacity as compared with that of either the far surface of the plate or of a specified artificial defect.

Considerable attention has been given to the inspection of thin-wall tubing, used in the manufacture of fuel elements. The present trend is to supplement or replace the eddy current test by an ultrasonic test which, although slower, is more specific for flaw detection.[82]

Examples of inspection, up to 1959, of nuclear chemical plant and Calder type reactors have been described,[83, 84] with particular reference to leak testing of vessels by soap bubble tests, and by infra-red gas analysis methods, leak testing of fuel element cans by the helium mass spectrometer, and radiographic inspection of welds in pressure vessels and fuel elements.

Subsequently there have been developments in inspection methods applied to massive butt and nozzle welds. In the first reactors of the U.K. civil power programme the pressure vessels were made of plate of up to three inches in thickness and the butt welds in these vessels were radiographed with conventional transportable X-ray sets of up to 400 kV. However, for radiography of welds in the thicker plate used for the Windscale A.G.R., and for some of the later C.E.G.B. reactors, radiation of higher penetrating power was required. Among the high voltage X-ray generators and gamma ray sources available, kilocurie Cobalt 60 gamma ray sources[85] were the first to be used on construction sites. Subsequently a prototype mobile 4·3 MeV linear accelerator for radiography was developed and successfully used for the inspection of the Trawsfynydd pressure vessels.[86]

Even with the best techniques and equipment the resolution of defects on a radiograph decreases as weld thickness increases. For this reason, and on account of the cost of installation and the radiation hazards associated with high energy radiography, ultrasonic methods are being developed. Improvements have been mainly in mechanization of scanning and recording[87] and in the use of multiple probes to detect the shape and orientation of defects.

Critical radiographic inspection of the fillet welding of nozzles into pressure vessels for fuel element charge/discharge pipes is not feasible. These welds, which may be full or partial penetration types, can be inspected ultrasonically from the inside surface of the nozzle pipe.[88]

Conclusions

In the heat transfer field, major developments have been concerned with forced convection from finned surfaces and the use of roughened surfaces.

There has been significant development of circulators for gases, liquids and liquid metals. Where contamination and radiation are involved, gas-bearing-mounted compressors and canned rotor pumps are of particular importance. These developments are also finding increasing application in fields outside atomic energy.

Control and refuelling equipment are examples of developments in atomic energy which are in the main conventional but in which the necessity for remote handling and reliability over a long period has significantly influenced development. Remote handling equipment has increased in flexibility and ruggedness.

Graphite and pressure vessel steels have been critical materials in the development of gas-cooled reactors in the U.K. The physical properties at high temperatures and in the presence of radiation have been extensively studied, and have been the principal influences on the engineering design of moderator and pressure vessel structures. A new development, certain to be of major importance, is the use of pre-stressed concrete for reactor pressure vessels.

Atomic energy has provided the incentive for the engineering development of several metals, including niobium, zirconium and beryllium. This has involved their manufacture on a large scale, to very high purity, and fabrication to very fine limits. Advances made in the technology of these metals have found application in many other fields, for instance in rocket technology. Fabrication techniques for these and more conventional materials, in particular welding, have received a tremendous impetus from the special demands of nuclear power and the results have been applied throughout mechanical engineering. Inspection techniques have received a similar impetus.

In ancillary equipment, the most significant developments have been in valves and joints. The development of larger

Dr H. Kronberger *obtained his B.Sc.(Hons) degree at King's College, Durham, and for his Ph.D. he worked at Birmingham University. He joined the war-time atomic energy team (Tube Alloys) under the late Sir Francis Simon at Oxford and worked on the separation of uranium isotopes. At Harwell, between 1946 and 1951 he continued this work with ultra-centrifuge and distillation techniques. Subsequently, he became head of the Capenhurst Laboratories and responsible for the development of the gaseous diffusion plant. In 1956 he was appointed Chief Physicist of the U.K.A.E.A.'s Industrial Group at Risley, and subsequently Director of Research and Development. From 1960, he was Deputy Managing Director (Development) of the Reactor Group of the United Kingdom Atomic Energy Authority. He died in 1970.*

Goliath cranes[89] has been a governing factor in reactor construction. The advanced state of vacuum technology and the industry that has grown up round it owes much to the early days of atomic energy diffusion plants.

Acknowledgments

I wish to make grateful acknowledgment to Mr V. Y. Labaton of Reactor Group Headquarters for his help in the preparation of this article, and the large number of specialists on whose advice I was able to draw.

REFERENCES

1. CETINKALE, T. N. and FISHENDEN, M., 'Thermal conductance of metal surfaces in contact'. Institution of Mechanical Engineers and American Society of Mechanical Engineers. General discussion on Heat Transfer, London, 1951. *Proceedings* pp. 271–275.
2. SANDERSON, P. D., 'Heat transfer from the uranium fuel to the magnox can in a gas-cooled reactor'. *In:* American Society of Mechanical Engineers and others. International developments in heat transfer, 1961. pp. 53–64.
3. UNITED KINGDOM ATOMIC ENERGY AUTHORITY, Unpublished work.
4. BRITISH NUCLEAR ENERGY CONFERENCE, 'Symposium on the use of secondary surfaces for heat transfer with clean gases'. 1961 October *British Nuclear Energy Conference J.*, vol. 6 pp. 275–417.
5. CENTRAL ELECTRICITY RESEARCH LABORATORIES, Unpublished work.
6. WALKER, V., 'The improvement of fuel element heat transfer by surface roughening.' 1961 April *Nuclear Energy*, volume 6 pp. 144–148.
7. DE BORTOLI, R. A. *et al*, 'Forced convection heat transfer burnout studies for water in rectangular channels and round tubes at pressures above 500 p.s.i.a.' 1958 *WAPD-188.*
8. FIRSTENBERG, H., *et al.*, 'Compilation of experimental forced-convection, quality burnout data with calculated Reynolds numbers' 1960 *NDA-2131-16.*
9. ROBERTS, H. A., and BOWRING, R. W., 'Boiling effects in liquid-cooled reactors'. 1959 February *Nucl. Pwr*, vol. 4, pp. 69–73 and 1959 March *Nucl. Pwr*, vol. 4 pp. 96–101, 118.
10. VOHR, J. H., 'Flow patterns of two-phase flow: a survey of literature', 1960 TID-11514.
11. BURGREEN, D. *et al*, 'Vibration of rods induced by water in paralled flow'. 1958 July *Amer. Soc. Mech. Engrs Trans.*, vol. 80 pp. 991–1003.
12. WHITLEY, S., 'Gas bearings for compressors in gas-cooled reactors'. *In:* Progress in nuclear energy. Series 4. Technology, engineering and safety vol. 3. Oxford, Pergamon, 1960 pp. 3–28.
13. WHITLEY, S., and WILLIAMS, L. G., 'Principles of gas lubricated shaft seals'. 1962 June *Journal of Mechanical Engineering Science* vol. 4 pp. 177–187.
14. BOWDEN, A. T., and MARTIN, G. H., 'Calder Works nuclear power plant: design of important plant items'. 1957 April *Brit. Nucl. En. Conf. J.*, vol. 2 pp. 156–167.
15. 'Pumps for water-cooled power reactors'. 1961 July *Nucleonics*, vol. 19. pp. 55–61.
16. BLAKE, L. R., 'Conduction and induction pumps for liquid metals'. 1957 February *Instn Elect. Engrs, Proc. pt. A.* vol. 104, pp. 49–67.
17. FENNEMORE, A. S., 'Linear induction pumps for liquid metals'. 1957, 17th May *The Engineer*, vol. 203, pp. 752–755.
18. WATT, D. A., 'The design of electromagnetic pumps for liquid metals'. 1959 April *Instn Elect. Engrs, Proc. pt. A.*, vol. 106, pp. 94–103.
19. BARNES, A. H., 'Pumping of liquid metals'. 1st United National International Conference on the Peaceful Uses of Atomic Energy, Geneva 1955. *Proceedings* vol. 9, pp. 259–264.
20. DENT, K. H., 'On load refuelling problems'. 1961 August *Nucl. Pwr*, vol. 6, pp. 69–74.
21. DENT, K. H., 'Improvements in or relating to heterogeneous nuclear reactors'. *Brit. Pat. 835, 764.* Appl. 14th August 1956. Published 25th May, 1960.
22. HACKNEY, S. and PACKMAN, G., 'Improvements in or relating to nuclear reactor refuelling devices'. *British Patent 847, 636.* Application 1st July 1957–13th March, 1958. Published 14th September 1960.
23. TINDALE, J., 'Control mechanisms for nuclear reactors'. 1960 November *Control*, vol. 3, pp. 89–92 and 1960 December *Control*, vol. 3 pp. 100–105.
24. WRIGHT, A. D. and LEBEAU, K. F., 'Control rod systems for AGR' 1961 April *Nucl. Engrg*, vol. 6 pp. 159–161.
25. FLETCHER, S. E. D. and BIRKENSHAW, D., 'Improvements in or relating to safety release mechanisms'. *Brit. Pat. 872, 092.* Appl. 8th September 1958. Published 5th July 1961.
26. UNITED KINGDOM ATOMIC ENERGY AUTHORITY, Unpublished work.
27. UNITED KINGDOM ATOMIC ENERGY AUTHORITY, Unpublished work.
28. FITCH, S. H. *et al*, 'Integral safety devices for reactors'. 2nd United Nations International Conference on the peaceful uses of Atomic Energy, Geneva. 1958. *Proceedings* vol. 11, pp. 186–192.
29. UNITED KINGDOM ATOMIC ENERGY AUTHORITY, Unpublished work.
30. HICKS, R., 'Japan's first nuclear power station. 2. Earthquake problems'. 1960 March *Nucl. Pwr*, vol. 5 pp. 108–110.
31. NOONE, M. J. and BISHOP, R. F., Pressure vessels for gas-cooled graphite-moderated ractors'. 1961 *Proc. Instn mech. Engrs Lond.*, vol. 175 pp. 471–495.
32. ROBERTSON, J. M. and NICHOLS, R. W., 'High temperature mechanical properties of steels used in gas-cooled reactor pressure vessels'. British Nuclear Energy Conference Symposium on steels for Reactor Pressure Circuits, London, 1960. *Report.* pp. 14–39.
33. WOOD, D. S. and SWIFT, P. B., 'The creep properties of a silicon-killed carbon-manganese pressure vessel steel'. ibid. pp. 111–126.
34. KEHOE, R. B. and NICHOLS, R. W., 'Brittle fractures and reactor pressure circuits'. 1951 March *Nuclear Engineering*, vol. 6 pp. 112–116.
35. ROBERTSON, T. S., 'Propagation of brittle fracture in steel'. 1953 December *Iron steel Inst. J.* vol. 175, pp. 361–374.

36. WINDER, B., 'A 4000 ton crack testing machine'. 1962 February *Nuc . Engrg* vol. 7. pp. 61–63.
37. MACKENZIE, I. M., 'Progress in development of some improved steels for gas-cooled reactor pressure vessels'. British Nuclear Energy Conference Symposium on steels for reactor pressure circuits, London, 1960. *Report* pp. 437–462.
38. PEARSON, T. F. and DE LIPPA, M. Z., 'The development of a high-tensile plate steel for use at elevated temperatures and pressures'. ibid. pp. 463–516.
39. RUSSELL, J. E., 'Niobium in carbon and low-alloy weldable steels'. ibid. pp. 534–548.
40. FAWCETT, S., 'The call for development in pressure vessels and containment buildings for nuclear reactors'. 1st symposium on Nuclear Reactor Containment Buildings and Pressure Vessels, Glasgow, 1960 *Proceedings*, pp. 3–18.
41. DAVIES, D. K., 'Design and manufacture of reactor vessels'. 1961 March *Metal Progr.* vol. 79, pp. 101–104, 148–154.
42. BRADLEY, N. and STANNERS, W., 'Pressure tube reactors'. British Nuclear Energy Society Symposium on Pressure Tube Water Reactors, Risley, 1962. *Proceedings*. To be published.
43. ZICK, L. P., 'Design of steel containment vessels in the U.S.A.' 1st Symposium on Nuclear Reactor Containment Buildings and Pressure Vessels, Gasgow, 1960. *Proceedings*, pp. 91–113.
44. WATERS, T. C., 'Reinforced concrete as a material for containment'. ibid. pp. 50–60.
45. DAVIDSON, I., 'Some contributions from nuclear power to engineering practice'. 1960 October *Instn Civ. Engrs Proceedings* vol. 17, pp. 121–136.
46. BAKER, A. L. L. *et al*, 'The design, construction, and testing of a prestressed concrete reactor pressure vessel model'. 1961 December *Instn Civ. Engrs Proc.* vol. 55, pp. 565–586.
47. BELLIER, J. and TOURASSE, M., 'Marcoule Nuclear Centre: caissons in prestressed concrete of the reactors G2 and G3'. 1959 July/August *Inst. Tech. Batinent Travaux Publics Ann.* vol. 12, Supplt.
48. 'Handling radioactive materials: manipulators'. 1962 June *Nucl. Engrg*, vol. 7, pp. 228–237.
49. LUSTMAN, B. and KERZE, F., '*The Metallurgy of Zirconium*'. New York, McGraw Hill, 1955 (National Nuclear Energy Series, Div 7, vol. 4).
50. WILLIAMS, L. R. and HEAL, T. J., 'The consolidation, fabrication and properties of niobium'. 3rd Plansee Seminar, Reutte, 1958, *Proceedings*, pp. 350–370.
51. UNITED KINGDOM ATOMIC ENERGY AUTHORITY, Unpublished work.
52. INSTITUTE OF METALS, Conference on the Metallurgy of Beryllium, London, 1961. *Proceedings*. To be published.
53. PIGOTT, A., 'The Machining of Uranium, Thorium, Niobium, Vanadium, Zirconium and Beryllium' Institution of Mechanical Engineers. Conference on Technology of Engineering Manufacture, 1958. *Proceedings*, pp. 365–373.
54. DICKINSON, F. S., 'Welding in nuclear power stations'. 1961 October *Nucl. Pwr.*, vol. 6. pp. 66–69.
55. TAYLOR, A. F., 'Welding problems associated with nuclear fuel elements'. 1960 October *Brit. Weld. J.*, vol. 7, pp. 615–622.
56. 'New welding techniques speed fabrication'. 1962 July *Nucl. Pwr*, vol. 7, pp. 44–45.
57. WILLIAMS, A. E., 'Experience obtained of Springfields on the application of vacuum to fuel element production processes'. Joint British Committee on Vacuum Science and Technology Symposium on User Experience of Large-scale Industrial Vacuum Plant, London 1961. *Proceedings*, pp. 35–41.
58. MILLS, L. E., 'Zircaloy welding techniques developed for plutonium recycle program UO₂ fuel element fabrication'. 1961 February *Weld., J.*, vol. 40, pp. 141–151.
59 BROWN, G. *et al*, 'Calder Works nuclear power plant: design and construction of the pressure vessel'. 1957 April *Brit. Nucl. En. Conf. J.*, vol. 2, pp. 132–145.
60. 'Trawsfynydd power station'. 1961 September *Weld. and Metals Fab.* vol. 29, pp. 350–369.
61. HORSFIELD, A. M., 'Application of electro-slag welding'. 1960 May *Brit. Weld. J.*, vol. 7, pp. 337–341.
62. DICKINSON, F. S. and WATKINS, B. 'Use of welding in chemical plants in the U.K.A.E.A.'. 1960 October *Brit. Weld. J.*, vol. 7, pp. 643–651.
63. UNITED KINGDOM ATOMIC ENERGY AUTHORITY, Unpublished work.
64. IGGULDEN, D., 'Valve types and applications'. 1960 September *Nucl. Pwr*, vol. 5, pp. 93–95.
65. BLAKEBOROUGH, C. D., 'Gas valves for nuclear power stations'. ibid. pp. 96–99.
66. GRUENWALD, K. H., 'Valves for water-cooled power reactors'. 1961 July *Nucleonics*, vol. 19, pp. 64–69.
67. STEWART, D. G., 'Leakage from carbon dioxide cooled ractors'. 1962 July *Nucl. Pwr.*, pp. 53–56.
68. UNITED KINGDOM ATOMIC ENERGY AUTHORITY, Unpublished work.
69. WOLFE, W. A., 'Burst Tests Help Set Design Stress for N.P.D. Pressure Tubes'. 1960 October *Nucleonics*, vol. 18, pp. 96–100.
70. LONG, E. and RODWELL, W., 'Improvements in or relating to an assembly comprising a heat source and associated apparatus'. *Brit. Pat. 897, 445.* Appl. 11th August 1958. Published 30th May 1962.
71. HEATH, H. H. and PHILLIPS, K. F., 'Unlubricated metals as bearings for carbon-dioxide-cooled nuclear reactors'. Institution of Mechanical Engineers Symposium on Non-conventional Lubricants and Bearing Materials such as used in Nuclear Engineering, London, 1962. *Proceedings*. To be published. *Also* other papers from this symposium.
72. CORNELIUS, D. F. and ROBERTS, W. H., 'Friction and wear of metals in gases up to 600°C.' 1961 April *Amer. Soc. Lub. Engrs Trans.* vol. 4, pp. 20–32.
73. ARCHARD, J. F., 'Single contacts and multiple encounters'. 1961 August *J. Appl. Phys.*, vol. 32, pp. 1420–1425.
74. JERMAN, R. B. *et al*, 'Evaluation of material wear and self-welding in sodium-cooled reactor systems. 1959 June *J. Basic Engrg*, vol. 81, pp. 213–225.
75. KISSEL, J. W. *et al*, 'Frictional behaviour of sodium lubricated materials in controlled high temperature environment.' American Society of Mechanical Engineers Spring Lubrication Symposium, Miami Beach, 1961. *Paper 61-Lub S-16*.
76. KRONBERGER, H., 'Vacuum techniques in the atomic energy industry'. 1958 *Proc. Instn mech. Engrs, Lond.*, vol. 172, pp. 113–131.
77. UNITED KINGDOM ATOMIC ENERGY AUTHORITY, Unpublished work.
78. MUNDAY, G. L., 'Ultra high vacuum technology'. 1959 June *Nucl. Instrum. and Methods*, vol. 4, pp. 367–375.
79. SANTELER, D. J., 'Application of vacuum process evaluation to the study of outgassing'. 1961 February *Vacuum*, vol. 11, pp. 1–9.
80. AMERICAN INSTITUTE OF CHEMICAL ENGINEERS and others, 'Symposium on nondestructive tests in the field of nuclear energy'. Philadelphia, A.S.T.M.. 1958.
81. HOLLIDAY, W. C. and NOONE, M. J., 'Metallurgical problems associated with the construction of gas-cooled reactor pressure vessels'. British Nuclear Energy Conference Symposium on Steels for Reactor Pressure Circuits, London, 1960. *Report*, pp. 207–239.
82. OLIVER, R. B. *et al*, 'Immersed ultrasonic inspection of pipe and tubing'. 1957 May/June *Nondest, Test*, vol. 15, pp. 140–144.
83. DICKINSON, F. S. and KEEFE, E. J., 'Non-destructive testing for nuclear plant'. 1958 May *Nucl. Pwr.*, vol. 3 pp. 224–226.
84. OWEN, SIR LEONARD, 'Welding and the nuclear power programme'. 1959 May *Brit. Weld. J.*, vol. 6, pp. 197–204.
85. FORBES, R. S., 'Radiographic inspection'. 1959 December *Nucl. Engrg.*, vol. 4 pp. 442–445.
86. HANSTOCK, R. F., 'Electron accelerators for site radiography'. 1961 February *Nucl. Pwr*, vol. 6, pp. 69–71.
87. CARSON, H. L., 'Ultrasonics and the examination of welds'. 1961 November *Weld and Metal Fab.*, vol. 29, pp. 437–445.
88. HANSTOCK, R. F. and LUMB, R. F., 'Ultrasonic inspection of nozzle welds'. 1962 January *Nucl. Pwr*, vol. 7, pp. 77–79.
89. LANDON GOODMAN, L., 'Mechanical Handling'. 1962 July *Chartered Mechanical Engineer*, p. 50.

Steam Generator Engineering

by P. B. Silk, BSc

Whatever the heat-source, be it solid fuel, oil or nuclear fission, steam remains the only practical medium for converting heat into the enormous blocks of power demanded by the modern community. In the last few years steam pressures and the sizes of boilers have increased at an almost revolutionary rate; with the re-introduction of reheat and with supercritical pressure, overall efficiencies are reaching 40 per cent, a figure undreamt of twenty or thirty years ago.

At the present time much thought and effort is being put into developing methods for the direct conversion of heat into electrical energy.

The most promising of these are the magneto-hydro-dynamic, the thermoelectric and thermionic generators, and the fuel cell. The main problems with these are either low efficiency or the lack of suitable materials to withstand the high temperatures involved. Such methods, however, are still in the laboratory stage and it is generally accepted that it may be many years before they become practical propositions for the large-scale, commercial production of power.

In the British Isles, therefore, where available water power accounts for less than one per cent of our requirements, steam remains the only satisfactory medium for producing the enormous blocks of power demanded by the modern industrial community.

Even in the nuclear field, for the nine atomic power stations now operating or under construction, steam is still the medium for transforming into mechanical work the heat produced by the reactors.

It might be asked 'What of the internal combustion engine. or gas turbines?'

The answer is twofold. In the first place this country's basic fuel reserves are in the coalfields and though attempts have been made to operate both diesels and gas turbines on pulverized coal, neither has been conspicuously successful. Secondly, while both machines cater very successfully for low power ranges, neither can touch for size the modern steam turbo-alternator set. For example, what gas turbine or diesel can produce in a single unit the 750 000 h.p. of the steam sets of the Thorpe Marsh Power Station of the Central Electricity Generating Board?

Even among those connected with the design, manufacture and use of steam-raising equipment, few realize the tremendous strides made in the last few years. Undoubtedly the most spectacular advance has been in the size of steam generators. In 1955–56 the largest boilers commissioned by the C.E.G.B.

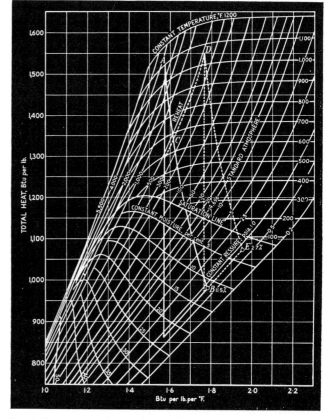

Fig. 1. The total heat-entropy or Mollier diagram for steam shows that in the higher pressure ranges there is a decrease of total heat, if temperature remains constant

had outputs of 550 000 lb of steam per hour.[1] Today there are under construction units capable of producing $3\frac{3}{4}$ million lb per hour, a sevenfold increase. Steam pressures have also risen considerably. While most new power stations were then operating at 900 lb/in² gauge pressure, with one or two at 1500 p.s.i.g., the C.E.G.B. has recently placed orders for two boilers to work at 3650 p.s.i.g. some 460 lb/in² above the critical pressure. Such increases are revolutionary by any engineering standards, and it says much for the enterprise and courage of both users and manufacturers that they have been accepted.

From the point of view of overall thermal efficiency, temperature is of major importance; it is unfortunate, therefore, that metallurgical considerations have prevented anything other than a very moderate increase from, say, 900°F to 1100 in steam temperatures over the period under review. Temperatures up to 1200°F have been tried but the complications involved, particularly in the welding of pipes have, in the last year or so, actually led to the more moderate figure of about 1055°F being retained.

What then is the philosophy behind these recent advances in steam pressures, temperatures and boiler outputs? To answer this some consideration must be given to the fundamental factors governing the steam cycle.

The largest single loss in a power station is that due to the heat thrown away in the condenser cooling water. With moderate temperatures and pressures and a simple cycle, this loss may amount to 60 per cent or more of the heat in the fuel. Engineers' endeavours have therefore been directed for some time towards reducing this tremendous source of inefficiency.

The main lines of attack have been directed to increasing the initial heat in the steam, reheating and heating of the boiler feed by steam bled from various stages of the turbine. The first two methods result in a reduction of the condenser loss as a percentage of the heat put into the steam. The third reduces the actual amount of steam and heat, rejected thus to the condenser.

Increases in temperature and pressure

While there appears to be almost no limit to steam pressure, commercial and technical aspects of metallurgy at present set a ceiling of about 1100°F to temperature.

An inspection of the Mollier diagram in Fig. 1 or of the Steam Tables shows that while, at constant pressure, there is always an increase in total heat with increase in temperature, the same does not necessarily apply with increase in pressure at constant temperature; in fact, in the higher ranges, there is a decrease in total heat.

What matters, however, is the available adiabatic or isentropic heat drop and while it is true that this is increased by raising the pressure, such increase in the simple cycle is only at the expense of extra wetness in the lower stages of expansion in the turbine, as shown in Fig. 2. If the percentage moisture is too high, blade erosion becomes serious and turbine efficiency drops.

With limits to both temperature and moisture content, there is, in the simple cycle, little virtue in going beyond about 2000 lb/in² which was, in general, reached about five years ago.

This barrier to higher efficiency can be surmounted by adopting the somewhat more complicated reheat cycle. Appearing in a British patent of 1905, reheat was first used 40 years ago on the North East coast but at that time it suffered from two defects. First, steam pressures were low so that, after any appreciable expansion in the turbine, the high specific volume called for large and expensive pipework and valves to carry the steam from the turbine to the boilers and back. Secondly, boilers, at that time, were, with some justification, considered less reliable than the turbines they served and it was accepted practice to have two or more boilers per turbine, plus a certain number as stand-by. This arrangement entailed complicated pipework if reheating was used. Today, however, largely due to the excellent work of the Boiler Availability Committee,[2] boilers match turbines in reliability. The unit system of one boiler per turbine permits simpler pipework and today's higher pressures reduce it to manageable sizes, making reheating a practicable proposition. With very high pressures and particularly with supercritical pressures, more than one stage of reheating may be justified.

Further reference to the Mollier diagram of Fig. 1 shows an important feature of reheating. The expansion line *AB* is given for a simple cycle from 2000 p.s.i.a. down to 28½ in. mercury vacuum, resulting in a moisture content of 11·5 per cent. If, however, one stage of reheat is adopted from *C* to *D*, with expansion down to the same vacuum, the final point *E* shows a moisture content of only 2·7 per cent. Reheating always brings the final condition point further over to the right-hand side of the diagram and this means drier steam.

In 1957–58 only one reheat set (Ferrybridge B) was commissioned for the C.E.G.B. By 1962 reheat was a standard requirement and the plant on order or planned for new stations or sections included over 50 boilers, every one of which for reheat.

Regenerative feed heating

Like reheating, regenerative or bleed heating originated over forty years ago in this country but not until the middle twenties did it become general practice. With the moderate steam conditions then prevailing, it seldom went to more than three or four stages. With today's high pressures seven stages are common, while for the latest supercritical pressure cycle now under construction eight stages are justified.

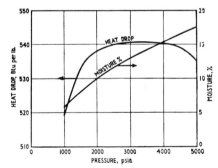

Fig. 2. Increase of pressure in the simple cycle causes greater wetness in the later turbine expansion stages. Turbine efficiency of 82%, 1100°F initial temperature, and no reheat are assumed in this graph of heat drop plotted against pressure

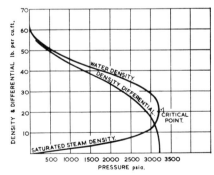

Fig. 3. Densities of saturated steam and water, and density differentials

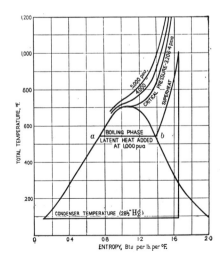

Fig. 4. The temperature-entropy diagram

▶

Table 1—Recent Advances in Unit Size

Output (MW)	12·5–52·5	60–75	100–120	275–375	500–550
up to 1957 (% of units)	40	55	5		
up to 1962 (% of units)			57	24	19

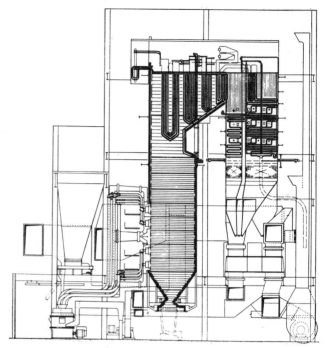

Fig. 5. The supercritical boiler at Drakeclow 'C' station supplies 2·5m lb/h at 3650 p.s.i.g. and 1110°F, with preheating to 1055°F

The design of the feed heating system has become a highly complicated matter and, in fact, the cycle of boiler, turbine, reheat and feed system has to be treated as a whole since all parts are interdependent. For example, the pressure at which reheat takes place affects the design of both the turbine and the boiler; the bleed points for feed heating to give the maximum gain in overall efficiency cannot be determined without reference to the convenience of the turbine designer and many other questions arise, such as the practical or economic features of steam or electric drive for the feed pumps. There are so many variables in the cycle that even with the help of computers many months of intense study are required to arrive at an acceptable compromise; for compromise it must be, between the theoretically desirable and the practically possible.

Like many types of engineering equipment, turbo-alternator sets have been increasing in size and boilers naturally have kept pace. When, however, the unit system of one boiler per set became established, boilers had to be doubled or trebled in size almost overnight.

But apart from this the size of sets for the C.E.G.B. has had, in recent years, more than a normal rate of growth, as shown in Table 1.

To produce from a single boiler the quantity of steam required for a 550 MW turbo-alternator set calls for a unit of considerable proportions. Two such units each has an output of $3\frac{3}{4}$ million lb/h of steam at 2400 p.s.i.g. and 1055°F, with reheat back to 1055°F.

One of these units has a single furnace, operates on natural circulation and some idea of its size may be gleaned from the fact that its drum is 120 ft long, 5 ft 6 in. in diameter and $5\frac{3}{8}$ in. thick. Without its internal fittings this drum weighs 250 tons and is positioned 160 ft above ground level. drum will weigh 250 tons and will be positioned 160 ft above ground level.

The second unit for exactly the same steam conditions and output is of somewhat different design. It has two furnaces, each split into two by a central dividing wall and the water circulation around the various heating surfaces is assisted by pumps incorporated in the circuit.

Supercritical pressure

The practical full-scale application of supercritical pressure is such a landmark in steam engineering that it warrants special consideration.[3]

If a closed vessel, partially filled with water, be heated then, as the temperature and pressure rise, the density of the steam increases and that of the water decreases. Continuing, there comes a point where the densities of steam and water are the same, (see Fig. 3). This is the critical point and corresponds with 3206·2 lbf/in²g and 705·4°F.

Fig. 4 shows the process on a temperature-entropy diagram. The line *a–b* represents the addition, during the

boiling phase, of latent heat without increase in temperature. At the critical point there is no boiling, no surface of demarcation between liquid and vapour and the two substances are indistinguishable. Further application of heat beyond the critical point continues to increase the pressure and temperature and supercritical steam can be superheated like any other vapour. In fact, superheating is more than ever necessary as the critical point is so far to the left-hand on the Mollier diagram. In addition, of course, one stage at least of reheating is essential.

These characteristics of supercritical steam have two important results in regard to boiler design. In a sub-critical boiler, the main function of the drum is to provide a place in which steam and water can be separated. If these are indistinguishable under supercritical conditions the drum may be omitted. Secondly, natural circulation depends on the difference in density between the water/steam mixture in the generating tubes exposed to the furnace heat and the water only in the unheated downcomer tubes outside the furnace. As the critical point is approached, this density differential becomes less and less (see Fig. 3) so that natural circulation is no longer possible and recourse must be had to assisted or forced circulation. It is interesting to note, however, that with skilled design it is possible to obtain satisfactory natural circulation with a pressure as near to the critical as 2735 lbf/in²g which is the drum pressure, corresponding with a boiler stop valve pressure of 2400 lbf/in²g.

Fig. 5 shows the general arrangement of a supercritical unit ordered by the C.E.G.B. for Drakelow 'C' Power Station. It was a 375 MW unit having an output of $2\frac{1}{2}$ million lb/h. It is designed to operate at both sub- and supercritical pressure within the range 2000/3650 p.s.i.g. and the steam temperature will be 1110°F. There is one stage of reheating up to 1055°F at 713 lbf/in²g.

The Field steam cycle

Attempts have been made to circumvent the efficiency limitation inherent in the Rankine cycle. For example, in 1950, the late J. F. Field proposed the application of gas turbine techniques to the use of steam, on the basis that, in the higher ranges of temperature, the gas turbine mechanism, involving a compressor, is more effective than the Rankine cycle. While, in the lower ranges, the reverse is the case; a suitable combination of the two, shown in Fig. 6, appeared to offer a better efficiency than either separately.[4, 5]

About 85 per cent of the exhaust from a high pressure turbine would go, with appreciable back-pressure and super-heat, to a compressor. The remaining 15 per cent would pass through a low pressure turbine and be condensed. A regenerator between the H.P. turbine and the compressor would transfer some of the heat from the H.P. exhaust to the compressed steam and the condensate from the L.P. turbine would be injected into the suction of the compressor, the proportions being arranged so that the delivery from the compressor was just saturated. From the compressor the steam would go to an externally fired superheater and back to the H.P. turbine.

At the time, considerable discussion and controversy took place as regards the practicability of the cycle and there is no doubt that large scale research would have been required to make it a workable proposition. Again, although there would be an apparent saving in capital cost on the boiler proper, the superheater would have to be very generously designed and of expensive materials while to produce 100 MW net output the turbines and compressor together would require a capacity of nearly three times this amount.

When first proposed, the Field cycle aimed at raising the then possible overall efficiency of 30 per cent up to 40 per cent but subsequent improvements in the conventional cycle have so far closed the gap as to make any improvement merely marginal. Nevertheless, the search for new ideas continues and considerable effort is still being put into investigating the practical possibilities of the Field and similar cycles.[1]

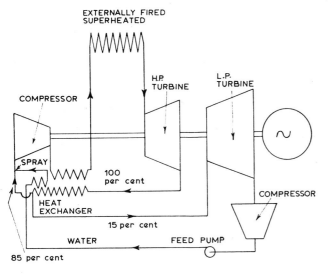

Fig. 6. The Field cycle combines gas and steam turbines in an attempt to beat the efficiency limitations of the Rankine cycle

Dual circulation boilers

Where make-up is high as, for instance, in installations requiring process steam, difficulties can arise from the deposition on turbine blading of solids, particularly silica, carried over in vapour form. This occurs at around 520°F but the concentration is not serious under about 700 lbf/in²g. Above this pressure, however, the concentration rises rapidly and had it not been for the highly efficient de-mineralizing equipment now available, central power stations, although having small make-up, would have experienced the same difficulties due to the much higher pressures which are now common.

For industrial installations, however, where the cost and complication of special water treatment plant is not justified another solution is possible.[6, 7] This consists of designing the boiler with two distinct circulatory systems. Turbine steam is taken, *via* the superheater, from the primary circuit and make-up to this circuit includes steam from the secondary circuit. This make-up steam is partially or wholly condensed on entry to the primary drum by mixing with the main feed water. The primary drum is set higher than the secondary so that blow-down takes place from the primary to the secondary drum and thence to waste. With this arrangement, solids tend to concentrate in the secondary circuit. leaving the turbine steam from the primary circuit comparatively pure. The success of the operation is, of course, dependent on the efficiency of the drum internal fittings which are designed to perform the usual functions of separating the steam from the water on arrival from the generating tubes and of drying the steam before its despatch to the superheater.

Boiler control

The modern highly rated boiler-turbine unit with its complicated reheat and feed-heating system calls for the control of a great number of variables, particularly when starting up from either hot or cold conditions and when changing load or shutting down. The ideal and most economical way to operate a conventional steam power station is, of course, to keep it on full load continuously but with more and more atomic power stations feeding into the grid and because these latter have to be operated as nearly as possible at 100 per cent load factor, it is inevitable that many, even of the latest steam stations, will eventually have to work on a two-shift basis. All recent steam stations, therefore, have had features built into them to enable this to be done and the C.E.G.B. have instituted very thorough tests to investigate the techniques of quick shut-down and start-up of the modern large units.

Information was required regarding the conditions for a hot start after an eight hour shut-down (corresponding to 2-shift working) and a warm start after a 36-hour shut-down (corresponding to week-end working).

When running-up a unit, two conflicting requirements have to be met. The rate of temperature-rise in the turbine must be rigidly controlled if shaft distortion and stresses on the casings are to be avoided; on the other hand, if too little steam goes through the boiler superheater, metal temperatures in it may rise beyond the maximum permissible. This un-balance had generally been allowed for by diverting considerable quantities of steam to the condenser. These tests, however, have shown that, with adequate instrumentation, the modern unit can safely be run-up to full load in three quarters of the time previously required and, further, that the quantity diverted can be reduced to about one fifth of what was previously necessary.

Fig. 9. This Goliath crane, specially developed for the construction of nuclear power stations, will lift 400 tons through 200 ft

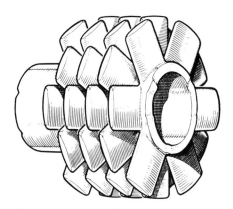

Fig. 8. Typical studded tube designed to increase heat transfer

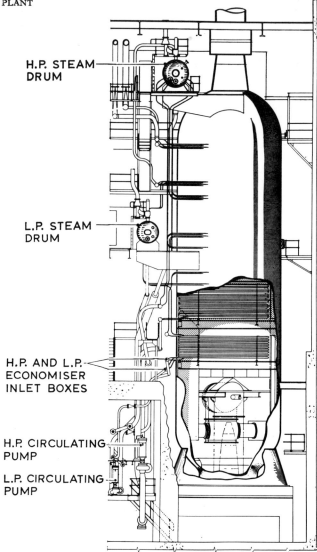

H.P. STEAM DRUM

L.P. STEAM DRUM

H.P. AND L.P. ECONOMISER INLET BOXES

H.P. CIRCULATING PUMP

L.P. CIRCULATING PUMP

Fig. 7. The layout of the boiler units at Sizewell is typical of all the other British nuclear power stations

Such comprehensive tests are also of importance in justifying features already incorporated into the units and in pointing to other desirable features for the future. For example, the modern boiler is so large that, from the point of view of ease of operation, as many controls as possible should be remotely operated. This, in a sense, is no new feature since semi- or fully-automatic control during running has been available for many years. But this is a very different thing from the overall unit control during the transient conditions of starting up in conjunction with a turbo-alternator set which has not yet been synchronized.

The latter condition requires a carefully thought out and delicately balanced programme but, thanks to the modern computer, even this operation can be made automatic[8–12] and it is interesting to note that the first application in Europe of digital-computer control of a power station boiler was incorporated on one of the 200 MW units by the C.E.G.B. at their West Thurrock Station. Standard programmes were written for starting from hot or cold, for normal running and for emergency or normal shut-down.

Each programme is fed into the computer by means of plug-boards which are easy to modify by re-arranging the plugs. The system is thus flexible and, when the best sequence has been established by experience, permanent programmes will be set up.

In addition to controlling the normal boiler auxiliary equipment, the computer also automatically operates the soot blowers when required, according to the changes in heat transfer, as calculated by the computer. The unit overall efficiency is also calculated by the computer at frequent intervals and any operations performed are automatically typed out with their times so that there will be a complete record for study and comparison.

Steam for nuclear power

It has been noted that, even with the advent of nuclear power, steam is still the medium employed for transforming into mechanical work the heat generated in the reactors. The design of the steam generators and, in fact, of the whole thermodynamic cycle, to meet the requirements set by the conditions in the reactors posed some interesting problems. Calder Hall has become the prototype for subsequent British stations and, as such, may be taken as an example for examination of some of the problems involved. The main limitation was one of temperature, both of the fuel and of the magnox cans in which the fuel rods were sheathed. At high temperatures natural uranium distorts and magnox loses its strength. These two effects eventually lead to bursting of the cans. At Calder Hall the safe can temperature was fixed at 768°F corresponding to a maximum temperature of carbon dioxide coolant of 637°F. Compare this with the furnace of a normal coal-fired boiler where the combustion gases reach 2000°F–3000°F. The resulting moderate steam temperature of 595°F, if used in a simple cycle, would have produced a low overall efficiency. Somehow this efficiency had to be increased. Re-heating was considered but discarded for much the same reasons as prevented it becoming a practical proposition in the early days of conventional stations when pressures were low and the unit system of one boiler per turbine had not yet arrived. In addition, in nuclear stations the turbine room had to be located some distance from the reactor buildings, so that the pipework problem became even more acute.

Eventually, a solution was found which was both elegant and simple.[13] It was to generate steam at two pressures, 77 per cent of it at 200 lbf/in²g and 23 per cent of it at 50 lbf/in²g. This use of a dual-pressure cycle enabled the H.P. steam to be generated at a pressure higher than could be done with a single pressure cycle. It led to an increase in overall thermal efficiency and it could be used to control, in a simple way, the temperature conditions in the reactor. While Calder Hall was designed primarily for plutonium production, subsequent stations have been optimized for electric power output. One noticeable difference, for example, is that at Calder Hall the L.P. steam is generated at the temperature to which the H.P. steam has fallen after expansion, so that at the turbine they mix at approximately the same conditions of temperature and pressure. In subsequent stations, however, the L.P. steam is generated at approximately the same temperature as is the H.P. steam with the result that the mixture becomes slightly superheated and gives better exhaust conditions after final expansion down to vacuum. This dual-pressure system, with minor modifications, has been adopted as standard for all the C.E.G.B.'s nuclear power stations.

The design of the steam raising units[14] was fairly straightforward as temperatures and pressures were so moderate, but rigid precautions, in both design and manufacture, had to be taken to avoid any possibility of either leakage of the carbon dioxide coolant gas into the steam circuit or of contamination of the gas itself by steam or foreign matter.

Fig. 7 shows the layout of a steam raising unit for the nuclear station at Sizewell but is generally typical of all other British stations. Four such units serve each of the two reactors. Each has a full load output of 534 500 lb/h of steam at 720 p.s.i.a. drum pressure and 750°F together with 174 000 lb/h of steam at 321·5 p.s.i.a. and 753°F. Thus, for the 578·3 MW net output of the station, over 5½ million lb/h of steam are generated. The illustration shows the high and the low pressure drums; the circulating pumps will also be noted.

The fact that the heating medium is very clean carbon dioxide as opposed to the dirty, dust-laden flue gases of a conventional boiler, enables very efficient types of extended heating surfaces to be used.[15] For example, at Calder Hall,

Chapel Cross, Hinkley Point and Sizewell Stations multiple flat studs are welded to the outsides of the water tubes. This increases their effective gas-side heating surfaces, as compared with plain tubing, by factors ranging from 3·7 up to 8·9. Fig. 8 shows this form of construction and when it is considered that the 12 Hinkley Point units contain a total of some 235 miles of tubing with 350 million studs welded thereto, the manufacturing problems become apparent.

The atomic power programme has given rise to a number of interesting innovations in manufacturing and erection techniques.[16] To mention only two of them, there is the heavy welding on site of the large vessels required and, the use of Goliath cranes for erection.

Welding techniques

As regards welding, the major items involved are the cylindrical shells for the steam-raising units and the containment vessels for the reactors. Owing to their size and weight, they cannot be transported as complete units but only in sections. For their subsequent welding-up on site, therefore, very complete arrangements have to be made for supplying all the services normally found only in the maker's works. Buildings have to be provided since satisfactory welding cannot be carried out in the open and special arrangments are necessary for stress-relieving, examination and testing of all welds made. These vessels were spheres 67 ft. in diameter, welded up from plate sections 3 in. thick, hot pressed to the necessary spherical curvature. Their weight was about 400 tons each.

On the other hand, one of the manufacturers adopted the novel plan of completing the boiler shells in their works,

[Courtesy, B.A.C.]

Fig. 10. Combustion research at the British Coal Utilization Research Association's Leatherhead Laboratories

launching them and then towing them to site by sea. This had the advantage of obviating much of the elaborate arrangements on site for welding and stress-relieving.

The steam raising units at Hinkley Point required shells 21 ft 6 in. in diameter, 90 ft long, with a maximum wall thickness of $2\frac{3}{8}$ in.

These shells had to be up-ended, transported and placed in their final positions. It was decided to provide, purely as an erecting tool, a large crane, capable of carrying out this operation in a matter of an hour, as compared with almost a month previously required. Fig. 9 shows one of the Goliaths at Trawsfynydd. It will lift 400 tons, has a span of 250 ft and a clear lift of 200 ft. These cranes, of course, had many other uses such as, for instance, the lifting and positioning of the various sections of the 3 in. thick reactor containment vessel after these had been welded up.

Fig. 11. A special section of superheater is welded into the main run to test the effects of corrosion on various metals

While many new problems posed by atomic power stations have been solved by adapting existing techniques based on conventional steam power tsations, the traffic has not all been one-way. For example, the experience of heavy site-welding has led one manufacturer to apply the same methods to the very large boiler drum for the 550 MW boiler, illustrated in Fig. 3. Since the inland transport of a drum, 120 ft long and weighing nearly 250 tons, would impose almost insuperable difficulties, it will be sent in four sections and will be welded, stress relieved and tested on site. It is believed that this will be the first time in history that this procedure has been adopted for a high-pressure boiler drum.

Similarly, now that several Goliath cranes exist in the country, it would appear logical that they should be employed for the erection of conventional power stations.

Mention has been made of the rapid strides in steam engineering in the last few years. Such advances were not made blindly but were rendered possible only by continuous, unremitting research.

Natural circulation

For example, as boiler pressures approach the critical, the density differential between steam and water, on which natural circulation depends, becomes less and less. It is essential, therefore, to have as accurate a knowledge as possible of the resistance opposing circulation. Many theories had been attempted to explain natural circulation phenomena but little reliable information was available with which to test those theories. To fill this gap the Water-Tube Boilermakers' Association decided to sponsor a major research project at the Department of Engineering of Cambridge University and their report was issued in January 1960 after several years' work.[17]

The results confirm that above about 1250 lbf/in²g the behaviour of a two-phase, water/steam mixture can be described adequately by the known theory which assumes that the mixture is homogeneous. Below this pressure, however, the effect of slip between the steam bubbles and the water must be taken into account.

During these experiments a new technique was developed for measuring the apparent density of the mixture. A collimated gamma-ray beam was passed from side to side of the pipe and its attenuation measured by a scintillation counter. This information enabled an estimate to be made of the relative or slip velocity between the two phases and correction factors to be established for application to the acceleration and gravitational pressure drops as calculated by the homogenous theory. The research covered 1 in. and $1\frac{1}{2}$ in. vertical and horizontal pipes and a pressure range of 250 to 2100 lbf/in²g.

The Cambridge project is an example of a piece of research the specific aim of which was to assist manufacturers to design high-pressure boilers with more certainty and, therefore, more economically.

Corrosion and erosion

A far more general, continuing programme has for many years, however, been undertaken by the Boiler Availability Committee.[2] This is sponsored and financed mainly by the W.T.B.A., the C.E.G.B. and the South of Scotland Electricity Board. The work is of two kinds; full-scale field trials at power stations and small-scale laboratory research on combustors carried out by the British Coal Utilization Research Association at Leatherhead and Wolverhampton. Fig. 10 shows such combustors in use at Leatherhead.

Because of the one-time shortage of coal, certain power stations were laid down or adapted for oil burning. Since 1956, the amount of oil consumed has, in consequence, increased from less than $\frac{1}{2}$ per cent to a much higher percentage of the fuel used by the C.E.G.B. As anticipated, the high sulphur content inherent in oil has led to low-temperature corrosion at the back-ends, notably in air-heaters.

As regards solid fuels, East Midlands coals in particular have an appreciable chlorine content and this results in high-temperature corrosion of superheaters. Another aspect of pulverized fuel is high-temperature deposits, again on superheaters.

All these problems are the concern of the B.A.C. Some of their recent work has shown that metallic zinc additive to oil reduces corrosion at the low-temperature back-end but may increase it in the high-temperature superheater zone. Other additives such as ammonia gas and *Teramin* are being investigated. Sodium in residual oils may also cause trouble and a water washing treatment has been developed, reducing the sodium content from 120 to 20 parts per million. Further work on sodium release has also been undertaken, this time in connection with pulverized fuel. It was found that particle size has a direct bearing on the matter; reducing this size from 38 per cent through 300 B.S.S. mesh to 75 per cent through the same mesh, lowers the alkali release from 30 to 20 per cent.

Another interesting kind of research project was carried out jointly by the C.E.G.B. and a boiler manufacturer. At two pulverized-fuel-fired power stations in N.W. England, the metal wastage of superheaters was investigated at elevated temperatures and under actual service conditions. In each case a special section of superheater, shown in Fig. 11, was welded into the main run and metal temperatures were controlled by a valve varying the steam flow through the section. One section was made up of seven different types of steel, the other of ten. At one station the fuel contained 0·43 per cent chlorine and 2·52 per cent sulphur and at the other the respective figures were 0·27 and 1·9 per cent. The trials ran for 8580 hours in one case and 4300 in the other.

These investigations were important because, while mild steel and low alloy steels are satisfactory for steam temperatures up to 950°F, most of the more recent stations are designed for 1050°F or more. Alloys had been produced[18-23] to resist creep and oxidation at these higher temperatures but their development had taken place under the somewhat artificial conditions in a laboratory where possible attack by bonded deposits could not easily be simulated.

Measurements were made of both the internal wastage due to steam, and the external wastage due to flue gases and deposits. The results showed very clearly that the superiority of austenitic materials became more marked with increase in chromium content.

Heat transfer

The advent of nuclear power has, of course, necessitated a tremendous programme of research not only by such authorities as the U.K.A.E.A. and the C.E.G.B. but also by individual manufacturing firms. A recent example may be quoted from the British Nuclear Energy Conference's symposium on *The Use of Secondary Surfaces for Heat Transfer with Clean Gases*. At this meeting four of the papers[24-27] dealt with heat transfer and pressure drop characteristics of cross-flow finned and studded tubes and Fig. 12 indicates one of the elaborate and costly rigs used. With this arrangement air or carbon dioxide under pressures up to 250 lbf/in²g and temperatures up to 320°F could be circulated by a fan the maximum capacity of which was 515 000 lb/h (at 200 lbf/in²g). The importance of work of this kind can be judged by the great advances made during the comparatively short period between Calder Hall and, say, the most recently designed nuclear power station at Sizewell. The steam-raising units at the two stations will be about the same height but although the diameter of those at Sizewell will be only 30 per cent greater than at Calder Hall, they will absorb, in the steam they generate, no less than six times the number of B.Th.U.'s per hour. Part of this increase is, admittedly, due to the temperature, pressure and velocity of the carbon dioxide being higher but the more efficient type and arrangement of the heating surfaces developed as a result of research work, is undoubtedly responsible for much of the economy.

Codes of practice

As regards pressure vessels in particular, the codes of practice, standard specifications, etc., in use in the various countries not only differ considerably but in no case are they based on a truly rational approach to the subject.[28]

Many codes today are oversimplified. They assume certain minimum values for a by no means comprehensive range of physical properties of the material to be employed. The temperature-dependent values of these properties are then modified by applying an arbitrary factor of safety to cover a wide variety of contingencies, some of which are, in fact, nothing more than bad features of design which have become accepted as inevitable liabilities. There have been few failures but the factors chosen, while high enough to cover ignorance, tend to handicap good design.

But what is the alternative? Two proposals have been advanced.[29] The first is an amended procedure which could be applied right away, as it requires little more knowledge than is at present available; the second is a target for the future as it calls for the accumulation of much more data and information but its usefulness at the present moment is that it indicates the directions in which future research should go.

Under the first amended procedure, the approach to the problem would be to establish an upper limit for the design stress under ideal conditions and to work backwards for each departure

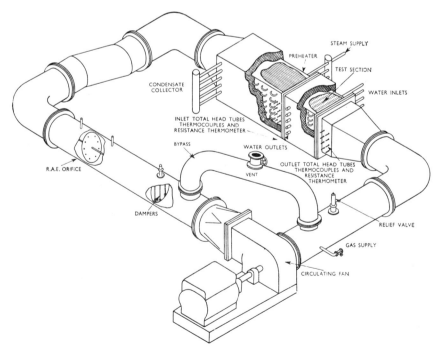

Fig. 12. One of the elaborate heat transfer rigs for testing pressure drop in tubes

from the ideal. Such departures would, for example, be represented by the addition of components such as nozzles, hand-hold fittings, etc. Each departure would be further conditioned by the possible methods of failure and by a factor to represent deficiencies in manufacture.

Under the second amended procedure, the use of a design stress would in effect be eliminated; direct design would be based on a comprehensive list of failure contingencies, critical factors and material properties. A start has already been made by the International Institute of Welding's Pressure Vessels Commision to this end by calling on various countries to contribute theses on designing against all known failure contingencies; it is quite certain that, as a result of this work, many gaps in our present knowledge will be disclosed.

Mr P. B. Silk *was educated at Oundle and, after serving with the R.N.A.S. and R.A.F., graduated with first class honours in Mechanical Engineering at Manchester University. In 1922 he joined the Power Station Design Dept of English Electric and in 1928 formed the Ash Co. (London) Ltd, which specialized in the Hydrojet ash-sluicing system. In 1938 he became Manager of the Crane, Conveyor and Structural Departments of Babcock and Wilcox Ltd. For some years he was Chairman of the Association of Crane Makers and President of the European Federation for Mechanical Handling (FEM).*

Combustion mechanisms

An altogether different type of research, also of an international character, was started some 12 years ago and is now being carried out in Holland under the auspices of the International Flame Research Foundation. Its sponsors are Great Britain, U.S.A., and the members of the European Coal and Steel Community.

The main object of these researches is to gain a better knowledge of what happens in the flame under various conditions of combustion.

Between 1949 and 1954 six performance trials were carried out on oil and gas flames in a special tunnel furnace in the Royal Netherlands Steel Works at Ijmuiden.[30-33] At the end of 1955 a second furnace was commissioned specifically for burning pulverized fuel and since then a series of tests have been made and reported.[34-38]

Fundamental work of this kind must be based on general considerations rather than on the evaluation of particular items of equipment. For example, the burners used were of a simple type designed by the Foundation's staff and were not proprietory articles. From a severely practical point of view, therefore, it may be difficult at first sight to see in what way researches of this kind can benefit industry but the published results can be interpreted by each firm according to its own needs and supplemented by further research on its own types of plant.

It should not be thought that the work of the I.F.R.F. referred to above is the only research carried out in the field of combustion. Investigations of this kind continue, not only in Britain but all over the world. It would be well-nigh impossible to include a complete list of references to papers on the subject but a short selection is appended.[39-43]

Two of these,[42, 43] deal with the use of models in connection with the design of boiler furnaces. In Britain, increasing use is being made of this technique to establish the best positions for the burners, the superheater and reheater surfaces and the general shape of the furnace itself. With the modern very wide furnace it is easy to run into difficulties because of the non-uniform gas velocity past the heating surfaces; this can lead to local overheating, slag-formation, and dead pockets which accumulated dust and loss of overall efficiency.

For example, a three-dimensional, perspex model was made of one half of a very large unit recently ordered by the C.E.G.B. The fluid used was water containing plastic tracer particles. The movements of these particles, when photographed, make them appear as short lines which give an indication of the flow pattern. By a suitable choice of velocity in relation to the Reynolds number, a reasonably accurate representation of gas flow conditions can be obtained. The main limitation of the method is, of course, that it is impossible to reproduce the varying specific volume of the gas as its temperature changes during its passage through the boiler; nevertheless very valuable information can be obtained at a reasonable expenditure of time and money.

Future trends

When writing of future developments, an engineer has a natural tendency to be conservative and to think that present practice represents about the limit to which design can go.

Thirty years ago the everyday use of 500 MW sets with boilers for 4 million lb/h of steam at 2400 lb/in² or more would have appeared to the writer as pure fantasy. But, one by one, the barriers have been broken and while, admittedly, the present important limitation on temperature seems difficult to surmount, there are ways of getting round it and it should not be long before we see the 1000 MW unit, possibly at supercritical pressure, with two or three stages of reheat.

In general, it will be many years before steam is superseded as the best medium for transforming heat-energy into mechanical work. The gas turbine has undoubtedly a greater part to play than at present and its use may possibly come in a combined cycle with the steam turbine. But much development work is still required to give it the necessary life between overhauls and to produce it in the sizes to meet present and future demands.

Many fascinating possibilities present themselves such as magneto-hydrodynamic generation, binary fluid and various combined cycles, but in the present state of the art any observations would be in the realm of crystal-gazing. The potentialities of nuclear fission or fusion, however, appear almost unlimited. In all these cases steam will probably remain their handmaiden for many years. In the natural uranium, gas-cooled, graphite-moderated reactor at present standardized for British nuclear power stations, permissible temperatures are low, allowing only moderate steam conditions.

There is little doubt, however, that in a few years steam temperatures and pressures will have caught up with those prevailing in conventional power stations.

While essentially safe and of predictable performance, the present type of nuclear station suffers from the disadvantage of large physical size and therefore of high capital cost. There are, however, so many possible variations in regard to the type of fuel, the method of cooling and the nature of the moderator that the future will most certainly see much more compact and cheaper reactors being built.

REFERENCES

1. C.E.G.B. Annual Reports and Accounts 1955–56 to 1960–61.
2. Boiler Availability Committee. Technical Papers 1–5 1947/1961 and Bulletins I to XI which include summaries of some 170 papers.
3. SILK, P. B., 'Critical and Supercritical Steam Pressures'. *Power and Works Engineering*. 1960 December.
4. FIELD, J. F., 1950 *Proc. Instn mech. Engrs, Lond.*, vol. 162 p. 209.
5. HORLOCK, J. H., 'The Thermodynamic Efficiency of the Field Cycle' *A.S.M.E.* paper no. 57-A-44.
6. PADDON Row, R. H., *J. Inst. Fuel* vol. 34 no. 243, pp. 143–152.
7. VERITY. C. H. E., 'Dual Circulation Boilers for Use with 100% treated Feed Water', THE CHARTERED MECHANICAL ENGINEER. 1961 February, pp. 124–125.
8. KESSLER, G. W., 'The application of Computers to automatic boiler operation'. *A.S.M.E.* paper no. 60-PWR-1.
9. CHADWICK, W. L., 'The general philosophy and objectives of Power Plant Automation'. *American Institute of Electrical Engineers*, paper no. 60-1164.
10. KESSLER, G. W., 'Power Plant Automation—Boiler Operation'. *Mechanical Engineering*, 1960 November.
11. NOYES, E. G., 'Power Plant Automation—Turbine Generators'. ibid.
12. HESS, W. T., 'Power Plant Automation—Economics'. ibid.
13. WOOTON, W. R., 'Calder Hall—Steam cycle Analysis'. 1957 April. British Nuclear Energy Conference Journal.
14. MORRIS, H., and WOOTON, W. R., 'Design and Construction of Heat Exchangers'. ibid.
15. ZOLLER, R. E., 'Extended heat exchange surface for marine and nuclear boilers'. 1958 *Trans. N.E. Cst Instn Engrs and Shipb.*, vol. 74.
16. 'Shop and Site Welding for Hinkley Point Nuclear Power Station'. 1960 November, *Welding and Metal Fabrication*.
17. HAYWOOD, R. W., KNIGHT, G. A., MIDDLETON, G. E., and THOM, J. R. S. 'Experimental Study of the flow conditions and pressure drop of steam–water mixtures at high pressures in heated and unheated tubes'. 1961 *Proc. Instn mech. Engrs, Lond.*
18. SMITH, A. I., 'Mechanical Properties of Materials at High Temperatures'. 1961 THE CHARTERED MECHANICAL ENGINEER, pp. 278–285.
19. JOHNSON, A. E., HENDERSON, J., and KHAN, B. 'The behaviour of metallic, thick-walled cylindrical vessels or tubes subject to high internal or external pressures at elevated temperatures'. N.E.L. Report A B Div. No. 1/59.
20. ARMSTRONG, D. J., and MURRAY, D., 'Techniques for creep testing at high temperatures'. 1956 *Journal of the Less-common Metals, Amsterdam*, 1(2), pp. 125–131.
21. SMITH, A. I., JENKINSON, E. A., DAY, M. F., and ARMSTRONG, D. J., 'Creep stress relaxation and metallurgical properties of steels for steam power plant operating with steam above 950°F'. 1957 *Proc. Instn mech. Engrs, Lond.*, 171(34), pp. 918–942.
22. SMITH, A. I., ARMSTRONG, D. J., and TREMAIN, G. R., 'Stress relaxation properties of steels for steam power plant components'. N.E.L.

Report A B Div. No. 13/60.
23. HIMS, H. E. C., and FREDERICK, S. H., 'Materials for advanced steam conditions and their influence on operation of marine turbines and boilers'. 1961 *Trans Inst. mar. Engrs.*
24. WORLEY, N. G., and ROSS, W., 'Heat transfer and pressure loss characteristics of cross-flow tubular arrangements with studded surfaces'. 9th Nov. 1960 *I. Mech. E. Symposium on the use of Secondary Surfaces for Heat Transfer with Clean Gases.*
25. LYMER, A., and RIDAL, B. F., 'Finned Tubes in cross-flow of gas'. ibid.
26. ROUNDTHWAITE, C., and CHERRETT, N., 'Heat transfer and pressure drop performance of helically finned tubes in staggered cross-flow'. ibid.
27. DAVIES, F. V., and LIS, J., 'Heat transfer and pressure drop characteristics of concentric arrangements of helical coils'. ibid.
28. CARLSON, W. B., 'Pressure Vessel Design'. *I. Mech. E.*, Scottish Branch paper, 1957 November.
29. CARLSON, W. B., 'Pressure Vessel Design Requirements in the future'. International Institute of Welding presented to A.W.S. 42nd Annual Meeting. New York. April 1961.
30. I.F.R.F., International Flame Research Committee. *Journal Inst. Fuel.* 1951. vol. 24, supp. p. 1.
31. I.F.R.F., *Journal Inst. Fuel.* 1952. vol. 25, supp. p. 17.
32. I.F.R.F., Reports of Burner Trials at Ijmuiden. *Journal Inst. Fuel.* 1953. vol. 26, p. 189.
33. I.F.R.F., 'The Radiation from Turbulent Jet Diffusion Flames of Liquid Fuel/Coke-oven Gas Mixtures'. *Journal Inst. Fuel.* 1956. vol. 29, p. 23.
34. THURLOW, G. G., 'Summary of work of I.F.R.F.'. *Journal Inst. Fuel.* 1960. vol. 33, p. 235.
35. TISSANDIER, G., 'Cold aerodynamic trials on a fifth-scale model of the Ijmuiden P.F. Furnace'. ibid.
36. HUBBARD, E. H., 'The first performance trial and first combustion mechanism trial with P.F.'. ibid.
37. ALPERN, B., COURBON, P., PLATEAU, J., and TISSANDIER, G., 'Microscopic examination of samples taken from a pulverized fuel flame'. ibid.
38. THURLOW, G. G., 'The work at Ijmuiden on Pulverized Fuel Flames up to mid-1960'. ibid. 34. 244.
39. ROSIN, P. O., 'The mechanism of combustion of pulverized fuel'. *Journal Inst. Fuel.* Aug. 1958, p. 346.
40. BAYLESS, A., 'The effect of particle size on firing pulverized solid fuels in boilers'. *A.S.M.E.* 57-A-276.
41. HOLDEN, C., and THRING, M. W., 'Liquid fuel firing—Major developments 1948–1959'. *Inst. Fuel* 1959.
42. CURTIS, R. W., and JOHNSON, L. E., 'Use of flow models for boiler furnace design'. *A.S.M.E.* 58-A-120.
43. PUTNAM, A. A., and UNGAR, E. W., 'Basic principles of combustion model research'. *A.S.M.E.* 58-A-73.

Gas Turbine Technology

by R. H. Weir, CB, BSc, FRAeS

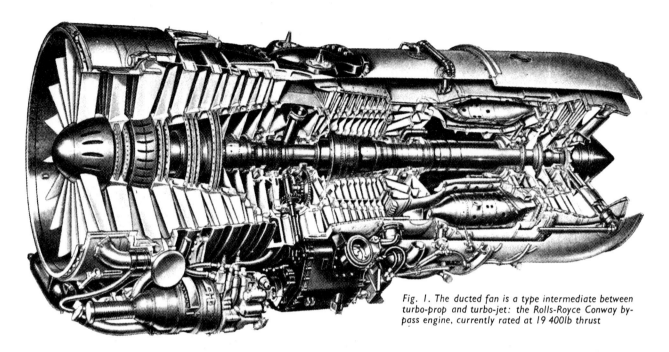

Fig. 1. The ducted fan is a type intermediate between turbo-prop and turbo-jet: the Rolls-Royce Conway by-pass engine, currently rated at 19 400lb thrust

At the time of its war-time debut in the form of the jet engine, the gas turbine promised a revolution, not only in aeronautics but also in prime-mover development generally. The first promise has been amply justified, the second development has been somewhat slower of fulfilment; but there are now prospects of further spectacular advances and other applications are multiplying rapidly, as a result of the research and development work described here.

The roots of gas turbine technology are diverse and buried deep in history. Yet to many people the gas turbine is synonymous with the jet engine which, in the course of a decade or two, has revolutionized the pattern of military and civil aviation, and offers the prospect of further spectacular advances in the next decade or two.

Military demands for high-speed fighting and transport aircraft gave a tremendous impetus to the aircraft gas turbine and sustained research and development effort at a high level for many years. Although military support has diminished somewhat in recent years, a counterbalancing influence has emerged in the rapidly growing market for gas turbines in civil aviation, which has led, in many cases, to the development of engines specifically for civil use. As a result, the aircraft gas turbine has demonstrated high standards of achievement for the designers of land and marine gas turbines to emulate, and through its development a vast amount of technical design data and know-how for use in land and marine gas turbines and related branches of mechanical engineering has become available. At the same time, the stringent aero-engine requirements for minimum weight and bulk, coupled with reliability, have opened up new areas of materials research and stimulated existing ones, and have led to many improvements in manufacturing techniques and machine tools, to the ultimate benefit of mechanical engineering in many industries.

When the results of research and development specifically connected with land and marine gas turbines are added, the total represents a vast and growing fund of information, applicable, with local adjustment, in many branches of mechanical engineering. This kind of hidden asset is particularly important in a country like Great Britain, in which the standard of living is dependent on the export of manufactured goods of high quality, based on native ingenuity.

Tangible proof of this is provided by aero-engine exports, which have been expanding continuously since the war, the bulk gas turbine engines and, as a result, 60 per cent of the turbine-engined aircraft in operation or on order in the free world have British engines.

No excuse need be offered for concentrating on the aircraft gas turbine in this introduction, nor for devoting to it a large proportion of what follows. The reason is simple. Much more money has been spent on research and development of the aircraft engine than on its land and sea counterparts, and in research and development, as elsewhere, money translated into intelligent and whole-hearted effort is what counts. A more comprehensive treatment of gas turbines in general is the Gas Turbine Progress Report of the A.S.M.E.[1]

Aircraft gas turbine research may be broken down into general research objectives aimed at meeting the basic requirements for an aero-engine, namely, high power output per unit weight, per unit frontal area (and bulk); economy of fuel consumption; reliability; longevity; reasonable noise level and minimum first cost. These basic requirements are by no means mutually compatible and some compromise is necessary. Much of this design philosophy applies to all types of gas turbines, indeed to most mechanical engineering; but the aero-engine requirements for minimum weight, bulk and frontal area and associated reliability are more stringent than most. In consequence, they entail exhaustive and costly inspection at all stages of manufacture, coupled with extensive development processes[2] which are also costly but unavoidable, in order to establish the necessarily small design safety factors.

Power and efficiency parameters

In a practical, constant pressure, open-cycle gas turbine, the potential output increases as the cycle temperature ratio is raised and as the pressure ratio increases up to an optimum value.[3, 4] Fuel consumption decreases primarily with increase of pressure ratio (up to an optimum value) and, slightly, with increase of temperature ratio on the turbo-prop engine, with the propeller giving approximately constant propulsive efficiency up to moderate flight speeds; for a turbo-jet engine, the same statement is true only if the higher jet speed associated with the higher cycle temperature is matched by an appropriate increase in aircraft speed to maintain propulsive efficiency approximately constant.

The ducted fan is an intermediate case in which increased cycle temperature may be used to increase power output at constant flight speed without deleterious effect on the specific fuel consumption, because jet velocity (and propulsive efficiency) are maintained approximately constant by the increased proportion of 'cold' air fed to the propulsive jet by the fan. An interesting example of the type is the Rolls-Royce *Conway* by-pass engine shown in Fig. 1, the first ducted fan engine in operational service (April 1960) and the forerunner of many other ducted fans on both sides of the Atlantic.

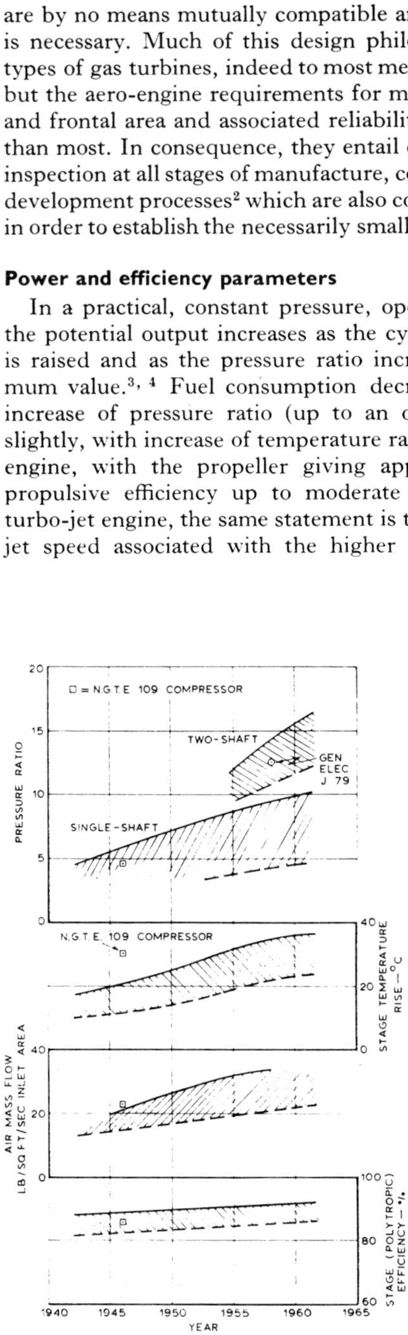

The power output of turbo-jet and ducted fan engines may be increased by burning fuel in the propulsive jet (reheat or afterburning), thereby increasing the jet velocity. The process is expensive in fuel consumption unless flight speed is raised to match the higher jet velocity. It lowers overall weight compared with the larger engine and can be used with advantage at low flight speeds, for short-term power boosting; for example, at take-off, if noise can be tolerated, or for short bursts of supersonic speed.

High output per unit weight and unit frontal area entails increased flow of air and involves lower hub/tip ratios in compressors and turbines and/or higher air and hot gas velocities through all components, and fuel consumption is likely to be adversely affected because of increased aerodynamic losses. On a given engine, airflow can be increased by injecting a refrigerant (usually a water/methanol mixture) into the compressor. Refrigerant consumption is high and this method of power boosting is only used at take-off, usually to restore the power lost at high ambient temperature.

Compressor development

The influence of the major parameters controlling the power output and efficiency of the aircraft gas turbine can be traced broadly by considering the published statistics throughout the years.[5, 6] Fig. 2 shows cycle pressure ratios increasing from about 4 : 1 with the original centrifugal compressor engines of the early 1940's to around 9 : 1 for single-shaft, axial engines in 1956. The emergence of twin-shaft L.P. and H.P. designs in the mid-1950's produced a sharp upward trend and by 1960 engines with pressure ratios of 15 : 1, such as the Rolls-Royce *Conway*, were in operational service. A proportionate increase in the number of axial compressor stages, and higher engine weights were avoided by increasing the loading per stage. Expressed in terms of the adiabatic temperature rise of compression per stage, the loading increased almost linearly, from about 20°C per stage average in 1950 to over 30°C in some current engines. This was achieved by refinement in the aerodynamic design of the compressor stages.

A notable exception to the general trend towards two shafts used with high pressure ratios, was the General Electric J.79 engine, with a pressure ratio of 12·5 : 1 on a single shaft, avoiding compressor/turbine matching problems over the engine speed range by incorporating six rows of variable stators in its compressor.

The increase of air mass flow per square foot of compressor inlet gross area has been more spectacular. Values of 35 lb/ft²/sec are common today, compared with an average of 20 lb/ft²/sec in 1950. Further progress in this direction is likely to be much slower, with a limit around 40 lb/ft² set by choking in the inlet guide vanes.

These developments would have been offset if accompanied by a loss of compressor efficiency but, in point of fact, a slow but steady rise of polytropic (or stage) efficiency is discernable, from an average of about 86 per cent in 1950 to around 89 per cent today; this is not always associated with the maximum blade loadings. A further small increase

Fig. 2. Axial compressor development over 20 years: cycle pressure ratios more than doubled for single-shaft engines

of average efficiency does not appear to be beyond the bounds of possibility for the future.

In Fig. 2, the points denoting the National Gas Turbine Establishment's 109 compressor of 1946[7] are worth noting as an early demonstration of what could be achieved by compressor design technique; and as an example of the necessary lead of the research unit over the fully-developed component in an engine.

Inducing more air into the axial compressor and raising it to higher pressure with reasonable efficiency in the minimum number of stages brought out many mechanical problems, due to incomplete knowledge of the detailed air flow.[6, 8, 9] Notably, the effect on the early blade rows of unsymmetrical pressure and velocity profiles (due to the aircraft intakes) was accentuated by the lower hub/tip ratios involved. Badly shaped surge lines were common and fatigue failures of blading occurred due to self-excited or externally-forced vibration. Practical mechanical solutions were found in all cases, but much research remains to be done to fill in many gaps in the knowledge of detailed air flow.

Combustion

In the absence of a unique criterion of merit linking the closely-integrated performance parameters (efficiency, pressure loss, stability limits, outlet temperature distribution, ignition characteristics and heat release) it is not easy to review briefly the progress in combustion. In practice this interplay between the operating variables means that individual designs can be trimmed to meet the needs of a particular application, but it follows that direct comparisons of isolated combustion chamber performance parameters can be misleading. Analyses made in recent years[10-13] into the performance of an idealized homogeneous combustion system in terms of reaction rate theory have successfully established the interaction of the main operating variables, i.e., mass flow, pressure, temperature and fuel/air ratio; but the application of such analyses to practical systems employing direct injection of fuel into the combustion space still remains very much an *ad hoc* process. The correlating parameter has been successfully used, however, to reduce the development testing time which is required to establish combustion chamber performance. Additionally, a scaling technique[14, 15] in which pressure is varied inversely as the linear scale, with gas temperatures and velocities remaining constant, has been shown to give similarity of combustion performance and a reasonably close approximation to chamber wall temperatures. The application of this technique in development presents some difficulty in accurately reproducing the full-size combustion chamber on a model scale, but it has possibilities when the test air quantities are limited.

The main aim of combustion chamber development has been to improve performance of existing designs so as to meet the demands for increased engine output. Increases in mass flow, often accompanied by a deterioration in the uniformity of the compressor outlet velocity profile, have required improvements in the entry aerodynamics to avoid excessive pressure loss through the chamber whilst increases in maximum cycle temperature have necessitated improvements to cooling air mixing arrangements so as to maintain the temperature spread at the chamber outlet within acceptable limits and avoid damage to the turbine due to local overheating.[16, 17] Altitude relighting characteristics have been improved, as have methods of flame tube cooling, and a considerable amount of work has been done to discover sources of incomplete combustion at altitude cruising.[18]

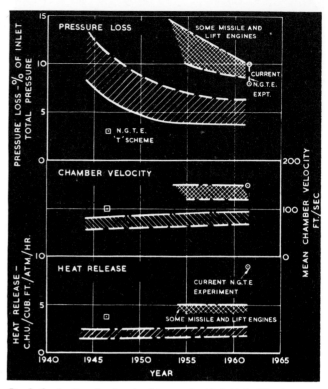

Fig. 3. Combustion chamber development: in this field the 'T' scheme, an offshoot of ram-jet research, established a lead in 1947

Bearing in mind earlier remarks about the pitfalls of characterizing combustion chamber performance in terms of any one parameter, it is nevertheless instructive to consider the extent to which the major design parameters have changed. The outstanding achievement has been the reduction of pressure loss, as shown in Fig. 3, from about 10 per cent of the inlet total pressure in 1945 to under 5 per cent on some modern engines; particularly as this has been accompanied by more uniform outlet temperature distribution. The mean air velocity through the combustion chamber has remained virtually unchanged at about 85 ft/sec (based on maximum cross-sectional area of air casing and air entry conditions). But chambers on low pressure ratio lifting or missile

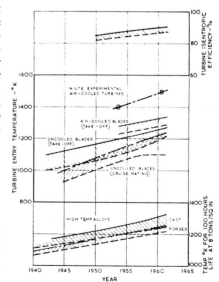

Fig. 4. Temperatures and efficiency: lower stage loading was a major factor in raising efficiencies

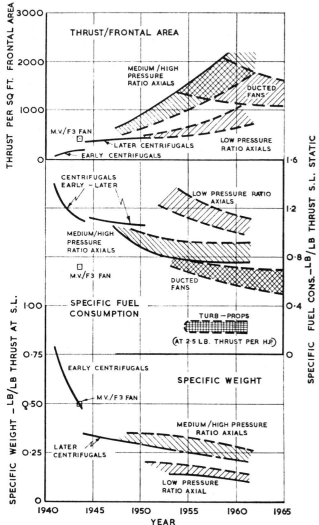

Fig. 5. Turbo-jet performance based on take-off thrust increased steadily over the years. A dramatic reduction of fuel consumption was combined with a lower specific weight of the engine

changes in temperature with the minimum of thermal fatigue, thermal distortion and spurious flow leakage.

In some early engines, turbines tended to be overloaded aerodynamically (in the interests of low weight) and turbine efficiencies were rather lower than had been expected. A later trend towards lower stage loadings was a major factor in initiating and maintaining a steady rise of isentropic efficiency, from about 85 per cent in the early 1950s. This can be seen in Fig. 4. Coupled with continuing refinements in aerodynamic design, this trend, has now resulted in efficiencies of 90 per cent and over. Weight penalties associated with more stages have been offset largely by improved mechanical design, and further refinements calculated to eliminate spurious leakage should yield an additional increase in efficiency.

The earliest engines had maximum turbine inlet temperatures of 1000° to 1100°K. Steady improvement in materials has yielded an increase of just under 10 degrees per year, producing engines which operate today at upwards of 1250°K with uncooled blading.

Since the 1939–45 war, the concept of blade cooling has been vigorously pursued in the research laboratories as a means of further uprating operating temperatures.[20, 21] Experimental air-cooled turbines were eventually demonstrated at 1400°K by N.G.T.E. in 1953 and work is currently proceeding at 1500°K. Fruits of this research began to materialize in the mid-1950s and British engines with air-cooled blading are now operating with turbine entry temperatures in the 1300° to 1350°K region. Such levels are still modest and substantial further improvements in both cooling and blade materials are to be expected over the next decade.

The mechanical behaviour of engineering materials under both steady and transient thermal stress in addition to external loads is very important in the design of turbine blades, discs, and casings. Research on this subject has been proceeding for several years[22, 23] and more reliable methods may supersede empirical methods to which designers are currently driven. The ultimate benefits of such research may spread far outside the field of gas turbine technology.

High temperature materials

In spite of periodic pessimistic opinions about absolute limiting temperatures for particular alloys, it will be observed from Fig. 4 that steady progress has been made[24] in the development of creep resisting alloys for rotor blading. In the late 1940s cast cobalt-based alloys[25] were developed with strength-to-rupture properties superior to nickel-based alloys; but since then, the latter, in the cast or wrought condition, have dominated the scene.[26] Fig. 4 also shows cast alloys maintaining a temperature advantage of 50°C over wrought alloys; but the better ductility and impact strength of the latter offset this advantage. Until recently British engine designers used only wrought alloys. The greater use of vacuum melting in the U.K. has altered the picture and it is probable that in future British designers will follow their American counterparts who have used both cast and wrought blades with success over a considerable period.

It would appear that not more than 50 to 100°C increase in temperature is to be expected from the conventional alloys of the type referred to in Fig. 4 and new alloys based on high melting point elements such as chromium, molybdenum, tungsten and niobium are being sought.[27] High tungsten content has the disadvantage of increasing weight, and niobium-based alloys currently appear more promising

engines have operated at up to 150 ft/sec. Heat release rates, similarly, show little change from a mean value of about 2 million C.H.U./hour. cubic foot. atmosphere; the exceptions again being special duty engines in which low pressure ratio has forced an increase to about 5 million C.H.U./hour. cubic foot. atmosphere.

Fig. 3 shows the lead established by the N.G.T.E. 'T' scheme combustion system of 1947,[19] an offshoot of ram-jet combustion research, which established a low limit of pressure loss without making any concession on chamber velocity or heat release rate. Current N.G.T.E. research, which is aimed at a significant increase in heat release rate and, therefore, reduced chamber dimensions, inevitably makes some concession in pressure loss because of the high chamber velocity used.

Turbine development

Turbines have been subjected to continuous pressure of development with a view to improving aerodynamic efficiency, increasing operating gas temperature and reducing weight, whilst retaining an ability to withstand rapid and repeated

if their oxidization resistance can be improved, or the blades can be suitably protected. There are many opportunities for the research metallurgist to exercise his ingenuity in this and allied fields. Although turbine blade cooling is available, it is always preferable to use uncooled blades, wherever possible, in the interests of overall performance.

Ceramics and cermets (ceramic–metal mixtures) are still being investigated[28] in search of a material with resistance to creep, fatigue, corrosion, erosion, thermal shock and impact which stands comparison with conventional alloys. It seems probable that most of these requirements will be met in the near future, except perhaps for impact strength. Ingenious design methods for minimizing the importance of impact properties would be a great advantage in gas turbine design.

Performance and weight

The upper limit of size of turbo-jet engines has increased steadily and engines available today range in maximum thrust from about 2500 lb to over 20 000 lb. Lack of space precludes any attempt to trace the individual influences on overall performance of the design factors and major component developments previously discussed, but well-defined trends in turbo-jet performance are discernible in Fig. 5.

Over the years, the thrust per square foot of engine intake gross area has risen steadily from about 1000 lb for medium pressure ratio axial turbo-jets of the early 1950s to over 2000 lb for some present-day engines of higher air flow, higher pressure ratio and higher turbine entry temperature. Values for low pressure ratio engines are about one-half, but these engines have compensating merits of simplicity, reduced length and low weight, which are particularly important in lifting engines; as well as lower first cost. The ducted fan also concedes thrust per square foot, to an extent depending on its bypass ratio, but has the compensating merits of lower fuel consumption and lower noise level.

As the achievement of minimum fuel consumption involves increased weight, it is particularly interesting to note that the dramatic reduction of fuel consumption, shown in Fig. 5, is actually accompanied by a reduction in engine specific weight.

Equally noteworthy is the long research lead established by the first ducted-fan engine, the Metropolitan-Vickers F.3 aft fan, a product of Government-sponsored research, which first ran in 1943. The ducted-fan engine gives promise of further reduction of specific fuel consumption in years to come. To show these future gains in perspective, turbo-props are included in Fig. 5 on a basis of fuel consumption per lb of propeller thrust.

Lift engines such as the Rolls-Royce RB.108, fitted in the Short S.C.1 V.T.O.L. aircraft, are setting the pace by a significant reduction of specific weight in low pressure ratio engines. Besides advancing the era of vertical take-off, such engines also provide a feedback of weight-saving ideas into the designs for other engines. For instance, short combustion chambers and welded-drums to replace individual compressor discs. There is evidence of this feedback in the equally striking reduction in the specific weight of propulsion engines which permits their use as dual-purpose engines, such as the Bristol Siddeley *Pegasus* ducted fan. This combines the lifting and propulsion roles with the aid of swivelling jets, as demonstrated in the Hawker P.1127 V.T.O.L. aircraft in 1960. Specialized lift engines with only one-half the specific weight of the RB.108 are promised in a few years' time and intense competition is building up between these and the combined lift and propulsion types. A practical aircraft including both concepts might well prove to be the ultimate solution.

In addition to component developments which have helped to reduce engine weight, there have been contributions from the materials field in the form of high strength/weight ratio alloys, based on magnesium and titanium. Titanium alloys have already made a significant contribution and will contribute more when their cost is reduced. For the future, resin-bonded glass fibre materials show promise of even higher strength/weight ratio but further research and development is required to increase their operating temperature range.

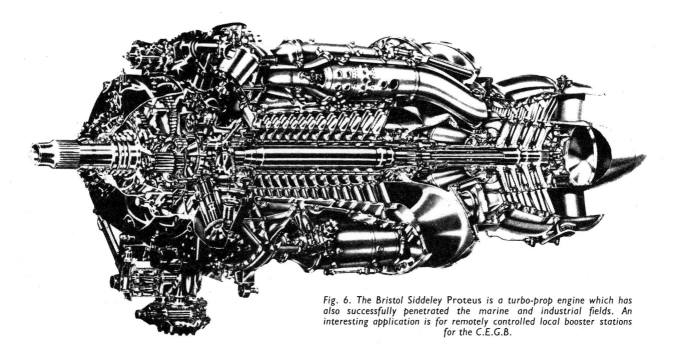

Fig. 6. The Bristol Siddeley Proteus is a turbo-prop engine which has also successfully penetrated the marine and industrial fields. An interesting application is for remotely controlled local booster stations for the C.E.G.B.

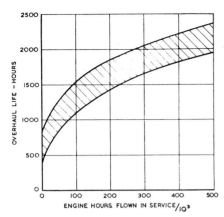

Fig. 7. The shaded area shows how the period between necessary overhauls increases with total engine hours flown by four-engined transport aircraft. Future turbines may start at 1500 hours' overhaul life

Turbo-props have benefited equally with turbo-jets from the component developments discussed earlier, but their gas flow rates are inevitably smaller than those of comparable turbo-jets and this has tended to restrict pressure ratio to somewhere between 5 : 1 and 8 : 1 so as to avoid very small axial compressor blading. One notable exception is the Rolls-Royce *Tyne* (pressure ratio 13·5 : 1) which develops 6000 h.p. in its latest form. Specific weights of engines in this range (which goes down to 500 h.p.) vary between 0·4 and 0·7 lb/s.h.p., and specific fuel consumptions from about 0·7 lb/s.h.p. for the 1000 s.h.p. engines to the 0·47 lb/s.h.p. achieved by the *Tyne*. The Bristol Siddeley *Proteus*, shown in Fig. 6, has been successful in penetrating the marine and industrial fields.

The prospects of the turbo-prop in its larger sizes became somewhat uncertain with the advent of the ducted fan, but the successful association of the shaft-power engine and the helicopter[29] provides considerable hope for the future. Any success in which the gas turbine shares will be well-earned, because of the significant contribution it makes to the helicopter, particularly the elimination of low frequency vibration, the reduction of noise level, and the flexibility provided by the separate power turbine.

Reliability and overhaul life

Whilst making the performance advances already described, the aircraft gas turbine has more than substantiated the claims made by its staunchest advocates in respect of overhaul life and reliability, by setting up standards superior to those achieved by fully developed piston engines in civil aviation. Overhaul lives over 2000 hours are now commonplace, and the Rolls-Royce *Dart*,[30] the first turbo-prop to enter airline service (in 1953), reached a new milestone in 1960 when it was approved for 3000 hours between overhauls in T.C.A. *Viscount* aircraft. Experience with aircraft gas turbines is now reckoned in tens of millions of engine flying hours; that is, thousands of years!

The shaded area between the curves of Fig. 7 is typical of the manner in which overhaul life has been built up in four-engined transport aircraft. The rate of build-up is, of course, dependent on the number of aircraft in use by a particular operator and their rate of utilization, and it has taken two to three years to achieve overhaul lives of the order of 2000 hours. Overhaul life increases more rapidly at the start of operational service, when the number of aircraft is usually small, and it would appear from this that future turbine engines should be capable of starting off with overhaul lives approaching 1500 hours.

Long overhaul life is of little value if it is associated with a high rate of unscheduled engine removals. Experience with

aircraft gas turbines to date has demonstrated that unscheduled engine removal rates as low as one per 4500 engine flight hours are achieved, coupled with overhaul lives approaching 3000 hours.

Fuels

Large-scale experiments have been conducted, mainly in the United States,[31] on high energy fuels for aircraft gas turbines or ramjets. Based on boron, aluminium or beryllium compounds or slurries containing pure metals to boost calorific values, many of these high energy fuels have toxic properties; in addition, they create deposition problems in the combustion zone, turbine and exhaust system and these factors, allied to their high cost, rule them out for civil use. Apart from these experiments, petroleum distillates with a freezing point of about − 60°C, similar to the four main specifications given in the Defence List,[32] have been used exclusively in aircraft gas turbines. Such fuels will continue in extensive use but the advent of sustained supersonic flight at Mach 2 to 3 may well influence aviation fuel specifications of the future, particularly if the fuel is used as a heat sink, in supersonic flight, as appears likely.

Many industrial gases and most natural gases have also been burned in gas turbines. Natural gas, which is clean and of high calorific value, and is usually available at a pressure high enough to inject into the combustion chamber without further compression, is probably the nearest approach to the ideal gaseous fuel. It is significant that a high proportion of industrial gas turbine sets have been sold for power generation in oilfields and are using natural gas. All gaseous fuel should be free from liquid fractions and its sulphur content should not exceed 3 per cent. This limit, however, may be extended in certain special cases.

In addition to the aviation distillate oils, ash-free diesel oil and gas oil are also acceptable fuels for a gas turbine, provided they are clean, relatively free from water and their sulphur content does not exceed 3 per cent. Sulphur-free and vanadium-free light crudes require only slight modification to turbine design. Heavier crudes and residual oils set a more difficult problem and require costlier modifications, for instance, a large external combustion chamber.

The considerable financial incentive to burn cheap residual fuel oils in industrial and marine gas turbines led to many attempts[33] to solve the problems of turbine fouling and the associated chemical attack on the hot blading. Only partial success can be claimed, in that the use of residual fuels in an open-cycle gas turbine entails close control of combustion conditions, the use of fuel additives, restriction of turbine entry temperature to a value below about 950°K, or combination of all three.

Burning solid fuel in a gas turbine frequently appears economically attractive in areas where the fuel is readily available, despite the higher capital cost of the more complex engine. As a result, interest in the solid-fuel burning gas turbine has been world-wide. In the U.K. development work was undertaken by some of the leading gas turbine companies under the Ministry of Power's sponsorship and with the assistance of research establishments such as the Fuel Research Station, the British Coal Utilization and Research Association and the National Gas Turbine Establishment. The objectives of this programme, which included both coal and peat burning, were described[34] in 1951 and progress was reported[35, 36] in 1955. The U.K. programme was subsequently terminated because interest grew in nuclear power and lessened in pithead electricity generation.

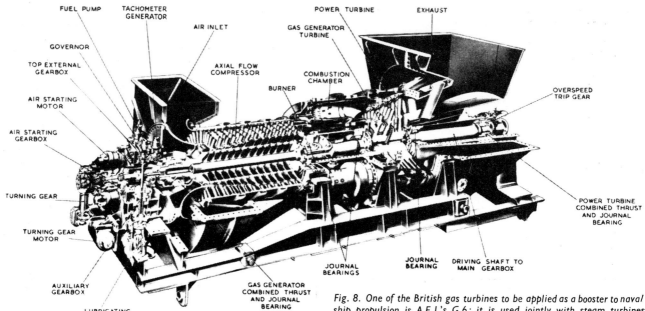

Fig. 8. One of the British gas turbines to be applied as a booster to naval ship propulsion is A.E.I.'s G.6; it is used jointly with steam turbines and develops 7500 s.h.p.

The open cycle gas turbine burning coal is a difficult proposition, requiring a special design of combustion chamber and an efficient dust separator to clean the gases before they pass through the turbine. Experience to date suggests that the first of these objectives can be achieved; it would also appear that turbine deposits may be controllable and that some degree of turbine blade erosion is acceptable with wide-chord blading. There is, however, some interaction between the two objectives in that a high efficiency combustion chamber tends to accumulate slag deposits; while designs which eliminate this undesirable feature tend towards lower combustion efficiency and an increase in the quantity of fly ash passed to the separator. The continuing interest abroad in the open cycle, coal-fired gas turbine, notably in the United States, U.S.S.R., South Africa and Australia,[37] provides support for the belief that the target, though difficult, is not technically impossible.

The marine gas turbine

The number of merchant ships propelled solely by gas turbines is small, probably because of the conservatism of ship owners and their mistrust of something new which might involve development expenditure. Engine powers range from 3500 to 12 000 s.h.p., including a few with free-piston gas generators. It is perhaps significant that only open-cycle turbines up to 12 000 s.h.p. have so far been developed for marine application. The possibilities for gas turbines of 20 000 to 30 000 s.h.p., probably closed cycle types, linked with a high-temperature, gas-cooled nuclear reactor, appear promising but the concept poses many problems which will require intensive development effort.

The application of gas turbine boost to naval ships, first demonstrated in a British gunboat in 1947, with the A.E.I. 2500 s.h.p. *Gatric* (G.1) engine, has now spread to ships of larger size and boost propulsion units are installed in high-speed craft of the U.S., German and Italian Navies, in addition to Royal Navy installations. Further considerable

numbers of these gas turbines are on order and under construction at present. In the U.K. the latest A.E.I. 'G' Series engine, the 7500 s.h.p. G.6 which is shown in Fig. 8, forms part of the propulsion machinery of two new classes of ship: general purpose frigates which started sea trials in 1961, and guided weapon destroyers. The frigate machinery, using a single propeller shaft, comprises a steam turbine for low power cruising and a G.6 boost gas turbine for higher speeds. The twin-propeller destroyer has a steam turbine and two G.6 gas turbines on each shaft.

Development of the A.E.I. 'G' Series of gas turbines[38] from the 2500 s.h.p. *Gatric* of 1947 to the 7500 h.p. G.6 of 1961 provides an interesting comparison with the progress already noted in the aircraft gas turbine field. Cycle pressure ratios of the 'G' Series have increased from 3·5 to 6·3 and turbine inlet temperature from about 1020°K to 1070°K. As a result, the h.p. per lb of air consumed has increased from just over 50 to about 70. From a starting point of 1·06 lb/s.h.p., specific fuel consumption of the G.6 has been brought down to as low as 0·68 lb/s.h.p. (a reduction of 36%) with a target figure of 0·72 lb/s.h.p. including inlet and exhaust duct losses. Specific weight remained constant at around 1·5 to 1·6 lb/s.h.p. for about 10 years, but has increased to about 5·5 lb/s.h.p. on the latest G.6 engine.

Whilst these achievements are, in general, less striking than the corresponding advances in the aircraft field, differences in design requirements must be borne in mind; for example, the G.6 is required to have a blade life at full load of 1000 hours, a much higher figure than that required from the aero-engine; and the warship engine has to withstand severe shock loads due to underwater explosions. The latter requirement is partly responsible for the higher weight of the G.6 engine, the remainder being accounted for by a conservative design, dictated by the necessity of getting the engine right, or very nearly so, first time and avoiding a development programme on anything approaching the scale of that usually undertaken on aero-engines. Thus, for example, plain bearings instead of the ball and roller bearings used on aero-engines are specified.

Further development of this type of turbine with the improved high-temperature materials now available is likely to be directed towards increased life at full load until a solution is found to the problem of chemical attack on the turbine blading.[39] This attack is due to sodium chloride from sea water ingested in rough weather and sodium sulphate from sulphur in the fuel. At present it limits turbine inlet temperature to about 1100°K.

The adaptation of aero-type gas turbines to marine use offers a tremendous reduction of weight: to at least one tenth, and possibly to one seventieth, and also of space requirements, compared with compression-ignition engines. But this is offset, to some extent, by the higher fuel consumption of the gas turbine.

Consideration by the Admiralty of the *Proteus* engine for marine use goes back to 1954. Three *Proteus* engines power the *Brave* class of Royal Navy fast patrol boats.[40] The first of these, *Brave Borderer*, was officially handed over to the Navy in January 1960, after extensive evaluation trials which started in late 1958 when the reliability of the *Proteus* had been proved in airline service. As ancillary power on the *Brave* class boats is provided by two Rover IS/60 gas turbines driving 40 kW generators, these craft are the first in the world to be powered exclusively by gas turbine engines. Reaching a top speed of over 50 knots, on trials in 1959, *Brave Borderer* became the fastest warship in the world. Two *Proteus* engines are also fitted in a smaller fast patrol boat prototype, *Ferocity*.

Hovercraft and hydrofoils

The gas turbine appears particularly suitable because of its high power/weight ratio, for powering two new forms of marine transport which have emerged as promising lines of development in recent years but have yet to prove themselves completely: the hovercraft which is equally at home over land, water or swamp; and the hydrofoil.

The possibilities of hovercraft were first demonstrated by Saunders-Roe with the SR-N1, using a 550 h.p. Alvis *Leonides* air-cooled piston engine to drive the fan which provides the cushion of air on which the vehicles rides.[41] The original SR-N1 was propelled by a small amount of air bled from the sustainer fan flow, but this system was replaced by a turbo-jet engine to extend the speed range of the vehicle. This initial attempt with a lightweight gas turbine was followed by the selection of shaft-power gas turbines for other hovercraft in 1961 and the association of the gas turbine with the hovercraft is now well under way.

Again, in hydrofoil research,[42] aircraft power plants, including gas turbines are strong challengers. Indeed, for the largest and fastest craft envisaged (500 tons and 100 knots) anything other than an aircraft-type turbine is inconceivable in view of the similarity between these boats, which support their weight on wings at high speed, and aircraft.

Industrial types

The industrial gas turbine has now established itself as a reliable power unit, having the advantages of compactness, cheapness when produced in reasonable numbers, and ease of running and maintenance, without the noise and regular complicated overhauls of a compression-ignition engine; or the complicated boiler, feed-heat and auxiliary systems of a steam turbine. Whilst development of the industrial gas

turbine may not have proceeded as quickly as the enthusiast would have wished, steady progress has nevertheless been made. A slower rate of progress than that achieved on the aircraft gas turbine is only to be expected, as the industrial engine in general is not made in sufficient numbers to justify embarking on the extensive and costly process of 'make and break' development commonly used on the aircraft types; indeed, many industrial gas turbines are built on a one-off basis, possibly as part of a larger engineering complex. In such circumstances they must work straight away. The designer of the industrial gas turbine usually has to aim at a total life of about 100 000 hours. Being denied, in many cases, the opportunity of determining safety factors during a development period, he has to adopt a cautious attitude in applying the new ideas thrown up by aircraft practice. An exception to this general rule might be a really efficient small gas turbine which, by justifying production on a scale equal to that of the automotive piston engine, would make the devotion of ample funds to its development a practicable proposition.

Voysey[43] suggests that the growing volume of gas turbine business is demonstrated by the fact that, of the 600 MW of Brown Boveri power generating plant installed since 1951, more than one-half have been installed since 1958. Nearer home, A.E.I. experience, on a much smaller scale, is similar, and the Ruston and Hornsby achievement with their T.A. turbine also suggests an accelerating rate of progress in recent times. Originally conceived about 1950 as a general purpose machine for stationary duty, this gas turbine has established itself as a reliable prime-mover in the oilfields and as a producer of power and process heat in industry during the last six years. Over 120 units are in service, having accumulated more than a million running hours; and about ten have still to be delivered. Designed originally for use either with or without a heat exchanger, the T.A. has an appropriately low pressure ratio (4 : 1), achieved in thirteen compressor stages (12·5°C per stage). At a turbine entry temperature of 1030°K, it yields a maximum of 1425 s.h.p. (60 h.p./lb air/sec), has a minimum specific fuel consumption of 0·83 lb/s.h.p. (thermal efficiency, 16·4%) and a specific weight of 9·5 lb/s.h.p., all without heat exchanger. A standard production T.A. turbine was installed in a matter of months to drive a generator in the Royal Navy's cruiser *Cumberland*[44] and completed successful sea trials during 1956. Besides demonstrating the suitability of the engine for marine and mobile applications on land, it led to further gas turbine development on behalf of the Admiralty.

The success of the T.A. turbine qualifies it as a reasonable yardstick for comparison of industrial engine performance parameters with those of established aero-engines; which gives some idea of the gap which exists between the two breeds of gas turbine. This gap has encouraged firms with established engines in production to seek applications in the industrial field. For instance, the first remotely-controlled power station to go into operation in Great Britain, at Princetown, Devon, in 1959, has a *Proteus* engine, coupled to a 3200 KVA alternator, to meet peak demands of very short duration. The *Proteus* develops 4150 s.h.p. and has a specific fuel consumption of 0·59 lb/s.h.p./hr on light diesel oil, thermal efficiency being 23 per cent. Remote control is effected over the public telephone line from Bristol, 100 miles distant, and up to forty functions such as starting, stopping, power control, etc., may be performed. A second set is now in operation at Lynton, Devon, and a third is destined for Porlock Somerset.[45]

The value of such sets will be appreciated if it is considered that in 1960 some 3 per cent of the total C.E.G.B. base-load plant was in commission for about 2600 hours, but only generated for 350 hours, with an output of 0·1 per cent of the C.E.G.B. total, at an average thermal efficiency of less than 10 per cent. Fig. 9 gives an idea of the potential field of application of gas turbine sets for 'peak-lopping'.[35]

A 15 MW generator based on the Bristol Siddeley *Olympus* turbo-jet, the Vulcan bomber engine, has been ordered by the C.E.G.B. for Hans Hall, Birmingham.[16] The initial rating of 15 MW corresponds to 21 000 s.h.p., at 25 per cent thermal efficiency. With other aero-engine firms interested, and a number of engines in current production available, an increase in the number of such applications is to be expected in future years.

Locomotive applications

From the beginning, the gas turbine has not had an easy passage in attempting to enter the locomotive power field. Although the efficiency target set by the conventional locotive steam engine was not unduly high, the lower weight of the gas turbine was not an overriding advantage, and could even be an embarrassment. Furthermore, in a majority of cases, the gas turbine had to operate on coal or residual fuel oil, for local economic reasons. The outstanding success of the diesel over the steam locomotive in recent years has not improved the prospects of the locomotive gas turbine but it still has a number of advantages over the diesel which will become significant when a thermal efficiency about one half that of the diesel is achieved at 25 per cent load.

Against this background, the U.S.A. fleet of more than 50 General Electric locomotives of 4500 h.p. and 8500 h.p. operated on selected residual oil to a G.E. specification, stands out in sharp contrast to experience elsewhere in the world, which has been limited to isolated prototypes enjoying varying degrees of success.[43]

Small gas turbines

The gas turbine, passing less than 10 lb/sec of air (or below 750 s.h.p.) justifies separate consideration because of its almost universal application potential and the challenging and exciting nature of development in the automotive power range, towards the bottom end of the scale. In terms of present day average road speeds, the overhaul lives of 2000 to 3000 hours achieved by aircraft gas turbines correspond to around 100 000 miles and bear comparison with the best achievements in automotive engineering—a happy augury for the automotive gas turbine.

Consideration of small gas turbines goes back to the mid-1940s on both sides of the Atlantic, with many notable firms involved.[47–57] Thousands of units have been supplied all over the world for a variety of purposes, including aircraft auxiliary power, air bleed and generator sets, marine auxiliary power, water pumping sets, exhaust heat processes and instructional units. Although the first gas turbine car, a Rover, appeared at Silverstone in 1949 and was followed by other automotive installations in the United States and the U.K., the gas turbine still has some considerable way to go before a real breakthrough into this field can be claimed. The pioneers, however, are still active in this field, as is shown by their latest engine in Fig. 10.

A rough idea of the range of small engines currently available and the standard of performance so far achieved is obtained from Fig. 11. The apparent scale effect on specific fuel consumption is illusory; being due to the fact that the

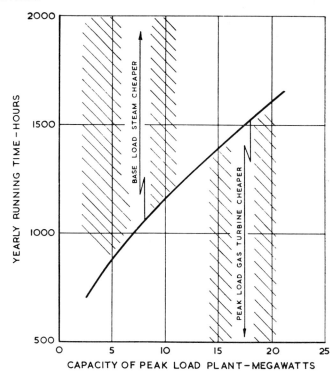

Fig. 9. The approximate economic limit of peak load gas turbine generating plant: cost is assumed to be £7 2s./kW maximum demand, plus 0.41d/kWh running charge

smaller engines have, in general, a lower pressure ratio than the larger units. The absence of scale effect is demonstrated by comparing the engines with heat exchanger in the 150–300 h.p. range with the higher pressure ratio, 200–400 h.p. engines. The smaller engine, almost inevitably, has a slightly higher specific weight and centrifugal compressors are used without exception, generally operating at a pressure ratio of about 4 : 1 to match the optimum simple cycle with heat exchanger, at a turbine entry temperature approaching 1200°K.

Specific fuel consumptions of the order of 0·55 lb/s.h.p. have been achieved at full power. Such a consumption rate begins to be competitive with the petrol engine, provided it is maintained at part-load conditions. This may be achieved by maintaining pressure ratio and turbine entry temperature at reasonably high levels at part-load, which entails some variable engine geometry to control the air mass flow, independently of engine speed; a feature announced only by Chrysler[47] so far. The alternative approach, favoured by Ford in America,[48] is to design for the very high pressure ratio of 16 : 1 at full load, at the expense of having to complicate the basic design.

The free-power gas turbine has a torque curve which matches vehicle requirements more closely than that of any piston engine without torque converter. It has many other advantages over the piston engine, already demonstrated by the aircraft gas turbine, and a determined attack on the part-load specific consumption problem is fully justified by the ultimate benefits to be obtained. There can be no doubt that it will have to be truly competitive to break into the automotive field in a big way, as this will entail abandoning a large investment in piston-engine tooling.

A development not necessarily linked to the small gas

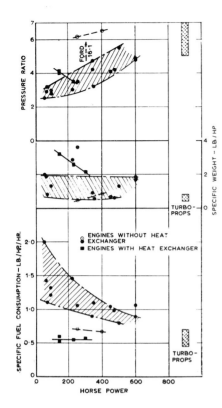

Fig. 10. The 25/140, now under development for automotive use, has two primary surface, contra-flow heat exchangers; overall measurements are only 29 in. high by 26 in. wide

Fig. 11. Standard performances of current small gas turbines: the apparent scale effect on specific fuel consumption is illusory. Turbo-prop pressure ratios continue

exchangers in series plus a power turbine should have a reasonably good performance and the successful development of a lightweight pressure exchanger might lead to the replacement of the conventional compressor and its driving turbine by relatively simple, mechanically-driven units in some gas turbine applications.

Special applications

A wide area of potential application of gas turbine type plant is surveyed by Hodge,[58] including gas turbines or components thereof, acting as ancillaries in industrial plant, e.g., blast furnace blowing; the gas turbine in combination with other cycle, steam or diesel; or used to supercharge a steam boiler combustion space; the application of gas turbine technology in refrigeration cycles, including those used for fractional distillation; and, finally, the gas turbine burning a particular fuel not economically usable by other means, such as coal mine upcast gas and sewage gases.

Integrated steam-gas turbine systems[59] and the association of the gas turbine with the diesel[60] offer high economy, usually coupled with increased complexity and first cost; and the potential customer is inclined to look hard at the cash price and may wonder about reliability and maintenance. In this connection, it is interesting to note that a Brown Boveri plant has accumulated more than 50 000 hours running whilst producing electricity and blast air, operating on blast furnace gas at a steel works in Luxembourg.

The air-bleed gas turbine which delivers its output in the form of pressure air tapped from the engine compressor has been developed and sold in considerable numbers, in small sizes; in spite of the limitation of delivery pressure to 40 lbf/in²g, imposed by the compressor of the basic gas turbine. It has always appeared that the market prospects for this type of gas turbine would be considerably enhanced by increasing the delivery pressure to 100 lbf/in²g. This is done in a portable turbo-compressor designed by Power Jets (R. and D.) Ltd and recently developed by Holman Bros Ltd in which air bled from the engine centrifugal compressor passes, by way of an intercooler, to a two-stage centrifugal compressor driven through step-up gearing by a separate power turbine.

During the last five years or so, much has been written about the potential role of the gas turbine in the nuclear field. From many papers on the subject a selection has been given in the references.[51-72]

turbine is the work undertaken by Power Jets (R. and D.) Ltd., to explore the potentialities of the pressure exchanger. This device, shown in Fig. 12, has a rotor with a large number of straight radial vanes, mounted in a cylindrical casing with ports in the end covers. Rotation of the rotor brings gases at different pressure levels into direct contact within the rotor channels and pressure waves traversing the length of the channels compress or expand the contents. During a compression, gas flows from the high-pressure port into the channels, and during an expansion process, gas flow is from the channels.

Compression efficiencies comparable with conventional compressors have been reported at the low pressure ratio of about 2·5 : 1 and, at a combustion temperature of 1250°K, it appears that rather more than one half of the air passing through the high pressure outlet port may be tapped off to do useful work in a turbine. Theoretically, two pressure

Fig. 12. The Power Jets (R and D) pressure exchanger

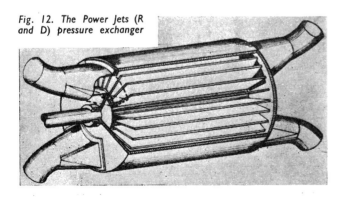

Much practical work remains to be done before a realistic assessment can be made, but it is already clear that the gas turbine offers one of the best possible means of exploiting the potentialities of high-temperature gas-cooled reactors. Conversely, the long-term prospects of the gas turbine at powers in excess of, say, 20 000 h.p. probably depend to a considerable extent on its adaptability to nuclear reactors.

As things appear at present, the closed cycle gas turbine is likely to be preferred to the open cycle, firstly for reasons of nuclear safety, and secondly because component sizes in the high-pressure closed cycle are moderate, even at powers of the order of 30 000 h.p. If air cannot be used as the working fluid because of oxidation of reactor materials or of induced nuclear reactivity, other gases, such as helium, carbon dioxide, neon or nitrogen, may provide feasible solutions. Of these, helium is currently favoured because of its extreme inertness, although it is not ideal as the working fluid in a gas turbine.

Interest has been renewed recently[73] in the possibilities of magnetoplasmadynamic electrical power generation, that is, the conversion of the kinetic energy of an ionized gas stream directly to electrical power, the subject of experiments by Karlowitz in the Westinghouse Research Laboratories in America in the 1930's. This renewal of interest is timely, in view of developments in the past ten years or so, in the physics of ionized gases, high temperature materials and space engineering. Success in this field would open a vast new area of research and development in which gas turbine technology would become involved.

Conclusion

Compared with an engine of the early 1950s, the modern high-pressure ratio axial turbojet develops the same thrust for rather less than half the frontal area, weighs 25 per cent less and consumes about 10 per cent less fuel. Further improvements of some 10 per cent in fuel economy are to be expected from ducted fan engines which will be operating in the early 1960s. Yet a further reduction of 10 per cent may well be obtained by 1970, the penalty being a larger diameter.

The low-pressure ratio axial type, designed and developed specifically as a lifting engine, now gives 8 lb thrust per lb of engine weight, and the promise of engines with twice this thrust/weight ratio in a few year's time considerably enhances the prospects of V.T.O.L. aircraft.

The high pressure ratio turbo-prop engine has reduced its specific fuel consumption to 0·47 lb/s.h.p., equivalent to 29 per cent thermal efficiency. With weights of the order of 0·4 lb/s.h.p., this represents an engineering achievement of first magnitude, by any standard. The future of the turbo-prop is seriously threatened by the ducted-fan engine, but shaft power engines will continue to find application in helicopters and small aircraft.

Advances by the aircraft gas turbine on all performance fronts have been accompanied by a demonstration of reliability and achieved overhaul life in civil aviation unsurpassed by any piston engine. Overhaul lives over 2000 hours are now commonplace; one engine is already approved for 3000 hours; and unscheduled engine removal rates as low as one per 4500 engine hours in flight have been achieved. Further research and experience will produce even better results.

As would be expected, the advances made by gas turbines outside the aircraft field have been much less spectacular. The research effort has been smaller, and development at the customer's expense is not popular. Steady progress has been made, however, despite setbacks, in burning the heavier

Robert Hendry Weir *was born in Glasgow in 1912 and went to Allan Glens School from 1923 to 1928. While gaining a B.Sc., with first class honours he also served an apprenticeship with Wm. Denny and Bros. He joined the R.A.E., Farnborough, in 1933, working in the Engine Department. He was technical officer in charge of world altitude record flights in 1936 and 1937, served at the Air Ministry, London, from 1939 to 1940, doing pressure cabin and engine development work and then went to the Aircraft and Armament Experimental Establishment, Boscombe Down, on special high-altitude flight and general engine liaison duties. From 1942 to* 1960 he worked in London in successive Ministries dealing with gas turbines. He became Director of Industrial Gas Turbines in 1950, Director of Engine Research and Development in 1952, and Director-General in 1953. From 1960 to 1970 he was Director of the National Gas Turbine Establishment, Farnborough, and since 1970 he has been Director of the National Engineering Laboratory, East Kilbride, Scotland.*

fuel oils and solid fuels, and the gas turbine is now an established and reliable power unit in the industrial field, with twenty years' development behind it and a promising future, judging from recent sales trends. The future may well show increased application of gas turbine technology in industrial processes, in combined steam/gas turbine cycles, and in self-contained, high-pressure air supply plant.

The number of ships propelled solely by gas turbines is small, including those with free-piston gas generators, but it is increasing and the gas turbine linked with a high-temperature gas-cooled nuclear reactor appears to offer a promising line of development for marine propulsion and a lead-in for the land-based power generating set on which the future of the large industrial gas turbine may depend.

The gas turbine has many advantages for vehicle propulsion, and the biggest challenge offered to it is penetration of the automotive field. The rewards are great but the competition, particularly from the diesel, is formidable. Recent work on the major problem of fuel consumption has yielded results at full power which are of the right order to begin competing with the petrol engine. The small gas turbine will have to be competitive in fuel consumption at part-load to replace the piston engine, which carries a large tooling investment; but the goal is demonstrably attainable, and the prize justifies the considerable effort involved.

Acknowledgment

The author gratefully acknowledges much help, generously given, by friends and colleagues in industry and at the N.G.T.E., during the preparation of this article.

REFERENCES

1. 1958 Gas Turbine Progress Report. July 1959. *Trans. ASME—Series A*—Journal of Engineering for Power. vol. 81 pp. 215–359.
2. LOVESEY, A. C., 'The Art of Developing Aero. Engines'. *Journal of the Roy. Ae. Soc.*, vol. 63, no. 584, pp. 429–449. Aug. 1959.
3. MORLEY, A. W., *Aircraft Propulsion—Theory and Performance.* Longmans, Green & Co. London 1953.
4. 'High Speed Aerodynamics and Jet Propulsion'. 1960 *Vol. XI. Design and Performance of Gas Turbine Power Plants* edited by W. R. Hawthorne and W. T. Olsen. Princeton University Press.

5. WILKINSON, P. H., *Aircraft Engines of the World*, 1960/61 Edition. Wilkinson, 724 15th Street, N.W., Washington 5, D.C.
6. CARTER, A. D. S. 'The Post War Development of the Axial Compressor in Great Britain'. Communication aux Journées Internationales de Sciences Aeronautiques, Paris. May, 1957.
7. CARTER, A. D. S., ANDREWS, S. J. and FIELDER, E. A., 'The Design and Testing of an Axial Compressor having a Mean Stage Temperature Rise of 30°C'. *Aeronautical Research Council*, R and M no. 2985 1953 Nov.
8. BLACKWELL, B. D., 'Some Investigations in the Field of Blade Engineering'. 1958 Sept. *Journal of the Roy. Ae. Soc.*, vol. 62, no. 573, pp. 633–646.
9. ARMSTRONG, E. K. and STEVENSON, R. E., 'Some Practical Aspects of Compressor Blade Vibration'. *Journal of the Roy. Ae. Soc.* Vol. 64, no. 591, pp. 117–130, March, 1960.
10. HERBERT, M. V., 'A Theoretical Analysis of Rate Controlled Systems'. *Combustion Researches and Reviews*. Butterworths, p. 76, 1957.
11. HERBERT, M. V., 'A Theoretical Analysis of Reaction Rate Controlled Systems'. 1960 8th International Symposium on Combustion.
12. JEFFS, R. A., 'The Flame Stability and Heat Release Rate of Can-type Combustion Chambers'. 1960 8th International Combustion Symposium.
13. MILLER, R. E., 'Some Factors Governing the Ignition Delay of a Gaseous Fuel'. 1958 Proc. 7th International Symposium on Combustion. Part 5, p. 417.
14. STEWART, D. G. 'Scaling of Gas Turbine Combustion Systems'. 1956 *Selected Combustion Problems, Vol. II.* AGARD Butterworths, London, p. 384.
15. HERBERT, M. V., and BAMFORD, J. A., 'Scale Effect in a Gas Turbine Combustion Chamber Fitted with a Swirl Atomiser'. 1957 Part I. *Combustion and Flame* p. 1360. 1961 March Part II. *Combustion and Flame* p. 5, 35.
16. HERBERT, M. V., 'Aerodynamic Influences on Flame Stability'. 1960 *Progress in Combustion Science and Technology*, vol. 1, p. 61. Pergamon, London.
17. LEWIS, W. G. E. and HERBERT, M. V., 'Aerodynamic Factors in Combustion Chambers'. 1957 *Engineering*, 184, 143 (2nd August, 1957).
18. LLOYD, P., 'Scientific Research in Combustion and its Application to Engineering Developments'. 1958 *New Scientist*, vol. 4, no. 93.
19. ASHWOOD, P. F., 'The T-scheme. A Low Pressure Loss Combustion Chamber System for Gas Turbine Engines'. 1947 *Flight*, vol. 52, p. 630.
20. SMITH, A. G. and PEARSON, R. D., 'The Cooled Gas Turbine'. 1950 *Proc. Instn mech. Engrs, Lond.*, vol. 163.
21. AINLEY, D. G., 'The High Temperature Turbo-Jet Engine. 1956 *Journal of the Roy. Ae. Soc.*
22. GLENNY, E. and TAYLOR, T. A., 'A Study of the Thermal Fatigue Behaviour of Metals: The Effect of Test Conditions on Nickel-based High Temperature Alloys'. 1959/60 *J. Inst. Met.* vol. 88, p. 449.
23. COX, M., and GLENNY, E. 'Thermal Fatigue Investigations'. 1960 26th August *The Engineer*.
24. TAYLOR, T. A., 'Aero-turbine Materials'. 1961 January *The Aeroplane and Astronautics*. Part I, vol. 100, no. 2571, pp. 90–93.
25. RICHMOND, F. M., 'High Temperature Materials'. Conference held in Cleveland, Ohio, April 16–17, 1957. John Wiley and Sons Inc., New York, 1959.
26. BETTERIDGE, W., *The Nimonic Alloys*. Edward Arnold, London 1959.
27. TAYLOR, T. A., 'Aero-turbine Materials'. 1961 February *The Aeroplane and Astronautics*. Part II, vol. 100, no. 2572, pp. 119–120.
28. GLENNY, E. and TAYLOR, T. A., 'The High Temperature Properties of Ceramics and Cermets'. 1958 *Powder Metallurgy*, No. 1/2, p. 189.
29. MORLEY, A. W., 'Gas Turbines for Helicopters'. 1958 Sept. *Journal of the Roy. Ae. Soc.*, vol. 62, no. 573, pp. 646–658.
30. ROSSITER, H. G., 'B.E.A.'s Experience with Propeller Turbine Engines'. 1960 June *ASME*, paper 60-AV-52.
31. WELLS, R. A., 'High Energy Aviation Fuels: Their Promises and Problems'. 1958 March *ASME*, paper 58-AV-26.
32. Petroleum Oils and Lubricants, Defence List, DL-2-A, H.M.S.O. Sept. 1960.
33. DARLING, R. F., 'Liquid Fuel and the Gas Turbine. A Ten-year Review'. 1959 Oct. *J. Inst. Fuel.*, vol. 32, pp. 475–484.
34. ROXBEE-COX, H., 'Some Fuel and Power Projects'. 38th Thomas Hawkesley Lecture. 1951 *Proc. Instn mech. Engrs, Lond.*
35. FITTON, A., and VOYSEY, R. G., 'Solid Fuel-fired gas turbines in Great Britain'. 1955 August *Engineering*.
36. VOYSEY, R. G., HURLEY, T. J. and BATTOCK, W., I.Mech.E. and ASME Joint Conference on Combustion, Oct. 1955.
37. LHUEDE, E. P. and ATKIN, M. L., 'Initial Running of Ruston TA. Turbine on Pulverised Coal'. 1960 Australian Dept. of Supply, Aeronautical Research Laboratories Mech. Eng. Note 240.
38. HARRIS, F. R., 'Gas Turbine for Naval Boost Propulsion (the A.E.I. G.6)'. 1961 March *ASME* Paper 61-GTP-5, presented to the Gas Turbine Power Conference of the ASME at Washington, D.C.
39. SHIRLEY, H. T., 'Effects of Sulphate-chloride Mixtures in Fuel-ash Corrosion of Steel and High Nickel Alloys'. Feb. 1956 *J. Iron St. Inst.*, vol. 182, part 2, p. 144.
40. MARKHAM, B. G., 'Marine Proteus Engines for the Brave Class Patrol Boats'. Oct. 1959 *Trans. Inst. Mar. Engrs*, vol. 71, no. 10, pp. 316–323.

also 'Development and Sea Trials of Marine Proteus Engines for Brave Class Fast Patrol Boats'. 1959 *ASME* Paper 59-A-273.
41. PUGH, A., Off the Ground: No. 2 of the Series. The Saunders-Roe Hovercraft, SR-N 1. March, 1961 *Flight*, vol. 79, no. 2716, pp. 399–400.
42. 'Hydrofoils: A Promise of High-Speed Water Travel'. July, 1961 *Product Engineering*, vol. 32, no. 29, pp. 15–17.
43. VOYSEY, R. G., Progress Review No. 50: Gas Turbines. Oct. 1961 *J. Inst. Fuel*, vol. 34, pp. 440–447.
44. KOHN, E. O., 'Marine and Mobile Applications of Industrial Gas Turbines'. 1959 *ASME* Paper 59-A-295 presented at meeting at Atlantic City, N.J. Nov.–Dec.
45. 'Electrical Power Generation by Remote Control: Bristol Siddeley 3-MW Plant for Peak and Emergency Service'. Jan. 1960 *Oil Engine and Gas Turbine*, vol. 27, no. 315, pp. 260–263.
46. 'Economic Use of Peak-load Generation Plant: Bristol Siddeley Offer Proven Aircraft-type Gas Turbine Equipment for This Role'. July, 1961 *Oil Engine and Gas Turbine*, vol. 29, no. 333, pp. 107–110.
47. LUDVIGSEN, K., *Car and Driver*, vol. 6, no. 12, pp. 82–84 and pp. 87–89. Chrysler's Turbine for Today. June, 1961.
48. SWATMAN, I. M. and MALOHN, D. A., 'An Advanced Automotive Gas Turbine Engine Concept.' *SAE* Preprint 187A. June, 1960.
49. WEAVING, J. H., 'Small Gas Turbines'. 1962 January *Proc. Instn mech. Engrs, Lond.* (A.D.).
50. 'Developed Design for Improved Performance: Electric Start Embodied in Budworth 90-h.p. Brill Mark II Design'. Dec. 1960 *Oil Engine and Gas Turbine*, vol. 28, no. 326, pp. 307–309.
51. 'Jupiter Marketed Under British Licence'. May 1960 *Oil Engine and Gas Turbine*, vol. 28, no. 319, pp. 37–38.
52. 'Some Applications of a Small Gas Turbine: Perkins-built 50 h.p. Unit Now in Production'. Oct. 1959 *Oil Engine and Gas Turbine*, vol, 27, no. 312, p. 148.
53. 'Small Power Unit Developments: Rover Single-shaft and Dual-shaft Designs of 90 and 140 h.p. August, 1960 *Oil Engine and Gas Turbine*, vol. 28, no. 322, pp. 150–151.
54. 'A New British Low-pressure Air Starting Trolley: Joint Venture by Auto Diesels and Standard Motor Co. Yields Industrial Equipment'. Jan. 1959 *Oil Engine and Gas Turbine*, vol. 26, no. 306, pp. 357–360.
55. 'Extension of American Power Range: Boeing 520 Series for Land, Sea and Air Applications'. April, 1960 *Oil Engine and Gas Turbine*, vol. 27, no. 318, pp. 384–386.
56. '225-b.h.p. Power Unit for Experimental Car: Progress in General Motors Automotive Gas Turbine Evaluation Programme'. Dec. 1959 *Oil Engine and Gas Turbine*, vol. 27, no. 314, pp. 219–224.
57. 'Aero. Engines 1961 (The Allison T.63, p. 88)', July, 1961 *Flight*, vol. 80, no. 2732, pp. 71–88.
58. HODGE, J., Gas Turbine Series, *Vol. 1—Cycles and Performance Estimation*. Butterworths Scientific Publications, London. 1955.
59. MILLS, R. G., 'Combined Steam Turbine–Gas Turbine Plant for Marine Use'. 1959 ASME Paper 59-A-237 for meeting Nov. 29th to Dec. 4th.
60. CHATTERTON, E., 'The Diesel Engine in Association with the Gas Turbine'. 1960 Proc. Instn mech. Engrs, Lond., vol. 174, no. 10, pp. 409–435.
61. GODWIN, R. P. and DENNISON, E. S., 'Nuclear Gas Turbine Plants'. Journal of Engineering for Power, 1959 July. *Trans. A.S.M.E.*, Series A, vol. 81, pp. 352–359. (One of a series of papers comprising '1958 Gas Turbine Progress Report').
62. BAXTER, A. D., 'Flying Under Nuclear Power'. 1961 Nov. *Aeronautics*, 1961 Sept. *Journal of the Roy. Ae. Soc.*
63. BAXTER, A. D., 'Nuclear Power in Flight'. 1961 April *De Havilland Gazette*, no. 122, pp. 64–67, 70.
64. 'Nuclear Energy Power Plants'. 1956 Jan. *Aeroplane*, pp. 48–49.
65. 'Nuclear Gas Turbine Marine Propulsion Unit: Some Considerations for a Single-loop Closed-cycle nuclear Gas Turbine Particularly for Merchant Ships'. 1957 April *Oil Engine and Gas Turbine*, vol. 24, no. 284, pp. 474–476.
66. PERKINS, J. E. B., 'The Gas Turbine in Atomic Energy'. 1956 July *Nuclear Power*, vol. 1, no. 3, pp. 108–113.
67. Nuclear Gas Turbines: Symposium Covering the Applications of Various Gas-turbine Power Systems to Nuclear Reactors. (First Annual Symposium on Nuclear Gas Turbine Power Plants', Gas Turbine Power Division of the American Society of Mechanical Engineers, Washington, pp. 699–702. Dec. 1955. 1956 Aug. *Mechanical Engineering*, vol. 78, no. 8.
68. BARNES, J. F., 'The Gas Turbine's Contribution to Nuclear Power'. 1961 Feb. *New Scientist*, vol. 9, no. 222, pp. 416–418.
69. DANIELS, F., 'Small Gas Cycle Reactor Offers Economic Promise'. 1954 March *Nucleonics*, vol. 14, no. 3, pp. 34–41.
70. SPILLMAN, W., 'Closed Circuit Gas Turbines in Aircraft Propulsion. (In German). 1960 Dec. *Neue Tech.* vol. 2, no. 12, pp. 41–50.
71. BERMAN, P. A., 'A Gas Turbine for a Helium-cooled Reactor'. pp. 293–305 of *Gas-cooled Reactors*. 1960 Symposium of the Franklin Institute and the American Nuclear Society, Philadelphia *J. Franklin Inst.* Monograph no. 7.
72. SHOULTS, D. R., 'Tests of a Direct Cycle nuclear Turbo-jet System'. 1958 Second U.N. Conference on the Peaceful Uses of Atomic Energy. Geneva, p. 11.
73. Second Symposium on Engineering Aspects of Magnetohydrodynamics at the University of Pennsylvania. 1961 March.

Petrol Engines For Automobile Use

Excluding one or two unorthodox schemes, there has been no major change in the basic form of the petrol engine for many years, but many changes of detail have been made possible by improved materials and production techniques; others were stimulated by changes in body design. Basic research is concentrated on combustion conditions and various attempts have been made to improve specific performance. There is new design interest in the use of light alloys.

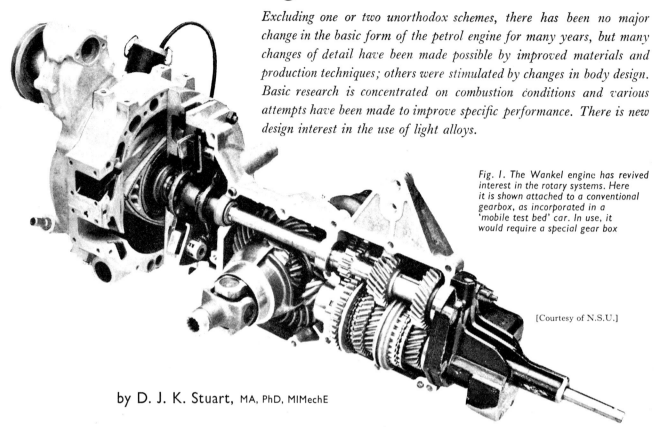

Fig. 1. The Wankel engine has revived interest in the rotary systems. Here it is shown attached to a conventional gearbox, as incorporated in a 'mobile test bed' car. In use, it would require a special gear box

[Courtesy of N.S.U.]

by D. J. K. Stuart, MA, PhD, MIMechE

Although specific power outputs of internal combustion engines in widespread use have increased steadily over the last thirty years, there has been remarkably little change in the basic layout of the I.C. engine during this period. In the case of the petrol engine, virtually no new forms have appeared, either in terms of cylinder arrangement or in combustion chamber and valve gear design. Though certain arrangements have gained in favour, to the extent that a relatively small number of them now constitute virtually the whole range of mass-produced engines.

There have been more significant changes in the diesel engine which has continued to gain ground in applications where operating economy is of prime importance. Both open and indirect chamber engines continue to flourish side by side, each predominant in its own particular sphere. Turbocharging is increasingly favoured in the competition to raise outputs.

Nevertheless, even in the petrol engine, there have been many changes in detail, often made possible by improved materials and manufacturing techniques; though in most cases the impetus to change comes from conditions imposed by the body designer and production engineer.

Rotary engines

The most publicized recent development in the field of I.C. engines has been the resurgence of interest in the rotary engine, represented by the N.S.U. *Wankel* unit,

shown in Fig. 1 attached to a Volkswagen gearbox. Since the first announcements made by the N.S.U. Company, a number of other firms have entered into agreements for development and subsequent manufacture. The progress of this type of unit has been periodically reported,[1-3] a recent review[4] giving an accurate picture of the present stage of the work and drawing particular attention to outstanding problems.

The main advantages of the rotary engine, i.e., freedom from reciprocating forces and a very high volume displacement for a given overall bulk, are well known and both these features are of particular significance in relation to present trends towards greater refinement of operation and more compact power unit.[5] The inherent disadvantages of the *Wankel* designs are the elongated form of combustion space with its high surface/volume ratio; the local concentration of heat in this area; the difficulty of applying satisfactory cooling in the heavily loaded bearing region; and the problem of maintaining adequate sealing whilst reducing wear to acceptable levels. The piston engine has reached its present advanced stage of design largely by progressive empirical work over many years, and it will be interesting to see the results obtained by the much more methodical and concentrated approach being employed on this new engine.

It is certain that the appearance of the rotary engine will give much impetus to the development of piston engines which, many authorities feel, will still be in an unassailable

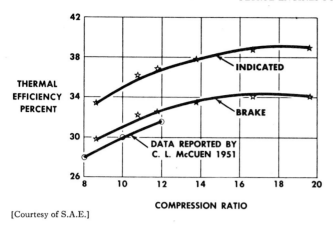

Fig. 2. Results of tests showing effect of varying compression ratio with maximum power setting at 2000 r.p.m.

pre-ignition[12] and increased stresses seem likely to restrict gains in output and efficiency from this source. The most realistic appraisal of the return offered by increased compression ratio has been obtained by Caris and Nelson,[8] in tests the results of which are shown in Fig. 2. Research into the basic combustion processes continues[13-15] but still falls a long way short of offering a complete theoretical analysis of combustion chamber performance and, in any case, seems unlikely to lead to more than marginal gains in practical design. There is, however, some divergence from this view: it is suggested that a more rapid rate of combustion should be achieved which would lead to some increase in the useful work done.[16]

There is little doubt that, in some conditions, the combustion process is protracted and even incomplete at the opening of the exhaust valve. This is clearly shown in photographs of combustion taken through a quartz window in the cylinder head (Fig. 3) and is even more graphically illustrated in slowed-down film sequences. This qualitative study gives no indication of the magnitude of the loss in useful work. As long as analysis is concentrated on study of the combustion process, quantative assessment will only become reliable when combustion theory is complete, and this is certainly not the case at present.

The situation in terms of basic thermodynamics is simple enough: at any stage of the process the gain in internal energy equals the heat supplied, minus external work done, minus heat lost in cooling.

This is a basic equation which the engineer can understand, and the physicist should not find too mundane. A good deal of use is made of the term 'heat release' which the engineer will interpret as the gain in internal energy plus external work done. This is also equal to heat supplied, minus heat lost in cooling. The surrounding walls of the cylinder and head are likely to extract heat from the gas at the maximum temperature stage, and return heat at the later stages of

position for road transport applications in the foreseeable future.[6, 7] Very high specific outputs, up to 135 b.h.p./litre, are now obtained from four-stroke petrol engines in racing cars and the seemingly modest output of approximately half this figure will be a realistic target for new designs of engines for mass-produced vehicles.

The technical means by which such progress could be made are well known, but a good deal of research and development is being aimed towards the commercial realization of the objectives.

Combustion in piston engines

Despite the promise of a limited return from increased compression ratios[8-10] practical problems such as rumble,[11]

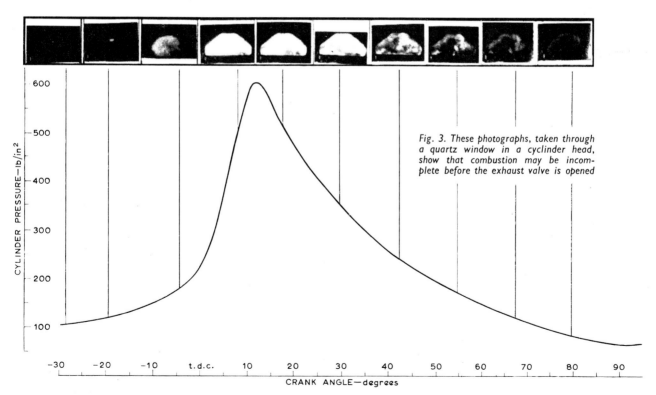

Fig. 3. These photographs, taken through a quartz window in a cyclinder head, show that combustion may be incomplete before the exhaust valve is opened

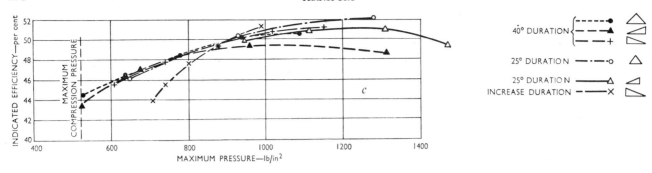

rig. 4. *Maximum cylcinder pressure, rather than the form of the heat release diagram, determines efficiency. The different diagrams are shown with compression ratio of 15/1*

expansion. 'Heat release' in itself is therefore no direct measure of combustion rate, and the alarming curves shown by Clarke[16] do not prove that the protracted combustion process is causing major losses in work done.

A simplified analysis has been carried out by Lyn[17] and although this study is primarily related to compression-ignition engines and, therefore, only includes calculations for high compression ratios, the general form of the results is obviously applicable to spark-ignition engines. Lyn investigated the effect of different forms of the heat release crank angle diagram, using mathematically convenient forms. His conclusions are extremely significant for they show that wide variations in the form, duration and timing of heat release only account for a variation of indicated thermal efficiency from 43·4% to 52·1%.

These values represent extremes, the corresponding theoretical maximum cylinder pressures being 520 lb/in² and 1270 lb/in². A more realistic comparison is shown in the variation of efficiency with duration of heat release for a linearly decreasing rate of heat release. Change of combustion period from 30 degrees to 70 degrees crank angle only reduces the efficiency from 51·4% to 43·9%, the corresponding maximum cylinder pressures being 980 and 700 lb/in². Advancing the ignition point in this latter case would increase the efficiency to approximately 46%, thereby further reducing the disparity.

Lyn's analysis is briefly summarized in Fig. 4, which shows that maximum cylinder pressure rather than any specific form of the heat release diagram is the most significant factor in determining efficiency. It is also shown that embarrassingly high rates of pressure rise (up to 100 lb/in²/ degree crank angle) can result if the period of combustion is unduly foreshortened. Such rapid rates approach those usually associated with various forms of uncontrolled combustion such as detonation, deposit ignition, etc, whereas 40 lb/in²/degree crank angle is the normally accepted optimum. Practical background to this study is given by Bowditch,[18] whose excellent quartz piston photographs indicate somewhat shorter normal combustion periods than those suggested by Clarke.[16]

Normal combustion is, however, a somewhat indeterminate condition in view of the cycle-to-cycle variations which occur in spark-ignition engines. Soltau[19] shows evidence of very large variations especially with weak mixtures which is illustrated in Fig. 5.

It is difficult to see how cooling can be controlled more effectively to reduce the sum total of direct loss and heat interchange between the high-temperature and low-tempera-

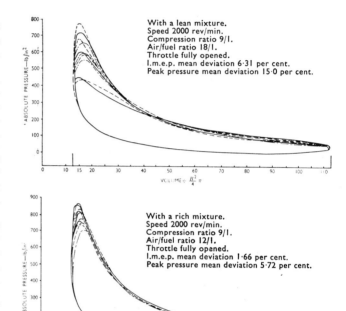

Fig. 5. How the indicator diagram varies, particularly with weak mixtures, in consecutive cycles of spark ignition engines

ture phases of the cycle. There are so many side-effects with any change in the cooling conditions that any advantage would be cancelled.

Recent experiments initiated by the author show that raising the coolant temperatures by a more 100°C does not significantly affect full-load efficiency, although there is some decrease in heat lost to coolant and increase in exhaust temperatures. Mechanical considerations prohibit the reduction in thermal inertia achieved in some of the more advanced designs of reciprocat' ig steam engine and inherent in the gas turbine cycle which is not subject to rapid fluctuations of temperature.

The extent to which mechanical considerations have actually outweighed thermodynamic factors is shown by the progressive move towards lower bore/stroke ratios in which the surface area/volume ratio of the combustion space is worsened and, for the same limits of temperature at the critical points, more heat is lost to the walls, with a corresponding loss of power. Fig. 6 shows this effect in a comparison of b.m.e.p. and heat losses in 850 c.c. and 950 c.c. engines having common bore and cylinder-head dimensions, the difference in swept volume being achieved by adjustment of strokes.

A recent incentive to combustion research has been the increasing concern over the content of unburnt hydrocarbons in vehicle exhausts.[20] The initial attacks on this problem resulted in the development of catalytic oxidizers[21] and after-burners[22] to be fitted in the exhaust system. Both these approaches present practical commercial difficulties and, despite the incentive provided by pending legislation in California, no commercially acceptable unit of adequate technical performance is yet available.

A more logical attack on the problem would be the improvement of combustion conditions to minimize the residual proportion of unburnt fractions.

Stratified charge engines

For a long time it has been recognized that stratification of the charge in the spark-ignition engine could lead to some considerable improvement in part-load efficiency. It is the difficulty of obtaining satisfactory ignition and the subsequent slow rate of burning which limit the minimum fuel/air ratio in conventional engines. In the case of the diesel, the whole combustion process is controlled by the form and rate of injection which reduces the delay and heat loss associated with vaporization in the cylinder. The heterogeneous mixture in the diesel cylinder includes areas which are approximately stoichiometric, yet still in the region of the flame front, so that a rapid combustion rate is maintained.

The reasons for the relative improvement in efficiency provided by stratified charge burning as opposed to throttling of the charge are not completely established on a qualitative basis. Baudry[23] calculates potential gain on the basis of the increases in the ratio of specific heats as the mixture is weakened but the reduction of heat transfer to the cylinder walls, resulting from lower cycle temperatures, is also a significant feature.[24, 25] There is also a reduction of throttling loss which normally increases the negative work loop.

A number of experimenters have achieved varying degrees

of success in carefully controlled laboratory engines and, from time to time, optimistic pronouncements are made. As recently as June of last year, an American authority predicted the general adoption of stratification within five to ten years, with overall fuel savings of up to 50%. There is certainly no diminuition of research effort in this direction, and both the antechamber and split charge within the main cylinder are being actively investigated.

When a separate chamber (similar to the antechamber in indirect injection engines) is used, it is fed with an approximately stoichiometric mixture to ensure a prompt ignition and initial combustion stage. The sudden expansion of the burning gases forces a flame front into the main body of the combustion space in which the mixture strength of the charge is progressively reduced as the load is decreased. A minimum size for the antechamber and a minimum mixture strength in the main charge are obvious limitations on the extent to which theoretical gains can be realized. These practical difficulties, coupled with losses inherent in the use of an antechamber, have led to disappointing practical results, despite apparently satisfactory functioning over a reasonable range of mixture-strength in the main chamber.

An alternative approach is to introduce two separate streams of charge through the inlet port. Some success with this system is reported by Baudry,[23] who indicates gains in specific fuel consumption, resulting from reduction of the mixture strength to approximately half the stroichiometric one (12%) in the optimum condition. It is surprising that the concentration of the rich mixture near the spark plug is maintained throughout the compression stroke.

As mentioned previously, some gain in performance and a general refinement in engine running could result from improved consistency of cycle operations, both between the various cylinders and within each individual cylinder.[19] Many forms of petrol injection have been proved feasible,[27] but the commercial impact is still relatively small. This is surely a matter of prime cost and, pending the development of a really cheap system, multi-choke carburettors have done much to equalize distributions in six and eight cylinder engines.

Ignition systems

New developments in ignition systems are centred on the elimination of the contact types in favour of a proximity-triggered relay.[28, 29] The mechanical superiority is clear and there should be a gain in consistency of firing, particularly at high speeds. Some further gains should be possible with a programme curve for ignition advance, in place of characteristics based on centrifugal-cum-manifold suction devices.

Low voltage ignition systems used in conjunction with surface discharge plugs have shown improved consistency, and recovery ability when plug fouling takes place. Otherwise,

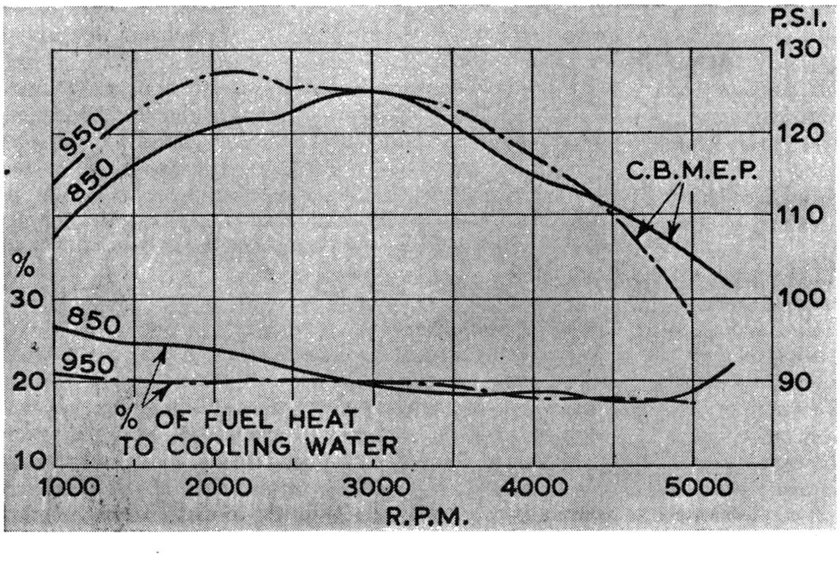

Fig. 6. A comparison of b.m.e.p. and heat losses for two engine sizes of equal bore but different stroke: mechanical considerations have outweighed thermodynamic factors in the progressive move towards lower bore/stroke ratios

their general performance barely matches that of conventional H.T. systems and the additional complications, particularly the screening required to reduce radio interference, do not, at this stage, seem justified.

The general question of efficiency should be viewed in its broadest perspective. Combustion research is only related to engine thermal efficiency whereas the ultimate criterion for the customer is m.p.g. This may properly be regarded as a matter of design, but the new concepts in passenger cars must modify the requirements for power units and should, in the long run, influence the emphasis placed upon the various avenues of applied research. The resurgence of the small car, aided by more compact and lighter power units, has done more to reduce average fuel consumption than all the more subtle improvements in basic engine efficiency in the last decade!

Research which has produced improved materials and better understanding of mechanical operations is of direct importance in raising the commercially attainable level of specific power output. Other specialized investigations have improved the smoothness of four-cylinder engines and those of 2 litres and above which, a few years ago, were unacceptable, could well make a large-scale come-back.

The new concept of the power unit is as a combination of engine and transmission and it is merely a question of time before research and development on automatic transmissions result in units which include large overdrive ratios and control engine conditions, so as to obtain a given power in the most efficient manner. Such developments could significantly offset the potential advantages of the stratified charge engine.

Research and design

The impact of research upon mechanical design is difficult to assess as new discoveries are gradually assimilated by the designer and lead to evolutionary rather than revolutionary changes. To be competitive, design must constantly improve and this makes more precise analytical study imperative.

There are many instances in which specialized research work has found practical application in engine design and has promoted further basic reappraisal. Dyke's carefully conducted experiments on the nature of piston ring operation[31] have had important consequences and have been followed by further studies.

Valve train dynamics is now regarded as an applied science in its own right and particularly notable in recent years were the contributions of Dudley,[32] Stoddart[33] and Nourse.[34]

Engine wear is a subject of intense current study and the use of the radioactive tracer technique has provided an important new tool for experimental investigations.[35-37] Studies of the rates of wear of piston rings and cylinder bores have led to the widespread introduction of the chromium-plated piston ring. The functioning of multi-grade oils has also been dramatically demonstrated by means of this technique.

The oil companies have long been in the forefront of I.C. engine research, both in their own laboratories, and by supporting work at other centres, notably the universities. There is need for still greater collaboration between engine manufacturers and the oil companies whose interests are so closely allied. To every lubrication problem there may be a number of possible solutions, but very probably one which is economically and technically sounder than the rest.

In the many instances in which the motor manufacturers rely solely on the technical resources of the oil companies to

Fig. 7. Polariscope equipment used for observing changing strain patterns in an aluminium block on dynamometer stand, using bi-refringent plastic coating on the diecast outer surfaces

sort out their lubrication difficulties, the optimum solution will rarely be found. Very few long-term projects or research programmes can be carried through in complete isolation. The lone research worker tends to become stale and to lose his perspective; a second, informed opinion is often invaluable and a certain amount of parallel research (of which there is all too little) can have stimulating effects.

The role of research in relation to design is changing rapidly because the designer has increasingly difficult questions to answer and because other research, unsponsored by design, is providing facts which cannot be ignored.

There are continual developments in new materials and production techniques, but these permeate slowly into production designs. Conservatism of outlook is a factor; but many of these new materials, though offering gains in cost, or functional merit, require very careful environmental study.[38] The designer and the chemist cannot provide this information and must look to the technologist who is capable of appreciating all sides of the problem to carry out the necessary exploratory investigations.

Typical of this general situation is the recent history of the aluminium cylinder block. To the casual observer, the U.S. motor industry may appear to be having second thoughts on light alloy in this application. The initial impetus, borne along on the weight-saving compact car campaign, appears to have spent itself. The more measured approach of the scientist shows no such violent ebb and flow of enthusiasm. He has recognized from the beginning that new materials require complete re-appraisal of the technical approach to them. For instance, die-casting techniques, which are essential to economical design, permit thinner sections to be used and scientific design must ensure that all the metal in the casting is there to perform a structural, as well as a filling-in function.

The use of light alloys

Noteworthy among the detailed investigations into the use of aluminium in engines was the work of the Ruston Hornsby team which has lead to the design of a highly successful range of air-cooled aluminium engines. A brief indication of the research and development behind these engines has been published,[39] and the application of photo-elastic techniques to study stress conditions in the American Motors designs represents applied research in the full sense. Fig. 7 shows the equipment used for this purpose. Die-cast aluminium cylinder-blocks have been introduced in some American engines.[40, 41] Current practice for aluminium cylinder bores is the cast-in dry liner of cast-iron, but alternative methods of solving this problem may result from current studies of various surface coatings on the aluminium.[42, 43]

The extreme hardness of hyper eutectic silicon-aluminium alloys and their lower coefficients of expansion have also evoked a good deal of interest. The addition of phosphorus to these alloys was found to assist in better nucleation of the silicon, and alloys containing up to 24% of silicon were at one stage used in the manufacture of experimental engines. Difficulties in machining and in the casting process itself have led to re-appraisal and it is now considered that 16%-silicon alloys are more practical materials for further investigation.

A number of small industrial engines have been made with pistons running direct in bores or aluminium containing no more than 12% silicon.

Yet another approach is that of R. C. Cross,[45] who reverses a currently accepted practice by incorporating deep bearing rings of steel or cast-iron in the piston to run against relatively soft aluminium bores.

The main justification for the aluminium cylinder block in the water-cooled engine is in the weight advantage but unless the engine weight is a fundamental factor in vehicle behaviour, (e.g., in a rear engine vehicle) no extra cost can be tolerated in order to effect such gains.

The cylinder-head is in a different category in that aluminium offers advantages resulting from its excellent heat-conductivity.

In thermodynamic terms, greater heat transfer from the combustion space is undesirable, but this is in practice more than offset by the better ability to prevent local hot spots.

Thus, a somewhat higher compression ratio can be

used with light alloys than with C.I., before uncontrolled combustion effects, such as detonation and pre-ignition, become troublesome. There is also some reduction of valve-seat temperature and a consequent improvement in valve life, despite the necessity to use separate valve-seat inserts

The aluminium cylinder head has long been accepted for high-output engines where some increase of cost is permissible, but current water-cooled head designs are not amenable to die casting techniques. A fabricated head of sandwich construction would greatly facilitate the casting process, and experiments are now being carried out with heads built up in this manner, the separate halves being joined together.[46]

At present, epoxy-resin adhesives are being used to join the halves, although the joint is only required to withstand leakage and to keep the head in one piece during assembly and servicing. The cylinder head studs are carried through all members of the sandwich and so carry all the loads arising during service.

Cooling and lubrication

In the petrol engine field, the water- versus air-cooling controversy remains unresolved.[47] Since water cooling offers much higher rates of heat transfer, there is less need for systematic study of temperature conditions in water-cooled designs.

More attention to such detail could, however, offer improvements, both in cost and performance. On the other hand, the absence of water jackets greatly simplifies the casting problems of the air-cooled engine. The objection to topping up and checking water level in water-cooled designs should soon disappear, as the separated reservoir system, recently introduced by Renault,[48] becomes more widespread. In view of the trend towards increased specific output, the facility of obtaining high local rates of cooling with liquid coolants seems to offer the more favourable long-term possibilities.

Bearing designs have become remarkably standardized and, research into basic oil-film conditions and into the behaviour of bearing metals themselves has led to interesting new developments,[49, 50] Fig. 8, for instance, shows a test rig, developed to simulate the operation of oil films in engine bearings.

However, there are so many design variations to cater for, that these developments will not result in any major change in fundamental thinking, at least in the immediate future. Although lubrication problems in the modern engine are being intensively studied and are consequently better understood, increasing performance is constantly required of the oil itself in order to extend the period between changes and to cope with increased loading; particularly where metallurgy and design have not succeeded in effecting the necessary improvement. Scuffing and pitting of tappets is a well-known instance.

A surprising amount is still left to the lubricating properties of oil, even in situations where pressure feed is available and the basic requirements for build-up of hydrodynamic films exist. Few engineers realize that special materials for journal bearings, with all their attendant limitations, are only necessary in order to compensate for imperfect functioning of the hydrodynamic film and to deal with inadequate filtration.

It is significant that the proportion of total heat transmitted to oil shows signs of increasing with engine power. Cooling

Fig. 8. Extensive research into the conditions within bearings has been carried out with equipment such as this rig for simulating the operation of oil films in engine bearings

is one of the main functions performed by the oil and some experiments have been carried out on fully oil-cooled engines. Nevertheless, water remains a most satisfactory coolant for vehicle engines.[51]

Future trends

Recent history has shown that some previously discarded forms of prime mover should be reconsidered in the light of advances in materials and design techniques. The gas turbine is the most striking example. A number of recent studies have been devoted to the Stirling cycle and these have led to practical experiments on a considerable scale.[55, 56]

High thermal efficiencies are certainly feasible and these engines are extremely quiet and smooth in operation, but it is difficult to obtain a high specific power output for a given weight and bulk, particularly, in smaller sizes. This remains, however, an interesting development.

Looking further to the future, the electro-chemical fuel cell offers a highly efficient and convenient method of conversion of energy to tractive effort.[57] However, much development must take place before a unit of high enough specific power is available for experimental installation in a passenger car.

The major long-term problem with the fuel cell is the commercial difficulty of eliminating impurities from the fuel. Very small quantities of such impurities cause rapid deterioration in the functioning and performance of the cell.

Despite limited horizons in terms of output and efficiency, there is still scope for considerable advances in the piston engine, and this will provide the necessary stimulus to promote continued and, in many cases, intensified research studies.

Alternative forms of prime mover will also continue to be the subject of extensive development but, if the piston engine continues to advance, it is unlikely to be ousted from its commanding position in the near future.

Dr D. J. K. Stuart, *at the time of writing this chapter, was Chief Research Engineer for B.M.C., an appointment he had held since 1957. In 1963 he joined H. W. Ward & Company and was subsequently elected to the Board as Technical Director. He is now Managing Director of Brockhouse Engineering Ltd, and he moved to this Company in 1967. Educated at Doncaster Grammar School, he entered Cambridge University in 1943 after obtaining an open scholarship in mathematics and gained first class honours in the Mechanical Sciences Tripos. He spent two years at Armstrong Siddeley Motors Ltd, on the development of gas turbine engines and returned to Cambridge in 1947 as a Research Student. Subsequently he was awarded a PhD for a thesis on Two Dimensional Flow in Aerofoil Cascades. On leaving Cambridge he joined Pilkington Brothers Ltd, as an experimental engineer, and a year later moved to the Austin Motor Co. Ltd, as a Senior Research Engineer in charge of investigations of automatic transmissions. He was appointed Deputy Chief Development Engineer in 1953.*

REFERENCES

1. FROEDE, W. G., 'The N.S.U.-Wankel Rotating Combustion Engine'. 1961 *S.A.E.* Internal Congress of Automobile Engineering, January.
2. BENTELE, M., 'Curtiss-Wright's Developments on Rotating Combustion Engines'. 1961 *S.A.E.* Internal Congress Auto. Engineering, January.
3. WANKEL, F. and FROEDE, W. G., 'Design and Present State of Development of a Trochoid Rotary Engine'. 1960 *M.T.Z.*, vol. 21, February.
4. 'Rotary Piston Engine Developments'. 1961 *The Engineer*, vol. 211, May.
5. SELF, K. W., 'The Future of Higher Horsepower Engines'. 1961 *S.A.E.* pre-print 384A, August.
6. DENT, R. E. 'The Engine Shape and Where to Put It'. 1958 *S.A.E.* reprint, November.
7. WINTERINGHAM, J. S., 'Potential Power Plants for Passenger Cars'. 1961 *Automotive Industries*, April 15th.
8. CARIS, D. F. and NELSON, E. E., 'A New Look at High Compression Engines'. 1958 *S.A.E.* reprint, June.
9. LOVELL, W. G., 'Some Chemistry of Future High-Compression Engines'. 1960 *Proc. Instn mech. Engrs, Lond.*, March.
10. MOLCHANOV, K. K., 'On the Problem of Gas Motion and Combustion in a Light Fuel Engine'. 1955 *Automobil'no-dorozhnogo Institu, Autotransizdat*, Moscow.
11. STEBAR, R. E., WIESE, W. M. and EVERETT, R. L., 'New Studies Provide More Information on Engine Rumble'. 1960 *General Motors Engineering Journal*, vol. 7, Jan.-March.
12. GOODGER, E. M., 'Abnormal Combustion'. 1960 *Automobile Engineer*, March.
13. EGERTON, Sir Alfred, SAUNDERS, O. A. and SPALDING, D. B., 'The Chemistry and Physics of Combustion'. 1955 *I.Mech.E. and A.S.M.E.* Joint Conference on Combustion, October.
14. VICH NIEVSKY, R., 'Combustion in Petrol Engines'. 1955 *I.Mech.E. and A.S.M.E.* Joint Conference on Combustion, October.
15. EGERTON, A. C. and LEFEBURE, A. H., 'Flame Propagation: The Effect of Pressure Variation on Burning Velocities'. 1954 *Proc. roy. Soc.* Ser. A 222.
16. CLARKE, J. H., 'Initiation and Some Controlling Parometers of Combustion in the Piston Engine'. 1961 *Proc. Instn mech. Engrs, Lond. (Automobile Division).*
17. LYN, W. T., 'Calculations on the Effect of Rate of Heat Release on the Shape of the Cylinder—Pressure Diagram and Cycle Efficiency'. 1960 *Proc. Instn mech. Engrs, Lond. (A.D.).*
18. BOWDITCH, F. W., 'Combustion Problems in Gasoline Engines'. 1960 8th F.I.S.I.T.A. International Conference.
19. SOLTAU, J. P., 'Cylinder Pressure Variations in Petrol Engines'. 1960 *Proc. Instn mech. Engrs, Lond. (A.D.),* December.
20. LARSON, G. P., CHIPMAN, J. C. and KAUPER, E. K., 'A Study of Distribution and Effects of Automotive Exhaust Gas in Los Angeles'. 1955 *S.A.E.* preprint, January.
21. HAMBLIN, R. J. and HAENSEL, V., 'The Catalytic Conversion of Automobile Exhaust Gases by Purzaust'. 1960 *American Inst., Chem. Eng.*, December.
22. FAITH, W. L., 'Status of Motor Vehicle Exhaust After Burners'. 1960 *American Petroleum Institute* preprint, May.
23. BAUDRY, J., 'A New I.F.P. Process for Engine Combustion: A Variable Air Fuel Ratio Spark Ignition Engine'. 1960 *S.A.E.* preprint.
24. CONTA, L. D., DURBETAKI, P. and BASCUNANA, J. L., 'Stratified Charge Operation of Spark Ignition Engines'. 1961 *S.A.E.* paper no. 375B.
25. CLEVELAND, A. E. and BISHOP, I. N., 'Several Possible Paths to Improved Part-Load Economy of Spark Ignition Engines'. 1960 *S.A.E* preprint, March 15-17.
26. CAY, E. J., 'Symposium on Petrol Injection: Studies of Automotive Petrol Injection in the U.S.A.'. 1958 *I.Mech.E.*, April.
27. BROCKHAUS, H., COWELL, T. F. and MASTERMAN, D. M. A., 'Injection Versus Carburation, A Comparison of Fuel Quality Requirements'. 8th F.I.S.I.T.A. International Conference.
28. WATSON, E. A., 'A Review of Problems and Developments in Quick Ignition Equipment in Piston Engines'. 1960 8th F.I.S.I.T.A. International Conference, May.
29. SPAULDING, G. E., 'Transfer Switch Ignition Systems'. 1960 *S.A.E.* preprint—Annual Meeting, January 11th-15th.
30. THE PONTIAC TEMPEST'. 1960 Detroit Section of S.A.E., Nov. 7th.
31. DYKES, P. DE K., 'An Investigation into the Mechanism of Oil Loss Past Pistons'. 1957 *Proc. Instn mech. Engrs, Lond.*, volume 171, page 413.
32. DUDLEY, W. M., 'New Methods in Valve Cam Design'. 1948 *Trans. Soc. Automot. Engrs, N.Y.*, vol. 2, no. 1, p. 19.
33. STODDARD, D. A., 'Polydyne Cam Design'. 1953 *Mach. Design*, nos. 1, 2, 3, pp. 121, 146, 149.
34. NOURSE, J. H., DENNIS, R. C. and WOOD, W. M., 1960 *S.A.E.* preprint, Summer Meeting, June 5-10, 34 pp.
35. AGIUS, P. J. and PEGG, R. E., 'Using a Radioactive Liner for Studying Engine Wear'. 1958 *Inst. Petrol. Rev.*, vol. 12, p. 337.
36. ABOWD, R., 'Probing the Causes of Piston Ring Wear by the Radiotracer Technique'. 1959 *S.A.E.* preprint, January.
37. CALOW, J. R. B. and ACTON, S. R., 'Symposium on Wear in the Gasoline Engine: Piston Ring Wear in the Gasoline Engine'. 1960 Shell Research Limited, October.
38. BONA, C. F., 'How to Select New Materials for the Automobile'. 1960 8th International F.I.S.I.T.A. Conference, May.

39. FEILDEN, G. B. R., 'A Critical Approach to Design in Mechanical Engineering'. 1959 Bulleid Memorial Lectures, Nottingham University.
40. MOELLER, E. G., WEERTMAN, W. L. and ERIKSEN, H. E., 'Chrysler Corporation Die-Cast Aluminium Slat Six Engine'. 1961 S.A.E. preprint 307C, January.
41. ADAMSON, J. F., BURKE, C. E., POTTER, D. V. and ZECHEL, W. J., 'The American Motors New Die Cast Aluminium Engine'. 1961 S.A.E. preprint 307A, January.
42. BAUER, A. F., 'Transplant Coated Aluminium Cylinder Bores, Physical Properties of this New Protective Coating in Comparison to Known Processes'. 1961 S.A.E. preprint 369C.
43. SMITH, N. W., 'A Profile of Aluminium Alloy Cylinder Bores'. 1961 S.A.E. preprint 369D.
44. SMITH, R. M., 'Hypereutectic Aluminium–Silicon Alloys'. 1958 S.A.E. preprint, Summer Meeting, June 8–13.
45. CROSS, R. C., An Address given at a meeting held on the 8th October, 1957, at the General Meeting of the Automobile Division of the Institution of Mechanical Engineers on 'Experiments with Internal Combustion Engines'.
46. BAUER, A. F., 'Aluminium Castings for Passenger Car Engines—A Comparison between U.S.A. and Europe'. 1960 S.A.E. preprint, 9th March.
47. LUDERITZ, W., 'Air Cooled Engines for Traction'. 1961 Diesel Engineers and Users Association Symposium, January.
48. THE RENAULT 4L, The Motor, 30th August. (Cooling System).
49. STERN, L., 'Progress in Engine Bearings'. 1960 Institution Automotive Aeronautical Engineers Journal, vol. 20, no. 3, April.
50. CHRISTOPHERSON, D. G., 'Developments in Diesel Engine Bearings'. 1960 British Power Engineering, vol. 1, no. 6, November.
51. ROSS, H. K., 'Heat Transfer Efficiency of Automotive Engine Coolants'. 1960 Coating and Chemical Lab. (U.S.A.). Report No. CCL 90, March.
52. MARKS, D. I. and REINERS, N. M., 'Forward Look on Engines'. 1958 S.A.E. preprint, August.
53. ROSEN, G. A., 'The Role of the Turbine in Future Vehicle Power-plants'. 1956 S.A.E. Buckendale Lecture, October.
54. ELTINGE, L., 'Turbo Auto's for the Future'. 1956 Petroleum Refiner, vol. 35, no. 7, July.
55. MEIJER, 'The Philips Hot-Gas Engine with Rhombic Drive Mechanism'. 1959 Philips Technical Review, May.
56. FLYNN, G., PERCIVAL, W. H. and HEFFNER, F. E., 1960 S.A.E. preprint. G.M.R. Starting Thermal Engine, January.
57. LIEBHAFSKY, H. A. and DOUGLAS, D. L., 'The Fuel Cell—Status and Background'. 1959 A.S.M.E. Paper No. 59, June.

Safety and Reliability in Cars

by R. J. Love, WhSch, MIMechE and
M. A. Macaulay, BSc, PhD, FIMechE

Reliability and safety are inter-related to the extent that a significant minority of accidents can be traced to vehicle defects. Both depend a great deal on the design of the car and this in turn relies heavily on suitable testing techniques. Of particular importance is the study of deceleration during impact; the aim being to protect the occupants by making the car body absorb the shock.

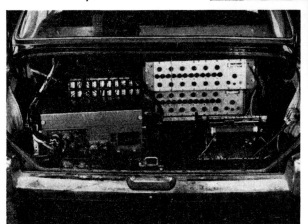

Fig. I. With the aid of magnetic tape equipment in the boot, strain gauge readings can be recorded for analysis by the laboratory equipment shown above. The resulting information indicates which components and features need particular attention

In order to keep this survey to a reasonable length and a coherent form, we shall restrict it to two aspects which are of topical interest. These are reliability, in the sense of the vehicle continuing to perform its designed function; and safety, in the sense of vehicle behaviour during a crash. The two are linked by the role of failure or maladjustment in causing accidents.

The extent to which the subject of reliability has come into prominence might suggest that there has not been a great need for reliability in the past or that engineering components have recently become very unreliable. Either conclusion would be unfortunate, because reliability has always been an objective of good engineering.

It would be more appropriate to say that design for reliability has recently been placed on a more rational basis and this is a natural development. As engineering devices become more complex and costly, unreliability of individual parts is of greater consequence. Also, there is the continual effort to reduce the weight and cost of components, involving higher operating stresses. Activity in the electronics and aerospace fields has given the subject an important stimulus and has indicated some lines of approach.[1]

There is also a great deal of interest in automobile safety at the moment; it is becoming the object of an increasing amount of legislation and manufacturers are concerned with present and future forms of this. However, in a democracy legislation is largely a result of public opinion.

The main reason for the increasing interest in safety seems to be that, with the increasing density of vehicles on the road, the accident rate, though not noticeably higher per vehicle-mile than it has been for a considerable time, is becoming higher overall.

Experience includes failure

An important fundamental factor in reliability concerns the divergence between the conditions which, for design purposes, are assumed to apply to components and those which actually prevail in service. Operating conditions such as loading are rarely known precisely, nor are there suitable analytical methods for assessing the durability of components or materials under complex conditions.

Consider, for example, the cam and follower. Usually, estimates of durability are based on the use of nominal peak stresses which are calculated from the curvatures of the cam and follower in contact at maximum lift, the load being the result of compression of the valve spring: no allowance is made for valve gear inertia, rubbing velocity, type of oil, temperature, etc. The problem is largely settled by using empirical design stresses, derived from past experience of similar components. Such experience must, however, include some failures in service before useful design stresses can be obtained from it. If economical designs are to be achieved, and if service is the only source of realistic experience, service failures are therefore inevitable.

To reduce the risks with new designs, extensive testing is carried out on prototypes and, later, on pre-production models. Whilst this is essential, complete success may not be readily achieved, because the test models are not likely to be exactly as the production models, test time may be restricted, and the number of test units does not constitute a sufficiently large sample in relation to subsequent production.

A suitable design having been established, it is then necessary of course, to ensure that the required quality of all components is consistently attained in production. Also, maintenance in service is important. Thus, almost the whole engineering activity—design, vehicle and component testing, manufacture, maintenance, service experience—is involved. A useful paper covering this whole field of activity in an automobile company and indicating how reliability can be achieved has been given by Simpson.[2]

Clearly the most important step that has been taken with regard to reliability is the introduction of reliability engineers and reliability departments. This has not been done in this country to the same extent as in America. Since so many engineering activities are associated with reliability, the responsibilities and functions of the reliability department must be carefully defined.[2-5]

Specifications and design

A programme for attaining reliability begins with specifications for operating requirements and cost, related to past experience with similar components. Crockett[6] illustrates the cost and reliability requirements by comparing different types of vehicle; cars must be produced at low cost and extended mileages are not required; from commercial vehicles very high mileages are expected and first cost is not as important as operating cost; military vehicles may need to run only very short mileages but very high reliability is required with a minimum of maintenance.

Information on materials and processes is continually increasing but the design technique itself does not appear to have changed significantly. Coutinho,[7] writing of engineering in general, says " . . . few practical techniques suitable for use in a design office are available . . . ".

As a result of instrumentation developed in recent years, considerable improvements have been made in measuring the loads on various components in service. Fig. 1 (bottom) shows instrumentation for recording strain gauge signals on magnetic tape in a test vehicle and Fig. 1 (top) shows tape records being replayed to a special purpose strain analyser in the laboratory.[8-10] Information of this type can, of course, be used to indicate the more important features which need attention.

In the long run, service load data, coupled with cumulative damage data for materials and components, should permit more realistic assessments of reliability at the design stage. Fatigue tests in progress at MIRA use service loading, random loading, 'programme' loading, etc.[11] Various other institutions are of course contributing information on this subject.

Proving ground tests

Developments in endurance testing are making an important contribution to reliability and this applies to both proving ground testing of vehicles and laboratory testing of components. An essential feature of proving ground testing is that a relationship must be established between test miles and miles in service.[12] A method introduced for this is a Weibull-Johnson plot of failure data.[2, 13-15]

On a special probability scale, percentage of parts failed approximates to a straight line relationship when plotted

against miles. This is illustrated in Fig. 2, which is given by Simpson.[2] One line is shown for proving ground data and another is shown for service data. A ratio of service/proving-ground miles, or a test severity ratio, is obtained as shown.

An undesirable feature of the test severity ratio is that it requires the occurrence of a significant amount of failure in service—the situation one is trying to avoid. A second point concerns the fact that, since the proving ground tests must be limited in number, the failure data will cover only the higher 'percentage failed' range, and it is necessary to extrapolate to the lower.

The straight-line character of the Weibull-Johnson plot, which makes it easy to extrapolate to low percentage-failed values, is the advantage claimed for this method but it is necessary to check its accuracy. Tallian,[16] examining the results of large numbers of tests on ball and roller bearings, found significant deviations from the Weibull distribution. Early failures, usually described as 'infant mortality', deviate from the Weibull plot,[17-19] and there is also the possibility of a change of slope where more than one mode of failure is involved.[20]

Proving-ground testing should not induce early failure by arbitarily increasing loads, because this can produce non-typical modes of failure.[12] Simpson[2] prefers the "more frequent application of conditions as severe as any encountered in service".

It would appear that the best principle is to employ a stress intensity as great as, or slightly greater than, the maximum likely in service, and to ensure that the vehicle can withstand significantly more applications of this stress than is likely to occur. The amount by which the test applications of stress are greater than those in service will depend partly on the margin of safety which is required and partly on the variability in strength of the production components to allow for the possibility that the test vehicles do not contain some of the weaker components which may be produced.

To decide on the most suitable form of test track and to establish the most suitable test procedure are clearly statistical problems. Important background data to assist with these points could usefully be obtained by the measurement of component loadings under a wide range of conditions in service; and by studying variability of component life in fatigue tests under realistic loads.

Laboratory testing

The importance of laboratory testing of components as an aid to reliability is obvious: where such testing is possible, it is much quicker and cheaper than proving-ground or service testing.[2, 5, 21, 22] Also, it can be used for developing the most economical designs for a wide range of components, whereas proving-ground testing will tend to confine attention to the discovery and correction of the weakest parts. Furthermore, the adequate testing of a number of items such as locks, doors, lighting, switches, etc, cannot be readily incorporated in the usual proving-ground test.

Until recently, almost all rig testing in the automobile industry has been very simple. Whilst it has been useful, it has in many cases served merely to indicate whether one design is better than another, rather than demonstrate whether a design is completely suitable for service. To improve accuracy when comparing components, and to provide a means of indicating the level of success in service, it is clearly necessary to introduce more realistic tests.

There is a definite move in this direction,[2, 15, 22-27,]

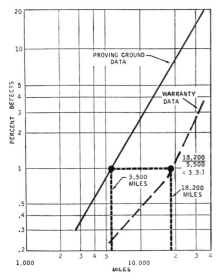

Fig. 2. On this special probability scale percentage defect graphs approximate to straight lines: proving ground data can be related to service experience

although the extent to which new methods have been applied in this country, is quite limited so far.

Programmed fatigue testing of components has been applied in the aircraft industry for many years. In Germany, it has also been used for many years on automobile components.[28] Beauvais and Sorenson[29] refer to laboratory tests in which completely realistic loads can be introduced by means of magnetic tape records taken in service and fed to servo-controlled hydraulic actuators. When a suitable test is employed, a test severity ratio can be derived in the same way as for proving-ground tests.[2, 20]

Whilst the programmed test constitutes an improvement in technique, there is very little data to demonstrate just how well it represents true service loading. There is a small amount of information to indicate that programmed loading affords a reasonably close representation of stationary random loading,[11, 30] but more work needs to be done to compare it with true service load histories. This is one of the objectives of the variable load fatigue testing in progress at MIRA.

Manufacture and maintenance

To attain reliability it is clearly necessary to keep variability of components to a minimum and quality control is therefore most important.[31] However, quality control is often associated with dimensional tolerances, surface hardnesses, *etc*, and some of the properties which can affect component durability may not be included. An important development contributing to reliability is the wider use of functional tests on specimen components.

Simpson[2] shows how a 'reliability requirement' in terms of the test which components must pass is drawn up and becomes part of the engineering specification. Suppliers of components are responsible for seeing that their products meet these requirements.[2, 32]

A difficulty pointed out by Wanley[33] is that, because jobs are becoming more repetitive, operators are less able to take a pride in them.

The need for a customer to have servicing carried out is an uncontrollable variable[32] and a possible source of un-reliability.[6] A significant step in this connection is the extension of servicing periods and introduction of 'lubed-for-life' components.

Information feedback

Since there can be no complete substitute for service operation, both in terms of the wide range of conditions and the large number of units involved, customer experience will continue to be a vital source of design information. Some years ago manufacturers' interest appeared to be confined almost entirely to information obtained from warranty data. The range of interest has clearly extended and warranty periods have also been increased.

Simpson[2] lists a number of sources of information, including warranty claims, questionnaires completed by owners, customers' letters, market research, public service and rental fleets. Nevitt,[15] and Hodkin and Nevitt[34] cite similar examples. Computer systems are used for recording and classifying data. The service data may indicate that a problem exists but not necessarily define the problem. Simpson explains that for such cases a 'parts recall procedure' can be used, faulty parts being returned by dealers for examination.

Corney[5] and Witts[35] show graphical representations of service data on failure rate, plotted against time, whereby

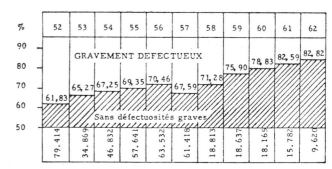

Fig. 3. *Percentage reject rates found during compulsory vehicle inspection in Belgium; the year is shown across the top*

the severity of such a problem can be judged as early as possible.

Published information indicating the level of success which can be achieved with reliability procedures of the kind referred to is lacking. However, Manoogian[32] shows that, with a particular model of car, the warranty expenses were markedly reduced over a period of four years, and this in spite of the fact that the warranty period had been increased from 3 to 24 months during this time.

Safety

The same engineering activities which are involved in ensuring reliability are needed in general to incorporate safety features in vehicles but the recent literature on the two subjects is quite different in character. The detailed results of reliability studies are kept as commercial secrets by the firms concerned. Accident studies and related work, on the other hand, are often carried out by groups not directly connected with motor vehicle manufacture. Their results are published freely but the application of their findings is not always straightforward and there is little published on the long-term use which can be made of the information.

For instance, there is little on how to make vehicles safer in the long term. But there is a great deal of public interest in automobile safety at present. This is evident in the appearance of publications such as Nader's *Unsafe at Any Speed*[36] and the steady increase in safety legislation. In this country we have, for instance, ever more frequent compulsory inspection and the compulsory fitting of safety belts to the front seats of all new cars. Abroad the best known examples are the new *US Federal Regulations on Safety*.[37]

The problems have been recognised for a long time and compulsory vehicle inspection was begun in some states of the USA in 1929 and the annual Stapp Safety Conferences have been running since 1955.

One possible adverse effect of longer servicing intervals is that, with a reduced need for maintenance, inspection is often apt to be less frequent. In countries where there is compulsory regular inspection of vehicles, this should be a less serious problem. Results from such inspections could also be used to amplify the kind of service feedback study outlined above which makers are beginning to carry out on the reliability of components.

From the point of view of the individual manufacturer who is trying to improve his product, published results from compulsory inspections suffer two drawbacks. The defects are not defined in sufficient detail and the vehicles are not separated by make. It might be difficult to justify the publication of results by make but there are precedents for telling a manufacturer confidentially how he stands in relation to the national average.

From an overall point of view, published results from compulsory inspections suffer from the lack of common standards. In 1962 Halleux[38] tried to compare defect rates from different sources but he found that the percentage of vehicles failing to pass the inspection without some rectification varied from 30% to 75%, depending on the country. Even within the USA the variation between States was almost as large. This does not appear to have improved greatly.

Despite these wide variations, the figures are sufficient to indicate that a surprising number of faults exist and they also indicate two distinct, but hardly surprising, trends. These are, firstly, that when compulsory inspection has been in force for some time, the reject rate falls; secondly, that older vehicles have more defects than newer ones. For instance, in Belgium in 1960, the reject rate for cars was 50% but by 1962 it had fallen to 30%.[38] A more detailed breakdown of the 1962 figures[39] shows that the 30% overall figure is the average of a reject rate running from 40% for ten year old cars to 20% for new ones. This is shown in Fig. 3.

Do defects cause accidents?

It is difficult to assess the effect of compulsory inspection on the accident rate. Preliminary work by the Road Research Laboratory before compulsory inspection was adopted in this country led to the comment that

the average death rate per vehicle-mile for all states in the USA which enforce vehicle inspection is lower than that for all those without it, but this is not conclusive proof of the effect of inspection, since the states which have vehicle inspection are probably those which are most interested in road and traffic safety and which make use of all kinds of other safety measures.[40]

The actual figures for average death rates are not quoted.

Attempts have been made to relate vehicle defects to accidents but the conclusions are even more varied than those on vehicle inspection. A small-scale RRL survey in this country[41] indicated that defects in vehicles were a major factor in 15% of the accidents studied. These were all day-time accidents and the comment was made that after dark the percentage would probably be higher since then there would also be lighting defects.

The Ministry of Transport records are quoted as reporting the presence of defects in about 3% of all accidents.[41] The conditions were different—the RRL figures were for all accidents whilst the Ministry of Transport's were for personal injury accidents only; and the RRL results were based on a small sample. Nevertheless, the difference in the percentages appears disproportionately high.

A similar difference of opinion is reported for Belgium.[42] The official statistics give the proportion of accidents caused by vehicle defects as just over 1%. Halleux, the Director of the Belgian Automobile Inspection Organisation, considers that, from an analysis of serious accidents which he himself has studied, 24% of them were caused or made worse by technical defects. A US authority is quoted as thinking that probably less than 5% of accidents are due to defects.[43]

There are wide differences in the terms of reference used. The difference between 'caused by' and 'present in' and the difference between 'accident' and 'serious accident' are only two of the most obvious but the disagreement still seems to be disproportionately large.

An overall picture does emerge, however. Most accidents are not caused by vehicle defects but the proportion, though fairly small, is probably high enough to cause some concern.

Safety belts

Work has been carried out in this country[41] and the USA[44] on studying vehicle accidents in fairly small areas. The investigators arranged with the local police to be called to accidents and made on-the-spot studies of them in much more detail than is normal. The results were analysed to give a breakdown of the different types of accident and the different types of injury. The conclusions arrived at differ in detail for the two countries but there is overall agreement. About 60 to 70% of accidents involve head-on impact—ie, within 45° of the direction of travel. About 10 to 15% involve side impact and about 5 to 10% involve rear impact. The main differences between the two countries appear to be that ejection from the vehicle and overturning are more common in the USA.

Most of the victims in the studies published have been drivers and front seat passengers. Investigation of a number of accidents in which there was a front seat passenger indicated that the passenger was slightly more liable to injury than the driver.[45] The overall injury rates without seat belts were about 85% for the passenger and 75% for the driver. Some results have also been published for rear seat passengers.[46] From a small number of fatal accidents it was concluded that about half the deaths could have been avoided if the rear seat passengers had worn seat belts. Most of the others were killed by the collapse of the side of the car.

For the front seat occupants in general there have been two different approaches which tried to assess the usefulness of safety belts. One of them was by Heulke and Gikas[44] of the University of Michigan who studied 139 fatal accidents and concluded that 37% of those killed could not have survived, even if they had worn belts. Again this was largely because of collapse of the car.

The other approach is reported by Lister and Neilson.[47] Several British seat belt manufacturers supply a blank accident report form with every belt they sell and the owner is asked to complete and return the form if he is involved in an accident. The completed forms are sent on to the Road Research Laboratory who analyse the forms. Some checks are made to ensure that the information in the forms is reasonably accurate.

In the published analysis, which contains no fatal accidents, it is estimated that the use of seat belts reduced serious injury by about 70%. The agreement with the American results is surprisingly close, especially if we remember that the American accidents, being all fatal, were presumably more severe. There seems to be little doubt that safety belts are the most important first step in reducing death and injury.

A survey in the London area in 1962 showed that 7% of cars were fitted with front-seat belts and that 70% of these in the driving position were worn, ie, about 5% of drivers. A further survey in much the same area in 1964 showed that 12% of cars had belts for the front seats and about 40% of these in the driving position were worn, ie, still about 5% of drivers.[47]

With legislation now making the fitting of safety belts in the front seats of new cars mandatory, the problem becomes one of persuading people to use their belts. Grime[48] comments on the protection of occupants without seat belts that, to be effective, protective devices must be close to the car occupant and this is the opposite of the requirement for a car equipped with belts.

In the latter case there should be as much room as possible in front of the occupant to allow the belt to extend.

Crash testing

The cases when seat belts are not effective are usually when the car collapses on to the occupant. Outside the car something can probably be done to soften the impact of the vehicle with roadside equipment. RRL results[49] show that 12% of the accidents in a sample studied involved collision with roadside furniture and a larger proportion of the people involved was injured than in any other kind of accident. It is suggested that collapsible roadside furniture would be of use.

Apart from this, however, we are left with the problem of how a vehicle deforms during an accident. It is almost impossible to discover after the event and this leads to the need for test crashes, an example of which is shown in Fig. 4, and their correlation with the results of observed accidents.

By means of suitable instrumentation and high-speed photography the sequence of events during a test crash can be studied and the final deformation of the car and the estimated injuries of the occupants can be related to those found in actual accidents. Examples of correlation between test crashes and accidents have been published.[50] Once a representative crash test has been established, it is possible to use it for improving vehicle behaviour. Such testing is being carried out in a number of places[51–54] and there are, for instance, SAE standards for various impact tests,[55] some of which are incorporated as test requirements in the new US Federal Legislation.

Apart from the structural integrity of the vehicle, crash tests are useful in studying what happens to the occupants. A man in a seat belt acts as a spring-mass system, the response of which to deceleration pulses (see Fig. 5) must be considered.[56–58] In addition, the man himself can be considered as a spring-mass system[59, 60] which can be simulated with more or less accuracy, in calculations or by means of anthropometric dummies, such as the one shown in Fig. 6.

The correct parameters for the spring-mass system and the injury-producing effects of sudden declerations can be established by tests with human volunteers, human cadavers and live animals.[61] This is all necessary to provide background knowledge before design and development work can be carried out to best advantage.

Design for safety

So far design and development work has been aimed at improving the chances of survival in present vehicles by making the passenger compartment as strong as possible, removing projections inside it and providing padding on all suitable surfaces. Seat belts and safety seats incorporating belts have also been studied.

The main difficulty with safety belts appears to be getting people to wear them and this has led to proposals for restraining devices, such as airbags which inflate automatically if an accident is imminent.[62] These appear to raise as many problems as they solve.

Work has also been aimed at preventing the intrusion of the steering column into the space occupied by the driver; at providing door locks which stay closed during impact; and reducing fire risks. A long-established controversy over whether windscreens should be made of toughened or laminated glass remains unresolved,[41] but this should become less important provided adequate safety belts are worn universally.

If these safety features can be incorporated successfully, the next stage will be to improve the behaviour of vehicles in accidents by obtaining optimum patterns of deceleration.

Establishing these patterns will mean obtaining better information on the injury-producing effect of sudden decelerations, the dynamic response of a man in a safety belt to different types of deceleration pulse and the mechanism by which the rapid deformation of the vehicle produces such pulses. Most of this information is not available at present.

An excellent survey of the present stage of thinking is given by Grime.[48]

A further problem which has not yet received a great deal of attention is that of reaching the best compromise design to cater for all the different types of accident which could occur. A preliminary theoretical article on the subject has been written by Miley.[63]

Some rough and ready feasibility studies have also been carried out[64] which indicate that cars which are much safer in accidents could probably be made and sold. But again these surveys in their present form raise as many problems as they answer.

REFERENCES

1. CHORAFAS, D. N., *Statistical Processes and Reliability Engineering*. 1960, Van Nostrand, 438 pp.
2. SIMPSON, B. H., 'Reliability and Maintainability: Part 2—Automotive: The Ford Reliability Program'. March 1966, *Mechanical Engineering*, vol. 88, no. 3, p. 46.
3. GRETZINGER, J. R., 'Chassis and Total Car Reliability'. 3rd October 1960. SAE Preprint No. S260.
4. SEHN, W. E., 'Initiating a Body Reliability Program'. 3rd October 1960, SAE Preprint No. S261.
5. CORNEY, C. T., 'The Achievement of Design Reliability'. June 1965, *Lucas Engineering Review*, vol. 2, no. 2, p. 38.
6. CROCKETT, C. V., 'Engineering for Reliability'. 4th April 1960, SAE Preprint No. S245.
7. COUTINHO, J. DE S., 'Reliability and Maintainability'. February 1966, *Mechanical Engineering*, vol. 88, no. 2, p. 22.
8. DRURY, C. G., and OVERTON, J. A., 'Vehicle Service Loads—Part 1; A Preliminary Study of Stress Level Counting'. February 1964, *MIRA Report* No. 1964/8.
9. ANDREW, S., and ATKINSON, M. R., 'The Effect of Vehicle Speed on Service Loads in Suspension Components'. November 1965, *MIRA Report* No. 1966/1.
10. ANDREW, S., ATKINSON M. R., and WHITTAKER, M. W., 'Vehicle Service Loads—Suspension Component Loads in a Front Wheel Drive Car'. October 1966, *MIRA Report* No. 1966/12.
11. BOOTH, R. T., and WRIGHT, D. H., 'A Ten-Station Machine for Variable Load Fatigue Tests'. October 1966, *MIRA Report* No. 1966/14.
12. MCCONNELL, W. A., 'How Good is Testing? A Correlation of Customer, Laboratory, and Proving Ground Experience'. 7th December 1959, *SAE Preprint* No. 5210.
13. WEIBULL, W., 'A Statistical Distribution Function of Wide Application'. 1951, *Journal of Applied Mechanics*, vol. 18, p. 293.

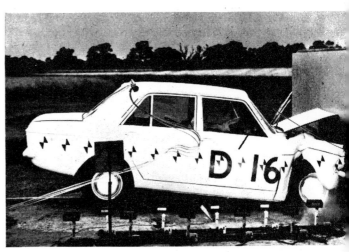

Fig. 4. Crash test of a Hillman 'Hunter' against a solid wall; note the dummy in the driver's seat

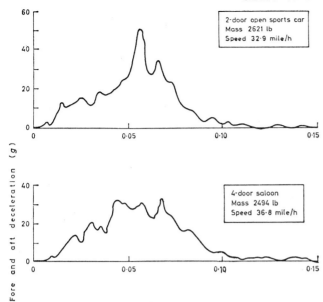

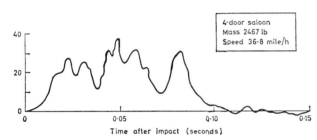

Fig. 5. *Deceleration pulses in British cars in head-on collisions with a concrete barrier*

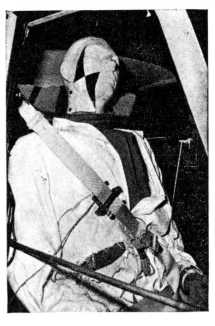

Fig. 6. Dummy used in simulating the effects of a crash on the human body; other tests are made with live volunteers, animals and even corpses

14. JOHNSON, L. G., '*The Statistical Treatment of Fatigue Experiments* 1964, Elsevier Publishing Co., London.
15. NEVITT, P. J., 'Reliability as a Design Criterion'. September 1965, Institution of Engineering Inspection, National Inspection Conference.
16. TALLIAN, T., 'Weibull Distribution of Rolling Contact Fatigue Life and Deviations Therefrom'. 17–19 October 1961, *ASLE Preprint* 61 LC-15.
17. GALLANT, R. A., 'Experimental and Statistical Analysis of Dome Lamp Quality: With Emphasis on the Weibull-Johnson Statistical Method'. June 1962, *SAE Preprint* No. 533C.
18. HOFWEBER, A. J., 'Body Component Reliability'. 3rd October 1960, *SAE Preprint* No. S262.
19. LUX, W. J., 'A Reliability Study of Diesel Engines'. April 1964, *SAE Preprint* S374.
20. VIGIER, M., 'The Interpretation of Endurance Tests by Means of Reliability Techniques'. February 1965, *Ingénieurs de l'Auto*, vol. 38, no. 2, p. 57: *MIRA Translation* No. 42/65. (See also Discussion of this Paper)
21. RODDEWIG, G. F., 'Experimental Bench Testing Techniques for Truck and Bus Components'. 22–24 October 1952, *SAE Preprint*.
22. CZARNECKI, R., 'Problems in the Choice of Equivalent Parameters for Loading a Motor Vehicle Body in Accelerated Endurance Testing'. August 1965, *Kraftfahrzeugtechnik*, vol. 15, no. 8, p. 290: *MIRA Translation* No. 3/66.
23. HARTMEYER, J. J., 'Fatigue Testing of Bodies in the Laboratory'. 6–8 October 1965, *ASBE Preprint*, Annual Tech. Convention.
24. SMIRNOV, G. A., and others 'Selection of a Programme for Testing Vehicle Final-Drive Gears and Half-Shafts on Rigs with Programmed Control'. April 1964, *Avtom. Prom.*, no. 4, p. 20.
25. CONOVER, J. C., JAECKEL, H. R., and KIPPOLA, W. J., 'Simulation of Field Loading in Fatigue Testing'. 10–14 January 1966, *SAE Paper, Automot. Engng. Congress.*
26. SJÖSTRÖM, S., 'Experimental Methods for Development of High Quality Commercial Vehicles for World-Wide Application'. 21 January 1965, *Inst. of Road Transport Engineers, Preprint*.
27. LOUCKES, T. N., and PORTER, C. L., 'A Summary of the Toronado Engineering Test Program'. Second Quarter 1966, *General Motors Engineering Journal*, vol. 13, no. 2, p. 29.

28. GASSNER, E., and SCHÜLTZ, W., 'Evaluating Vital Vehicle Components by Programme Fatigue Tests'. 1962, *FISITA*, I. Mech. E.
29. BEAUVAIS, R. Z., and SORENSON, G. R., 'Magnetic Tape and Servo-Hydraulics Applied to Truck Frame Testing'. 30th March–3rd April 1964, *SAE Paper*, Automobile Week.
30. KOWALEWSKI, J., 'On the Relationship between Component Life under Irregularly Fluctuating and Ordered Load Sequences—Part I.' September 1963. DVL Report No. 249. *MIRA Translation* No. 43/66.
31. PONTA, P. H., 'Process and Product Control'. 15th August 1964, *Automotive Industries*, p. 68.
32. MANOOGIAN, J. A., 'Reliability by Design—The Lincoln Continental Story'. 4th April 1962, *SAE Preprint* No. S331.
33. Discussion on 'Quality and Reliability of Cars'. February 1967, *Chartered Mechanical Engineer*, vol. 14, no. 2, pp. 84–88.
34. HODKIN, D., and NEVITT, P. J., 'The Significance of Customer Environment in Vehicle Testing'. 19–21 April 1966, *SAE Symposium, Environmental Engineering—Its Role in Society*, vol. 4, Paper 5.
35. WITTS, M. T., Contribution to Discussion on "Better Engineering with Simple Statistics'. October 1966, *Chartered Mechanical Engineer*, vol. 13, no. 9, p. 450.
36. NADER, R., *Unsafe at Any Speed*. 1965, Grossman, New York.
37. *Federal Register*, vol. 32, no. 23, Part II, 'Initial Federal Motor Vehicle Safety Standards'. 3rd February 1967, Washington, D.C.
38. HALLEUX, ALBERT, JR, 'Statistics Concerning Different Aspects of Technical Inspection of Motor Vehicles'. September 1963, Paper 7 Aspects Techniques de la Securite Routiere. Special Number on Technical Inspection. no. 15.
39. HORION, A., 'Belgian Statistics of Defects in 1962'. Paper 6 of Ref. 38.
40. Road Research Laboratory. *Research on Road Safety*. 1963, HMSO Chapter 12, 'Vehicle Inspection'.
41. Road Research Laboratory. 'Crash Injury Research'. Chapter 13 of Ref. 40.
42. HALLEUX, ALBERT, SR, 'Influence of Vehicle Condition on the Number and Importance of Accidents'. September 1963, *Journes International de l'Inspection Techniques des Vehicules Automobiles.*
43. LITTLE A. D., Inc. 'The State of the Art of Traffic Safety'. Part Two: Regulatory and Legal Factors, Chapter V 'Compulsory Vehicle Inspection' 1966.
44. HUELKE, D. F., and GIKAS, P. W., 'Causes of Deaths in Automobile Accidents'. April 1966, University of Michigan Medical School.
45. KIHLBERG, J. K., 'Driver and his Right Seat Passenger in Automobile Accidents'. 1966, Proceedings of the Ninth Stapp Car Crash Conference, Paper 21.
46. GIKAS, P. W., and HUELKE, D. F., 'Pathogenesis of Fatal Injuries to Rear Seat Occupants of Automobiles'. Paper 22 of Ref. 45.
47. LISTER, R. D., and NEILSON, I. D., *The Effectiveness of Safety Belts.* RRL Report No. 16, 1966.
48. GRIME, G., *Safety Cars* RRL Report No. 8, 1966.
49. STARKS, H. J. H., and MILLER, M. M., *Roadside Equipment and Accidents*, RRL Report No. 22, 1966.
50. NAHUM, A. M., SEVERY, D. M., and SIEGEL, A. W., 'Automobile Accidents Correlated with Collision Experiments: Head-On Collisons' Paper 18 of Ref. 45.
51. FIALA, E., and REIDELBACH, W., 'Methods and Facilities for Crash Tests'. July 1964, *ATZ*, vol. 66, no. 7.
52. FRANCHINI, E., 'Crash Testing Evaluation at Fiat'. January 1966, SAE Automotive Engineering Congress.

Mr R. J. Love *served an apprenticeship at HM Dockyard, Portsmouth, and attended the Royal Dockyard School. He obtained a Whitworth Scholarship and studied at King's College, London. He is now Group Research Head, Materials and Components, and Aerodynamics, at MIRA, having joined the Institution of Automobile Engineers Research Department (later to become MIRA) in 1943 as a research Engineer. His work has been associated with many aspects of the performance and durability of components, with particular emphasis on metal fatigue. He has written a number of papers on this subject.*

Dr. M. A. Macaulay *went to Glasgow University after service in the Army and graduated in Mechanical Engineering in 1952. He then spent three years with the North British Locomotive Co., working on the design and construction of diesel and electric locomotives. In 1955 he went to Bristol University as research assistant to Sir Alfred Pugeley and studied the effects of impact on simplified model railway coaches. In 1960 he joined the Motor Industry Research Association to start a new research group concerned mainly with predicting and measuring the loads and stresses in vehicles in service, and with vehicle crash testing.*

REFERENCES

53. HOHLIN, N. I., 'Studies of Three-Point Restraint Harness Systems in Full-Scale Barrier Crashes and Sled Runs'. 1966, Proceedings of 8th Stapp Car Crash Conference, p. 258.
54. STONEX, K. A., and SKEELS, P. C., 'A Summary of Crash Research Techniques Developed by the GM Proving Ground'. 4th Quarter 1963, *GM Engineering Journal.*
55. SAE Recommended Practice.
     J849 Rollover Tests.
     J850 Barrier Impact Tests.
56. KROELL, C. K., and PATRICK, L. M., 'A New Crash Simulater and Biomechanics Research Program', p. 185 of Ref. 53.
57. CICHOWSKI, W. G., 'A New Laboratory Device for Passenger Car Safety Studies'. March 1963, SAE National Automobile Meeting.
58. NEILSON, I. D., 'The Dynamics of Safety Belts in Motor Car Head-On Impacts'. September 1966, I. Mech. E. *Symposium on Ergonomics and Safety in Motor Car Design.*
59. McHENRY, R. R., 'CAL Computer Simulation of the Automobile Crash Victim in a Frontal Collision'. Cornell Aeronautical Laboratory Transportation Research Dept. Internal Report.
60. SEVERY, D. M., 'Human Simulations for Automotive Research'. 1965, SAE Special Publication No. SP-266.
61. PATRICK, L. M., 'Human Tolerance to Impact—Basis. for Safety Design'. 1965, SAE International Automotive Engineering Congress.
62. CLARK, C., and BLECHSCHMIDT, C., 'Human Transportation Fatalities and Protection Against Rear and Side Crash Loads by the Airstop Restraint'. Paper 3 of Ref. 45.
63. MILEY, R. C., 'The Synthesis of the Optimized Impact Environment'. Paper 25 of Ref. 45.
64. Republic Aviation Division: Fairchild Hiller Corp. 'Feasibility Study of New York State Safety Car Program'. State of New York Dept. of Motor Vehicles Final Report. August 1966.

Control Of Automobile Air Pollution

by K. C. Salooja, DSc, PhD

US legislation already prescribes limits for certain pollutants which will become more rigorous in the near future. Whether or not Europe follows suit, the developement of economical devices for reducing air pollution is of great concern to all engineers. The long-term solution appears to lie with redesigned carburettor and exhaust systems.

Table 1—Permitted Maximum Concentrations in Exhaust Gases

Pollutant	Typical average	Limits in US Standards	
		Current	Proposed for 1970
CO	(3·5% Vol.)	(1·5% Vol.)	23 g/vehicle mile (ca 1% by volume)
HC	900 ppm	275 ppm	2·2 g/vehicle mile (ca 180 ppm)

The important pollutants are carbon monoxide, unburned or partially burned hydrocarbons (HC) and oxides of nitrogen (NO_x).

While research continues to determine whether the highest concentrations of CO and NO_x encountered in areas of high traffic density are toxic or not, it is agreed that hydrocarbon concentrations present no significant health risk. However, under warm and sunny conditions, such as those in Los Angeles, hydrocarbons can present a health hazard indirectly by forming photochemical smog.

At present only the USA has enacted legislation limiting the emission of pollutants. The maxima permitted for exhaust emission during a specified test procedure[1] are compared with normal concentrations emitted in Table 1. Proposals for restricting NO_x have not yet been finalised.

Current standards are for American cars of engine displacements greater than 140 in³ (2295cc). For smaller cars, mostly European and Japanese, the standards are less severe since, in comparison with the larger cars, they produce less exhaust gas albeit with a higher concentration of pollutants. The current standards for cars with engine displacement of 50–100 in³ (820–1640 cc) are: CO, 2·3 % (volume) and HC 410 ppm, and for cars with engine displacement of 100–140 in³: CO, 2·0 % (volume) and HC 350 ppm. A similar allowance for engine size has been made in the proposed 1970 standards by specifying emission per vehicle mile.

Origin of pollutants

There are four sources of pollution from vehicles. The exhaust accounts for the major share with 65%, the crankcase for 20% and 9% and 6%, respectively, are due to carburettors and fuel tank.[2]

Of these four items engine exhaust poses the most difficult control problems. Emissions from the crankcase can be completely controlled by channelling the blowby gases through a valve into the inlet manifold of the engine and this is being done in most of the cars in current production.

The minor emissions from carburettors and fuel tank contain hydrocarbons only, and these can be controlled by absorbing the vapour on a bed of activated charcoal and later releasing them into the inlet manifold[3].

The pollutants in the exhaust are unavoidably formed during combustion, even in the most efficient engine; partly because, at the temperatures involved, the dissociation processes prevent the complete conversion of the mixture to CO_2 and H_2O; because at these temperatures oxides of nitrogen are formed; and also because the combustion process is quenched, when still incomplete, by the relatively cold cylinder surfaces. In practice, excessive emission also occurs because rich fuel-air mixtures are used, often poorly mixed at that.

Although attempts are being made to minimise these factors, over-rich mixtures are used deliberately for maximum power and to avoid rough running due to uneven distribution as between cylinders. In any case, carburettors are rarely capable of delivering exactly the required fuel-air mixture under all driving conditions. Poor mixing of fuel and air is difficult to avoid completely because gasolines have a rather wide boiling range and space limitations around the engine lead to inadequacies in the design of the induction system.

Changes in air-fuel ratio and ignition timing, and also in vehicle driving conditions, influence the emission of pollutants.

As shown in Fig 1, CO and HC emissions are substantially reduced while NO_x emission is increased by progressive reduction of fuel content from a rich condition to a lean ratio of about 17:1.[4] Beyond this, engine operation becomes

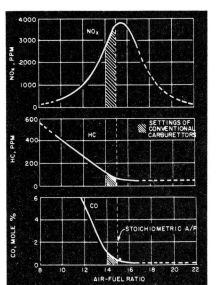

[Courtesy: SAE Journal]

Fig. 1 CO and HC emission is reduced while NO_x is increased by reducing the fuel content of the mixture. These data are based on a car speed of 60 mile/h; misfire regions are shown by broken lines

unsatisfactory; misfiring may occur and lead to an increase in the emission of HC and CO.

Changes in ignition timing affect HC and NO_x emissions have little influence on CO emission. Retarding, rather than advancing, the ignition reduces HC[5] as well as NO_x[6] emissions. Hydrocarbons are reduced because retarding ignition causes the exhaust gases to leave the cylinder at a higher temperature: combustion of hydrocarbons, which partly occurs in the exhaust manifold, now occurs more briskly. Oxides of nitrogen are reduced by virtue of the fact that retarding ignition lowers the peak cycle temperature.

Excessive retardation of ignition, however, can cause significant power loss.

As shown in Table 2, changes in driving conditions markedly influence pollutant emissions.[7] Not that any limitation can therefore be imposed on the driving modes because these must depend on circumstances; but this knowledge is useful for test purposes and for devising control mechanisms.

Engine design

Amongst engine design features, the fuel-air system has the greatest influence on the emission of pollutants. The other features which exercise some influence include combustion chamber surface/volume ratio, valve overlap, inlet-manifold air-pressure and coolant temperature.

The fuel-air delivery system is important because the fuel/air ratio has a dominant influence on pollutants and also because in current carburettors fuel metering and delivery and fuel-air mixing are often imperfect. Some improvement is possible by using fuel injection instead of carburation because injectors are capable of metering and delivering fuel more accurately and have the added advantage that charge stratification in the inlet manifold can be avoided by siting the injectors close to the cylinder inlet ports.

One severe drawback of carburation is that the large amount of fuel on the manifold walls precludes the possibility of effective fuel cut-off during deceleration, when the emission of hydrocarbons is exceptionally high. Again because of the manifold, carburettors are designed to deliver extra fuel during acceleration since the increase in inlet manifold pressure on opening the throttle causes some fuel to condense on the manifold walls and starve the engine.

Although these drawbacks can be completely avoided with fuel injection, carburettors are used far more extensively (and will probably continue to be) because they are far cheaper and simpler. However, attempts are being made to produce cheaper fuel injection equipment and also to reduce the shortcomings of carburettors.

Equal gas flow to all cylinders can be achieved by designing the manifold branches to be of equal lengths and ensuring that none involves an abrupt change in the direction of flow. However, space around the engine limits this approach.

One means of improving the homogeneity of the fuel-air mixture is heating the inlet manifold. This, however, entails certain disadvantages. It reduces the charge density and hence the power output, it increases NO_x emission and is also liable to increase knocking.

Variations in other engine features, such as combustion chamber surface/volume ratio,[8] valve-overlap,[5] inlet-manifold air-pressure[6] and coolant temperature,[6] are of very limited use. None of these influences the emission of CO. Those which reduce HC increase NO_x. A decrease in chamber surface/volume ratio reduces HC but involves a loss in engine efficiency due to the decreased compression ratio.

Table 2—How Driving Affects Pollution

Pollutant	Driving Mode			
	Idling	Acceleration	Cruising	Deceleration
HC (ppm)	300–1000	300–800	250–550	3000–12000
CO % (Vol.)	4–9	1–8	1–7	3–4
NO_x (ppm)	10–50	1000–4000	1000–3000	5–50

Table 3—Effects of Catalytic Converters on Emissions

	Arvin UOP	Grace-Norris	Walker Cyanamid
CO, % (Vol.)	0·62	1·19	0·73
HC, ppm	186	272	224

Fuel factors

Variation in the hydrocarbon composition of fuels within the normal boiling range has been found to influence the emission of HC only slightly[9] and has virtually no effect on NO_x.[7]

It has been thought for several years that while fuel composition does not have much influence on the total HC emission, it may affect its nature. Since the smog-forming propensities of individual hydrocarbons are known to differ markedly,[9] it was felt that fuel composition could influence smog formation.

However, recent investigations indicate[10] that variations in fuel composition, unlike those in engine operating conditions and design, have little influence on exhaust emission.

Lead-alkyls do not affect the emission of CO and NO_x. Whether or not they influence HC emission is still being hotly argued. Pahnke and Squire[11] have claimed, on the basis of their studies in 122 cars, each driven over some 20 000 miles, that lead-alkyls cause only a slight increase in HC. On the other hand, according to Gagliardi,[12] who carried out his studies in six cars, also over 20 000 miles, lead-alkyls increase HC (by ca. 45% after 2 400 miles) and this effect is significant at the 99·9% confidence level.

Even if eventually lead-alkyls are cleared of all blame for augmenting HC and/or being pollutants themselves, a major threat to their use could arise from a different direction, namely catalytic systems designed to destroy pollutants. The available systems (none of which is in commercial use as yet) are known to be prone to poisoning by the lead in the exhaust gas.

If the efforts to develop poison-resistant catalyst continue to fail, the use of lead-alkyls may have to be abandoned.

Minimisation of pollutants

In discussing the reduction of pollutants one has to consider first the targets—which pollutants have to be reduced, and by how much—and then the suitable means.

The current and proposed 1970 US legal limits are shown in Table 1. The expected targets for the late seventies, according to the Morse Committee,[13] are: CO, 0·5% V, HC, 50 ppm, and NO_x, 250 ppm. In the UK and on the Continent there is as yet no specific legislation for the control of air pollution by cars. The unpleasant Los Angeles-type smog, which triggered off the legislation in the USA, is

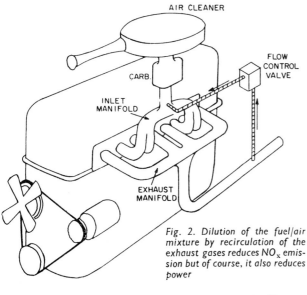

Fig. 2. Dilution of the fuel/air mixture by recirculation of the exhaust gases reduces NO$_x$ emission but of course, it also reduces power

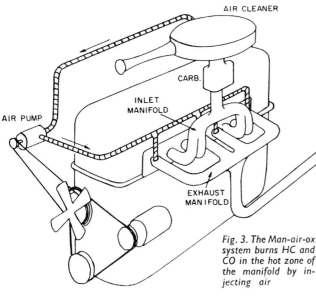

Fig. 3. The Man-air-ox system burns HC and CO in the hot zone of the manifold by injecting air

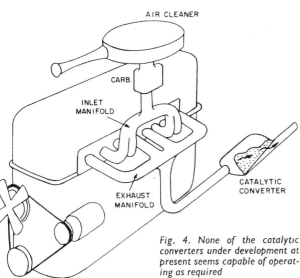

Fig. 4. None of the catalytic converters under development at present seems capable of operating as required

unusual in Europe, nor it is likely to occur, at least in North and West Europe, because the requisite weather conditions do not normally prevail.

Nevertheless car manufacturers are carrying out intensive investigations to minimise pollutants with the objective of meeting the US targets, since cars exported to the States have to conform to these. Resulting developments, particularly inexpensive ones, will also reduce air pollution in Europe and thus further reduce the need for legislation.

However, if legislation is considered necessary in future, it is likely to apply at first to CO only, the pollutant formed in the highest concentration.

The formation of CO is mainly affected by changes in the fuel/air ratio; a leaner mixture reduces the emission. No other factor is effective unless it indirectly influences fuel/air ratio. Since there is a practical limit to reducing the fuel content we have to rely on the methods available for reducing the emission of CO, once it is formed.

The formation of HC can be reduced either by using leaner mixtures or by reducing the quenching process. The latter could be done by modifying the engine design so that surface/volume ratio is reduced; or by increasing the temperature and pressure in the combustion chamber, eg, by increasing the coolant temperature and/or the manifold pressure.

Modifications to the engine design, like leaner mixtures, are limited by adverse results. Increased coolant temperature reduces HC emission but increases NO$_x$ emission, it also increases the knocking and/or pre-ignition of the fuel. Increased manifold pressure reduces HC emission slightly but this too increases the NO$_x$.

The formation of NO$_x$ can be reduced mainly by lowering the maximum temperature reached during the combustion cycle and/or reducing the availability of oxygen. The factors which affect the temperature include compression ratio, spark timing, and exhaust dilution and those which effect the availability of oxygen are mixture ratio, exhaust dilution and even the *Man-air-ox* system. Changes in the hydrocarbon composition of commercial gasolines do not appreciably influence NO$_x$.[7]

Exhaust-gas dilution, which reduces both the combustion cycle temperature and the availability of oxygen, has been shown to reduce NO$_x$ particularly effectively; an arrangement for this is shown in Fig. 2. Recirculation of about 15% exhaust-gas has reduced NO$_x$ emission by 85%[4] but reduces power by 16%. Recirculation of larger amounts would make the mixture difficult to ignite.

Destroying pollutants

There are three principal methods of reducing CO as well as hydrocarbon pollutants. The *Man-air-ox* system (Fig. 3) oxidises unburned HC and CO in the hot zone of the exhaust manifold by the injection of air close to the exhaust valves. This system, used in many current models in the USA is capable of markedly reducing CO and HC. A greater reduction, particularly in HC, is possible by maintaining the exhaust manifold at a higher temperature (by insulation) and/or by increasing the residence time of the gases in the manifold (by enlarging its size). Brownson and Stebar[14] found a 35% reduction of HC and a 50% reduction of CO due to air injection. With additional insulation and enlarged manifold a further reduction of 28% HC occurred, but no change in the CO content.

The hot gas leaving the exhaust manifold can also be passed through a bed of catalytist to promote the combustion of CO and HC as shown in Fig. 4. The effects of three catalytic devices[15] are shown in Table 3.[15] None of these systems seems as yet capable of satisfactory operation for 50 000 miles as required by the California Air Resources Board.

Direct-flame after-burners can also be fitted in the path of the exhaust gas. One such device is reported to reduce CO to 1·17% (by volume) and HC to 221 ppm.[15] Another device is claimed to reduce CO to 0.13% (by volume) and HC to 50 ppm.[16] Some practical problems with after-burners are that, for effective combustion, exhaust composition must not vary too much; and that special materials are needed to withstand the high temperatures involved. To maintain proper fuel/air composition some, 7% of additional fuel might be required.[16] Another disadvantage is that the high temperatures involved in combustion lead to the formation of some additional NO_x.

A series of modifications developed by Chrysler reduce emissions below the levels of the current US limits. These are designed partly to reduce the formation of HC and CO, and partly to promote their destruction. Optimum fuel/air mixture and spark-timing are used for all engine operating conditions by means of modifications to the carburettor and the distributor; a special distributor control valve is also incorporated.[17] Results show a reduction in CO and HC by 54% and 76% respectively.

The most successful method yet devised for decomposing NO_x is that of catalytic reduction in a reducing atmosphere. eg, one containing CO.

$$NO + CO \xrightarrow{\text{catalyst}} \tfrac{1}{2}N_2 + CO_2$$

$$NO_2 + 2CO \xrightarrow{\text{catalyst}} \tfrac{1}{2}N_2 + 2CO_2$$

Several catalysts, including copper oxide and copper oxide-silica[18] oxide-alumina,[19] and barium promoted copper chromite,[20] have been shown to be effective.

In general, increasing temperature promotes these reactions; on the other hand, the presence of oxygen beyond 2%, although it promotes the oxidation of CO, greatly reduces the decomposition of NO_x as shown in Fig. 5.

Meeting specific targets

Thus, when attempts are made to reduce CO and HC by weakening the mixture, nitrogen oxides are increased; and when, to reduce HC and NO_x, ignition timing is retarded and/or exhaust dilution is increased, engine power output and fuel economy suffer. There is no single approach which would reduce all the pollutants without, at the same time, adversely affecting the engine performance. A number of methods have to be used to achieve the desired reduction in pollutants with minimum penalties.

The most effective means of reducing the formation of CO is by operating the engine on the leanest practicable mixture; and the most economical method for reducing the emission of CO is by the *Man-air-ox* system. Further reduction, if required, can be brought about by means of a catalytic converter or an after-burner. Whatever means are adopted to reduce CO emission they would also reduce HC emission.

The most satisfactory methods for reducing both together, as applied in the current production cars in the USA are to minimise over-rich running, particularly in deceleration; retard ignition timing; and use manifold oxidation.

Catalytic converters and after-burners are not in use yet. They could reduce pollution further than is practised at present.

The reduction of NO_x at the same time as the other pollutants presents the biggest challenge at present since means of reducing it generally increase CO and HC and also reduce engine efficiency. Moreover, any NO_x formed cannot be destroyed as easily as CO and HC. Therefore, the best overall approach involves operating the engine to produce minimum NO. One course is to dilute the charge with exhaust gas but this necessarily involves a marked decrease in power output which is less likely to be acceptable for small (generally European) cars than for large cars. An alternative course, which involves the use of a rich fuel mixture, is preferable since decomposition of NO_x in the exhaust gas requires catalysis in a reducing atmosphere, such as is conveniently provided by the other pollutants.

The catalytic decomposition of NO_x also brings about a reduction in the CO content. If a further reduction in CO were desired, the exhaust gas would have to be treated in a catalytic converter or an after-burner.

Future developments

The most useful developments will be those which improve the performance of the fuel-air induction system, the exhaust manifold, catalytic converters and after-burners.

Development of the first is most needed at present because a direct and highly effective method for reducing CO is to operate engines on or near–stoichiometric fuel/air mixture.

This, however, is impracticable with the current carburettors because these are incapable of smooth operation on such lean mixtures. Some improvements in carburettors have been reported recently. [21, 22]

It is rather ironic that these desirable developments will become far less relevant when, as in the US in future, NO_x too will have to be reduced drastically. This is because, in all probability, the engines will then again have to be operated on rich fuel-air mixtures.

Considerable work, leading to the development of the *Man-air-ox* system, has shown the marked reduction in CO and HC due to combustion in the exhaust manifold. As indicated previously, there is considerable scope for further reduction of pollutants by improved design of that part.

The most urgently needed catalytic converter is one which can effectively decompose NO_x without requiring a strongly reducing atmosphere. The development of such a device is bound to receive much attention.

Catalytic converters for curbing CO and HC emissions are likely to be needed only when a reduction in these emittants

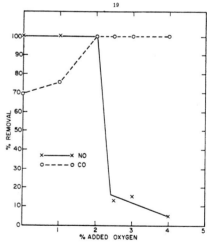

[Courtesy: Franklin Institute]

Fig. 5. Oxygen, while eliminating carbon monoxide, prevents the catalytic removal of nitrogen oxides so that whatever is done to reduce one pollutant tends to favour the other

Dr K. C. Salooja, *obtained his PhD at Imperial College, London, in 1954. His earlier studies leading to BSc in Chemistry, MSc in Chemical Engineering and BA in Economics were carried out at the University of Punjab in India. From 1955 to 1965 he worked at Shell's Thornton Research Centre and since then he has been at Esso's Research Centre at Abingdon. His research work has dealt largely with the mechanism of combustion, performance of fuels in different applications, and more recently with combustion-related atmospheric pollution problems. He has published numerous papers, and was awarded a DSc in 1968.*

is desired below the level attainable with other methods.

The ability of after-burners to reduce CO and HC to extremely small amounts, coupled with their advantage over catalytic converters in not being susceptible to poisoning by impurities in the exhaust gas, foreshadows considerable development for meeting the long-term targets.

In the light of the work reviewed here, the outlook for greatly reducing the current level of air pollution due to petrol-engined vehicles is very favourable. The crucial issue to be decided in the future will be what concentrations of pollutants in the atmosphere of busy places are injurious to health.

The greater the extent of the reduction desired, the higher will be the cost involved; but the technical problems involved for achieving a very low level of pollution are not insurmountable.

REFERENCES

1. 30th March 1966, *US Federal Register*, vol. 31, no. 61–2; 4th January 1968, *ibid*, vol. 33, no. 2–2.
2. MacGregor, J. R. 1966, *JSAE*, vol. 74(1), p. 84.
3. Clarke, P. J., Gerrard, J. E., Skarstrom, C. W., Vardi, J., and Wade, D. T. January 1967, *SAE Paper 670127*.
4. Kopa, R. D. January 1966, *SAE Paper 660114*.
5. Hagen, D. F., and Holiday, G. W. March 1962, *SAE Paper 486C*.
6. Huls, T. A., Myers, P. S., and Uyehara. June 1966, *SAE Paper 660405*.
7. Walker, J. A. September 1965, *British Technical Council of the Motor and Petroleum Industries Report BTC/65/F1*.
8. Scheffler, C. E. January 1966, *SAE Paper 660111*.
9. McReynolds, L. A., Alquist, H. E., and Wimer, D. B. May 1965, *SAE Paper 650525*.
10. Papa, L. J. May 1967, *SAE Paper 670494*.
11. Pahnke, A. J., and Squire, E. C. 1966, *Oil and Gas Journal*, vol. 64(50), p. 106.
12. Gagliardi, J. C. January 1967, *SAE Paper 670128*.
13. *The Automobile and Air Pollution: A Program for Progress, Part 1*. October 1967, Report to the Commerce Technical Advisory Board of the Department of Commerce (USA) by committee headed by R. S. Morse.
14. Brownson, D. A., and Stebar, R. F. May 1965, *SAE Paper 650526*.
15. Sweeney, M. P. 1965, *J. Air Poll. Control*, vol. 15, p. 13.
16. Schnabel, J. W., Yingst, J. E., Heinen, C. M., and Fagley, W. S. March 1962, *SAE Paper 486G*.
17. Beckman, E. W., Fagley, W. S., and Sarto, J. D. January 1966, *SAE Paper 660107*.
18. Sourirajan, S., and Blumenthal, J. L. 1961, *Air and Water Poll. Int. J.*, vol. 5, p. 24.
19. Baker, R. A., and Doerr, R. C. 1964, *J. Air Poll. Control Assn*, vol. 14, p 409.
20. Roth, J. F., and Doerr, R. C. 1961, *Ind. Eng. Chem.* vol. 53, p. 293.
21. Bartholomew, E. January 1966, *SAE Paper 660109*.
22. Lawrence, G., Buttivant, J., and O'Neill, C. G. May 1967, *SAE Paper 670484*.

The most prominent recent trend in marine propulsion has been the continuous encroachment of diesels on steam turbines—a trend which may yet be stopped in the larger vessels of the future. But the diesel itself is becoming ever more competitive; it burns low-grade fuel, lends itself to direct drive and may permit complete automation of the engine room.

Marine Engineering

by R. Cook, BSc, FIMechE

The last decade has been one of intensive development in marine engineering. It has seen the virtual disappearance of the naturally-aspirated direct-drive diesel engine as propelling machinery for new ships; the emergence of the supercharged, direct-drive diesel as the predominant prime mover; and the decline of the steam turbine for all but the very largest powers. It has seen the installation of gas turbine, free-piston gas-generator/gas turbine and nuclear-propelled machinery in merchant ships, the introduction in ocean-going vessels of a high degree of automatic and remote control of machinery and the widespread adoption of alternating current for ships' electrical supply.

The reasons behind all this activity are many. The growth of economic nationalism has intensified competition for the world's sea carrying trade (which is still growing) and the shipowner has been compelled to examine ever more closely all possible methods of increasing efficiency. The super-tanker, the large bulk-carrier and faster cargo-liners, all requiring more powerful propelling machinery, have appeared during this period. The spectacular development of the two-stroke diesel engine has enabled marine-engine builders to take advantage of this demand and, in doing so, largely to oust the steam turbine.

Again, full employment and rising standards of living have made it more difficult for the shipowners to recruit and to retain the skilled manpower necessary to operate the often very complex machinery of a modern ship and this has impelled them to examine the possibilities of crew reduction by the automatic or remote control of machinery.

The importance of the market for marine machinery can be gauged from the fact that over the last four years the total shaft horsepower of diesel and steam machinery installed annually in the world has averaged between 5 and 6 million which, very roughly, represents well over £100 m. This takes no account of the large amount of auxiliary machinery installed for lighting, pumping, cargo handling and other purposes, nor does it include naval machinery.

Diesel engines

Reference has already been made to the predominance of the slow-speed, direct-drive diesel engine, almost invariably of turbocharged, two-stroke, single-acting crosshead design.[1] A variety of scavenging systems are employed in conjunction with exhaust-gas driven turbochargers operating on either the pulse or constant-pressure systems. They are designed to operate on heavy fuels with viscosity of up to 3500

[Courtesy: W. Doxford and Sons (Engineers) Ltd]

Fig. 1. In the last decade the power per cylinder of marine diesels has, in some cases, increased by 150 per cent. The most recent large marine diesel engine of entirely British design: this turbocharged opposed-piston engine is designed for an output of 20 280 bhp at 115 rev/min; length 59ft 2in; height 33ft 6in; and its weight is 580 tons. The width over the test bed is 13ft. These 'J' type diesels are probably the lightest and most compact of their kind

Redwood No. 1 seconds at 100°F, which costs 40 per cent or so less than conventional marine diesel fuel. The power per single cylinder, offered by the nine principal manufacturers, ranges from 2100 to 2670 bhp, representing an increase over the last decade, in some cases, of 150 per cent. This has been accompanied by reductions in length and specific weight of about 40 per cent.

One of the most recent and largest engines of this type has a bore of 930 mm, a stroke of 1700 mm and develops 30 000 bhp in twelve cylinders at 112 rev/min at a bmep of 123 lb/in^2, on a fuel consumption of 0·34 lb/shp. It has an overall length of 75·2 ft, and weighs 1180 tons.

The only large marine diesel engines of entirely British design are Doxford's, the most recent of which is shown on the test bed in Fig 1. In 10-cylinder form this 'J' type, turbocharged, two-stroke opposed-piston engine is designed to develop 22 000 shp in service at 115 rev/min at a bmep of 125 lb/in^2; it has a length of 64·9 ft, an overall height of 33·5 ft and a weight of 640 tons and is probably the lightest and most compact engine of its type.[3]

As might be expected, the recent world-wide design and development activity in connection with direct-drive diesels has been accompanied by very considerable research effort. This has been mainly concerned with the problems raised by the up-rating of engines and by the use of cheap residual fuels. Among the former category are those concerned with scavenging and turbocharging, thermal loading of cylinder components, fuel injection systems and crankshaft stresses, while the use of residual fuels has led to considerable research in connection with cylinder wear and cylinder lubrication.

With the pulse system of exhaust-turbocharging now employed in the majority of direct-drive diesels, the pipework is kept as short, smooth and free from bends as possible in order to preserve and utilise the kinetic energy of the pressure pulses which arise from rapid port opening. The design of the exhaust system and the matching of turbine compresser and engine for pulse operation tended in the past to be an *ad hoc* affair, based upon accumulation of data from previous installations. In recent years much research has been carried out, both in this country and abroad, with a view to evolving more rational design procedures.[4] In this country the British Shipbuilding Research Association, in co-operation with British turbocharger manufacturers, has been carrying out a programme of work on turbocharger design since it was realised that future development of the two-stroke engine was dependent upon an increase in turbocharger efficiency and that the optimum design for constant pressure operation was not necessarily optimum under pulse flow.[5]

In the constant-pressure system, the exhaust gases are led to a receiver which feeds the turbine and the volume of pipework is large enough to maintain sensibly constant pressures at the turbine entry. This system offers better turbine efficiency, is less sensitive to exhaust pipe geometry and permits the fitting of gratings to protect the turbine from broken piston rings; it is, however, inferior to the pulse system in that it is not capable of supplying sufficient air at low loads and supplementary means have therefore to be adopted, usually in the form of scavenge pumps.

Engine makers have carried out a great deal of work with variants of these two basic systems.[6] It has been shown recently that, by combining both systems in a two-stage arrangement, a very marked advantage can be obtained,

particularly with certain numbers of cylinders; but of course at the expense of some complication.[7]

Reducing wear

So far as cylinder wear is concerned, probably the greatest advance has been the introduction by the oil companies of alkaline-additive lubricating oils which reduce the rate of wear when heavy residual fuels are burnt to approximately that formerly obtained with distillates. Work by Ricardo and Co. for the BSRA has shown that, with heavy fuels, wear is strongly influenced by sulphur content and that raising the temperature of the outer section of the liner has a greater effect in reducing wear than raising the temperature of the inner section where maximum wear occurs.

Further tests with a high-alkalinity lubricant showed a reduction of wear comparable to that in marine service but the rate of wear still responded to a change of jacket temperature. This suggests that wear remaining when using a high-alkalinity lubricant is still of a corrosive nature and could be further reduced by the employment of higher jacket temperatures or a higher level of alkalinity in the lubricating oil.

Other work by Ricardo and Co. for BSRA has led to the very important conclusion that centrifuging the fuel has little, if any, effect on the wear of liners, pistons and rings. Some interesting conclusions on wear have also been reached, following the fitting of a radio-active cast-iron piston ring to the 760 mm bore engine of a Swedish motor ship.[8]

Cases of excessive liner and ring wear are still being experienced and a number of makers have been experimenting with timing and distribution of the lubricant feed.

The rapid increase in engine ratings of recent years and the adoption of very large bores has focused the designers' attention on the thermal loading of cylinder components which was the subject of a recent symposium held by the Institution.[9] More recently, results have been given of a very exhaustive investigation on a 760 mm bore opposed-piston engine from which the effects of further up-rating can be predicted, thus assisting in the design of liner metal thickness and coolant-side conditions so as to minimise thermal loadings.[10] Recent unpublished work for the BSRA has shown the very great importance of certain thermal barriers to heat transfer and the marked influence which certain kinds of corrosion inhibitors exert in this connection.

Fuel injection equipment has been the subject of a good deal of experimenting, partly with a view to mechanical simplification and partly because of deficiencies of the conventional cam-operated jerk-pump. An interesting development in this connection is that of the BICERA hydraulically operated pump-injector system which promises an improved performance at the higher rates of injection which are now being called for.[11]

There has also been renewed interest in the gas-operated system which utilises the cylinder compression pressure and dispenses with the need for camshaft drive.[6]

A problem which has caused considerable concern in connection with both slow-speed and medium-speed diesel machinery is that of crankcase explosions. These have been shown to be due to the generation of condensed lubricating oil mist on an overheated surface. Two methods are in use for dealing with this situation. In one, explosion doors are provided which vent through flame arresters; while, in the other, reliance is placed upon the early detection and warning of oil-mist formation. The respective merits of these two systems for application to marine diesel engines have been the subject of recent research.[12]

Medium-speed diesels

In spite of considerable development during the past decade, the share of the marine market gained by medium-speed oil engines has been extremely modest and largely confined to the propulsion of very small vessels and larger vessels, such as car-ferries, in which the height of the engine was decisively limited. Strenuous efforts are now being made by the medium-speed engine manufacturers to increase their share of the market. The predominant type at present is the exhaust-turbocharged, trunk-piston, 4-stroke engine which is built in both in-line and V-form. The prototype of one of the latest British engines has a bore of 15 in, a stroke of 18 in and is rated at 203 lb/in^2 for continuous duty at 500 rev/min. In 16-cylinder V-form it gives 6660 bhp for continuous duty on heavy fuel. Fig. 2 shows this engine in 12-cylinder form.

With medium-speed machinery, gear transmission is essential, since propeller speeds of the order of 500 rev/min are precluded by cavitation and efficiency considerations. Multiple-engine arrangements provide flexible manoeuvring, low-power operation and maintenance and additional reliability, as well as reducing engine-room length. The single-engine installation, however, has the attraction of simplicity and low cost.

Attention now is being directed to the development of more powerful direct-reversing engines, capable of developing 8000 to 10 000 shp. A new two-stroke design, has a bore and stroke of $14\frac{1}{4}$ in and $18\frac{1}{2}$ in respectively, is built in both inline and 50° V-form and is initially rated at 150 lb/in^2 bmep.

Another new two-stroke design was recently introduced by Fairbanks-Morse. This is a direct-reversing, opposed-piston engine, operating at 400 rev/min and developing no less than 1000 bhp/cylinder. In this through-scavenge design the upper exhaust piston has a diameter and stroke equal to half that of the lower piston, the latter accounting for some 87·5 per cent of the power developed. The bore and end stroke of the lower piston are 20 in and $21\frac{1}{2}$ in, respectively. The engine operates at 130 lb/in^2 bmep and is to be made as a 6 or 9-cylinder in-line engine and as a 12-cylinder V-engine. For heavy fuel applications stationary rings, held in place by a cylinder sleeve, are used to prevent lubricating oil from the cylinder walls passing down into the crank chamber.

As with slow-speed diesels, research has been largely concerned with up-rating and the use of residual fuels. There is, however, one problem which has so far not appeared in the slow-speed engine; pitting of the water-side of the liner. This is due to vibration of the liner and the question of the most suitable counter-measures has been the subject of a great deal of research, both in this country and in America.[13]

With both slow and medium-speed diesels a stage is being rapidly approached when peak cylinder-pressures will set a limit to the degree of supercharging which can be employed, and hence to power output. Peak pressures can be reduced by reducing compression ratios, but there is a limit to this if starting and idling are not to suffer. One solution which has been investigated by BICERA is the employment of a variable compression piston; promising results have been reported.[14, 15]

Steam turbines

Progress with steam turbine machinery during the past decade, while not perhaps as spectacular as that with diesels, has nevertheless been considerable. Turbines, boilers, gearing and auxiliaries have all been the subject of much research and development. Some idea of progress may be gained from the fact that the 20 000 shp PAMETRADA-designed machinery for

Fig. 2. Medium-speed engines have only made very modest inroads into the marine market; this prototype Mirrlees National KV Major should help to remedy the situation

the 71 000 ton dw tanker *British Mariner*, which entered service in 1963, has been credited with an all-purpose fuel rate of 0·492 lb/shp/h with steam at 890°F and 560 lb/in^{2}g.[16] This may be compared with the design fuel consumption of 0·573 lb/shp/h of the 16 000 shp machinery of 42 000 ton dw tankers completed in 1959.[17] Some of this improvement must, of course, be ascribed to the increase in power.

Towards the end of 1963, PAMETRADA completed trials of their prototype I machinery which develops 22 000 shp and, with inlet steam at 1035°F and 800 lb/in^{2}g, has the most advanced steam conditions of any marine steam turbine yet developed. Designed for astern operation at 950°F it was run for two hours at full power at 1035°F without adverse effect.[16] Part of this plant is shown on the front cover of this journal.

More recently, PAMETRADA has, in co-operation with a panel of owners, turbine and boiler designers and other interested parties, prepared a proposal for a 30 000 shp installation with an all-purpose fuel consumption rate marginally less than 0·40 lb/shp/h. This design incorporates a three-cylinder reversing turbine, taking steam from a single main boiler at 1000°F, 1000 lb/in^{2}g with reheat to 1100°F. It incorporates a refined but not over-complex fuel system and engine-driven alternator and main feed-pump.

Such progress has not been obtained without considerable research on blade testing, partial admission losses, water catchment, thermal distortion of turbine cylinders, back-to-back tests of gearing, development of journal and thrust bearings and vibration of blades and discs.[18, 19]

Reducing corrosion in boilers

Boiler defects are the greatest single item prejudicing the reliability of turbine installations. Bonded deposits on the gas side of superheaters, corrosion of the supports and of soot blower elements, deterioration of brickwork, and corrosion and blockage of air heaters have necessitated extensive maintenance in many turbine-driven vessels. These troubles have been the subject of much research, both in this country and in America; based on its results, boiler manufacturers have made significant changes in their designs.[20]

Bonded deposits arise from the vanadium and sodium content of some of the cheaper fuels now being used. They have been greatly reduced by the use of long retractable, rack-type soot-blowers and by water-washing of the fuel to reduce sodium content. This inhibits the low-melting point sodium compounds which are the basis for deposits.

Water-washing of the fuel is, however, a delicate technique, considered by many to be unsuitable for use on board ship. More recently good results have been obtained simply by centrifuging the fuel.[21] The increasing use of steam-atomising burners has also been found to reduce deposits.

In many boilers now coming into use the spacing between superheater tubes has been greatly increased and furnace volumes are such as to ensure complete combustion of the fuel before the gases enter the tube banks. With these improvements there seems to be no reason why satisfactory operation should not be obtained on these inferior fuels with steam temperatures up to 1000°F. Higher temperatures appear unlikely because of the risk of corrosion when fuels containing vanadium and sodium are used. Considerable advances have been made in reducing brickwork maintenance, partly by increased application of water-walls and partly by use of plastic refractories.[22]

With some of the poorer fuels, blockage of air-heaters by corrosion products and deposits is serious when the metal temperature is below 220 F Improvement has been obtained in horizontal-tube heaters by the use of ferruled tubes and insulated tube plates at the inlet end.[20] Both these expedients are designed to overcome the cooling effect created by the entering air. More recently, encouraging results have been obtained with vitreous-enamelled tubes and it seems probable that this will permit air-heaters to be designed for higher efficiency than at present, without risk of corrosion.

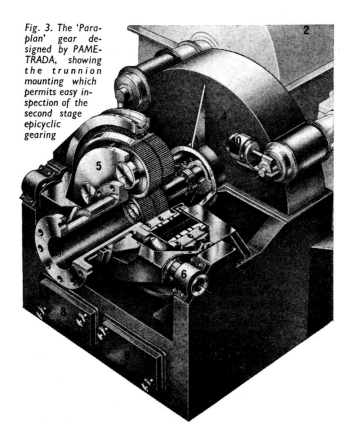

Fig. 3. The 'Paraplan' gear designed by PAMETRADA, showing the trunnion mounting which permits easy inspection of the second stage epicyclic gearing

Reducing the size of gearing

Of all the elements of a marine turbine, the reduction gear is the most expensive and bulkiest. It is not surprising, therefore, that with the direct-drive diesel offering such strong competition to the turbine, two famous organisations should propose to take advantage of recent progress to reduce the size and cost of this item. In the Stal-Laval design, a number of which are now in service, epicyclic gears are used for the first speed reduction and conventional parallel shaft gears for the second stage.[23] PAMETRADA, on the other hand, in their *Paraplan* design, have chosen conventional parallel shaft gear for the first stage and epicyclic for the second.[16]

These new types offer a reduction in cost estimated at approximately 20 per cent and a reduction in weight of approximately 40 per cent. They also permit a substantial saving in engine room space.

Fig. 3 shows the *Paraplan* gear; the trunnion mounting permits easy inspection of the second stage of reduction. A valuable feature is the ease with which propeller speeds of, say, 85 to 90 rev/min can be accommodated. This permits advantage to be taken of the 6 to 10 per cent increase in propeller efficiency at this speed, compared with conventional speeds around 110 to 115 rev/min.

Very significant advances in design and load capacity of parallel shaft gearing were reported at the Institution's International Conference on Gearing.[24,25]

Gas turbines not economical

The gas turbine received a tremendous boost immediately after the war as the result of its application to aircraft and its champions were not long in advocating its claims for marine propulsion. In naval service, where its lightness and ability to take load quickly are of supreme importance, it has achieved very considerable success, both for the propulsion of major warships and for high-speed craft.[26] Its application to merchant ships, however, has been attended by less happy results.

The first application to merchant-ship propulsion was the British experimental set installed in the Shell tanker *Auris* in 1952 which replaced one of four medium-speed oil engines connected to the propeller through electrical transmission. Some six years later the propelling machinery of this vessel was entirely replaced by a single gas turbine of 5500 bhp, driving the propeller through double-reduction helical gears, a hydraulic coupling, astern converter and friction clutch.[27]

The turbine was a two-shaft, open-cycle unit with inter-cooling between HP and LP compressors and with a vertical tubular heat exchanger preheating the high-pressure air before entry to the single, vertically mounted combustion chamber.

A built-in additive system was incorporated in the fuel supply to combat ash deposition. The technical performance of this installation, in spite of some teething troubles, was encouraging but the owners decided that it was uneconomic and, after six months' operation, the vessel was laid-up.

Towards the end of 1956 the liberty ship *John Sergeant* was re-engined by the US Maritime Administration with 6000 shp gas turbines. An open-cycle regenerative turbine of the two-shaft type was fitted, driving a variable-pitch propeller through conventional double-reduction gearing. This vessel was operated commercially over a period of three years during which time the machinery gave excellent service and achieved an all-purpose fuel rate of 0·52 lb/shp/h on low-vanadium Bunker C and distillate fuels (the latter being

160

required for starting and manoeuvring).[28] The rather elaborate system for fuel-washing and addition of magnesium-sulphate was found to require careful and continuous attention. The compressor and turbines were regularly water-washed after each long sea-passage.

Inspection at the end of the three-year period showed that no repairs or replacements were necessary in the high-temperature parts of the plant. Despite this favourable experience, the vessel was subsequently laid-up and it can only be concluded that, as in the case of the *Auris*, the machinery was considered to be uneconomic.

Over 40 vessels have been fitted in recent years with free-piston gas turbine propelling machinery, three of these in British yards for British owners, during 1959–60. A full account of the performance of two of these installations has been published.[29] The first comprised two GS-34 gasifiers of SIGMA design, delivering gas to a 2000 hp, reversible turbine, connected to the propeller through reduction gearing. The second comprised five GS-34 gasifiers, delivering gas to a non-reversing turbine of 4000 shp, connected through reduction gearing to a variable-pitch propeller. On trials these installations achieved fuel consumptions of 0·43 lb/shph and at a later stage, using a modified turbine, a 0·45 lb/shph was achieved in the second vessel when using 1000 Redwood No 1 second fuel.

Among the many troubles experienced with these three installations, probably the most serious were pressure-pulsations in the air-flow, giving rise to stability problems with multi-gasifier operation and to unpleasant low-frequency noise; and mismatching of turbine and gasifier, necessitating turbine replacement. To these were added a variety of troubles with combustion, lubrication and cylinder components, the successful solution of which would seem to call for further research and development.

Some three years earlier, in 1957, the US Maritime Administration equipped the *William Patterson* with six free-piston gasifiers which were feeding two turbines geared to a single shaft. After four voyages and various alterations, a consumption of 0·496 lb/shph was claimed.[30] Troubles similar to those referred to seem to have been experienced with this installation. In the last year or so little has been heard of this type of machinery and there is no doubt that shipowners in general do not favour it.

Nuclear propulsion costs too much

The world's first nuclear merchant ship, the *Savannah*, went to sea under nuclear power for the first time on 23rd March 1962, a day which may come to be considered a milestone in maritime history. This vessel was fitted with 22 000 shp turbine machinery, taking steam at 480 lb/in² and 467°F from a pressurised-water reactor. Its subsequent performance has amply demonstrated the practicability of this new form of ship propulsion.[31] A large number of technical papers have since appeared, covering every aspect of the design, construction, testing, manning and operation of this remarkable vessel.[32]

At no time, however, has it been suggested that the installation is economical. The capital cost of a nuclear reactor is many times that of the boilers which it replaces and the associated fuel cost is not low enough to counterbalance this.[33]

The problems involved in the application of nuclear power to marine propulsion have been studied intensively, both in this country and America, during the last 20 years. The

[Courtesy: Clarke Chapman and Co. Ltd]

Fig. 4. Gantry-type cranes are now being fitted to vessels to speed up the handling of dry bulk cargo

British shipbuilding industry has been to the forefront of those exploring these possibilities. Through the British Shipbuilding Research Association it has had a design-study team attached to the UK Atomic Energy Research Group since 1955. These studies have suggested that there does not appear to be any marine reactor in sight which promises to prove economical. In view of this, prospects of a British nuclear merchant ship appear remote unless the Government is prepared to bear the cost.

The case for nuclear propulsion is strongest where high powers and high utilisation factors are involved; this suggests that, if the present trend towards larger vessels and higher speeds continues, the day may not be too distant when nuclear propulsion will come into its own for restricted classes of merchant ship.

Propellers and transmissions

During the last few years there has been a rapid increase in the use of controllable-pitch propellers.[34] The great advantage they offer is improved manoeuvrability, especially with direct bridge control.

Their main application has, therefore, been in services such as harbour tugs, ferries and coastal passenger and cargo vessels. Application to larger vessels has been much slower, largely because of the capital cost (some three times that of the fixed propeller); the more expensive survey fees; a slight reduction in propeller efficiency due to increased size of boss; and an understandable reluctance on the part of shipowners to fit machinery which is inaccessible. Nevertheless, units of 23 000 shp are now coming into service and their use on larger vessels seems likely to increase with the trend towards bridge-control of machinery.

An alternative to the controllable-pitch propeller is a hydraulic transmission such as the one developed by PAMETRADA to meet the needs of gas turbine drive. This was fitted in prototype form on the gas turbine tanker *Auris*.[27] Like the controllable-pitch propeller, it could equally well be used with steam-turbine drive and thus eliminate the need for astern turbines.

The PAMETRADA transmission consists of an ahead-coupling and an astern convertor, the primary wheels of both units being driven by the engine and the secondary wheels attached to the primary pinion of a reduction gearbox, so that the propeller can be driven ahead or astern by admitting oil to the appropriate element without reversing the main machinery. A friction clutch is incorporated in the ahead-coupling, so that coupling losses can be avoided when the ship reaches the open sea.

Consideration is now being given to the application of this form of transmission to tugboats.

Electrical installations

A great deal of thought has been given in the last decade to the improvement of ship's electrical systems. The potential benefits of ac supply, particularly in regard to first cost and maintenance, have long been recognised, but it is only in recent years that the development of winches for operation on ac has reached a stage where the fitting of the system became practical. Since then ac supply has been used increasingly in all classes of vessel. Typically, a passenger vessel installation is one with generation and primary distribution at 440 V, three-phase, 60 c/s and a secondary distribution through transformers at 230 V for lighting and the smaller power consuming devices.[35] Further transformation is carried out from 230 V to 55 V for those items of portable equipment in which the hazard of shock is greatest. The non-earthed system is now generally considered to be the most desirable for ships' installation.

Consideration of the fault currents obtainable points to a total generating capacity of 4·25 MVA as the limit for the conventional 440 V, 3-phase system. In the 100 000 ton tanker *British Admiral*, which is due to enter service this year, a 3·3 KV system has been adopted, this being the first time that such a high voltage has been installed in any ship except one with turbo-electric drive.

Another recent development has been to drive the alternator from the propeller shaft. This can be so attractive economically as to justify the provision of quite expensive means for maintaining frequency within narrow limits whilst shaft speed varies. A number of such systems are now being fitted.

Since a dry-cargo ship spends about half of its time loading and unloading in port, improved methods of handling are obviously of major importance in any attempt to improve the

economics of operation. In the last few years notable progress has been made in this field.[36] The traditional derrick methods have been speeded up by improvement of the well-known Union Purchase system and by the introduction of swinging derricks.

The Union Purchase system owes its popularity to the fact that it requires only two winches and is capable of fast working. The swinging derrick systems employ three winches, are slower, but can place the load more accurately. There have also been developments in the design of heavy derricks, some of which can now handle loads of up to 180 tons.

Deck jib or gantry cranes are finding increasing use on cargo ships. Their main advantages over derricks are their faster working and accuracy in positioning the load; disadvantages are their higher initial cost and the fact that they cannot be adapted for lifting heavy loads. Savings of 30 per cent of time in port have been claimed for ships fitted with cranes. Gantry type cranes of various designs are also being fitted to dry-cargo vessels for the rapid handling of bulk cargoes. Fig. 4 shows a recent installation suitable for all types of dry bulk cargo.

In the larger tankers coming into service, 18 in and 20 in valves, used to maintain high loading and discharge rates, are mechanically operated. It is now becoming the practice in such ships to locate the controls for the cargo valves, together with the necessary remote-reading level-indicators, in a central control-room. Consideration is being given to the next logical step, namely, automatic control of the pumping system.

Automatic control

One of the most striking developments in ship propulsion over the last five or six years has been the adoption in many ships of a high degree of automatic control.[37] The driving force behind this has been the shipowners' desire to economise in engine room staff, particularly in view of the difficulties which they are experiencing in maintaining the number and quality of engineer officers and ratings required for the complicated machinery of the modern ship. Besides leading to a reduction in numbers, these measures have an important influence in helping to recruit and maintain staff, since the automation of such operations as cleaning of oil purifiers and soot-blowers renders working conditions more pleasant. It is usual, in these highly automatic installations, to group all the important controls, monitoring and recording devices in a separate control-room, situated inside the engine room because this further improves working conditions, both as regards noise and ventilation. A recent example of such a control room is shown in Fig. 5.

Experience with both motorships and steamships has shown that reductions in engine room staff up to about 50 per cent can be made by such means and shipowners are now engaged in assessing the economic aspects. This is not as simple as might at first be imagined, since many items on the credit side will require assessment over a long period if reliable conclusions are to be drawn.

British shipowners do not, at present, favour direct control of machinery from the bridge for normal merchant ships on long voyages although this may well prove to be the long-term outcome. Before it can be achieved, however, it will be necessary to determine whether direct access from control room to machinery-space is essential. At present few superintendents would be willing to forego this facility.

As an example of the degree to which automatic control is now being applied, some ships now building in this country are to be fitted with auxiliary generators which will be con-

Fig. 5. The high degree of automatic control in modern ships has resulted in improved working conditions, as in this air-conditioned and sound-insulated control room of the 'Southampton Castle'

trolled entirely automatically: instruments will sense the need to start the standby alternator sets which will then be brought into operation and put on the line automatically.

Vibration and smoke

Vibration in some form exists in all ships and whilst in most it is not unpleasant or troublesome, in some it causes acute personal discomfort and gives rise to local structural damage; the increasingly high power now being applied to a single screw and the increased dimensions of ships have rendered these problems more acute. A very great deal of research has been carried out to provide the designer with data which will enable him to ensure that the level of vibration does not exceed acceptable limits.

The two most important types of vibration encountered are main hull vibration and vibration of the propulsion system. The former may be excited either by the engine or by the propeller. In recent years great progress has been made in the prediction of the natural frequencies of the hull and it is now possible to predict with some confidence whether a given degree of machinery unbalance will give rise to an unacceptable degree of vibration. In the case of propeller excitation, the frequencies involved are higher and the problem more complex. Propeller excitation arises firstly from fluctuating torque and thrust due to varying flow conditions, these pulsations being transmitted to the hull through the shaft; and, secondly, the passage of the propeller blades causes variations in the hydrodynamic pressures on the stern plating which may set up vibration. The influence of the number of propeller blades, clearance between propeller and the ship's stern and rudder, and the shape of the stern have all been the subject of recent research.[38]

Torsional vibration of shafting systems is a phenomena which has been long known and understood; it is, therefore, only infrequently encountered in modern marine installations. In recent years, however, increasing attention has had to be paid to axial vibration of shafting systems. This form of vibration is particularly liable to occur in large ships and is just as likely to cause damage to the ship's structure as to the elements of the shafting system itself. Thanks to systematic investigation into the stiffness of thrust blocks, their seatings and associated hull structure, the designer can now take steps in the design stage to obviate the possibility of this trouble.[39]

Another problem which has been solved by the application of research is that of the descent of smoke from ship's funnels on to decks and accommodation. This problem has been accentuated in recent years by the fashion for short funnels.

Model tests in the wind tunnel have for some years been used with success to overcome this trouble but the results of recent research have led to the formulation of simple design rules which make it possible to dispense with wind tunnel tests in all but exceptional cases.[40]

To the naval architect and marine engineer accurate measurement of the power delivered to the propeller is vital. Due to difficulties associated with the marine environment, this quantity cannot be measured with a guaranteed accuracy greater than ±2 per cent although, in favourable circumstances, the error may be much less. This degree of accuracy can now be obtained more consistently with modern forms of torsionmeter. One noteworthy development has been the use of an ultrasonic technique to determine the modulus of rigidity of shafting.[41] This measurement can be made on board the ship and, when used in conjunction with optical

Mr R. Cook *took an honours degree in Marine and Mechanical Engineering at Armstrong College (now King's College) Newcastle in 1924. After drawing office and sea service experience he joined the Fuel Research Division of the DSIR in 1928. During the next 12 years he was engaged on a variety of researches into the uses of coal and its by-products in internal combustion engines and boilers. In 1940 he joined the Ministry of Supply where he became responsible for the direction of all research and development in connection with unexploded bomb disposal. In 1945 he was appointed Chief Marine Engineer to the British Ship Research Association. Mr Cook, who retired in 1966, is a past Chairman of Council, now a Vice-President and Hon Treasurer of the Institute of Marine Engineers and is the author of numerous papers.*

calibration of the torsionmeter, gives results at least as accurate as the older method of static calibration of meter and shaft, prior to installation, which was both expensive and inconvenient.

Future trends

Shipowners, in their quest for more efficient operation, carry out more research and development work on their vessels than is generally realised. A recent examination showed that British shipowners were spending very large sums on such projects.

Typical of the subjects receiving attention in this manner are corrosion control,[42] explosion hazards,[43] gas freeing[44] and tank cleaning processes in tankers, the automatic control of machinery[37] and the development of waste heat recovery systems.[45]

As far as propulsion machinery is concerned, the future, at least for the next decade or two, would seem to lie with diesel and steam turbine machinery. None of the alternative types of propulsion tried or suggested in recent years appears to have prospects of ultimate success on a large scale, save perhaps the gas turbine. The fundamental difficulty there is that, in the simple open-cycle form, the thermal efficiency is low so that fuel consumption is greater than that of the steam turbine and approaching double that of the diesel engine.

Moreover, the use of heavy fuels gives rise to difficulties, even with present gas temperatures of 1100°F. Development of a liquid-cooled rotor, capable of operating at, say, 2200°F on heavy fuel would go a long way to make this type of machinery competitive but the problems involved would necessitate considerable research.

The slow-speed, direct-drive diesel seems likely to maintain for some time its present supremacy for the propulsion of ocean-going vessels with powers of up to about 20 000 shp. For higher powers the steam turbine may well stage a comeback; much will depend upon the performance of the new diesel installations of 20 000 to 30 000 shp which are now coming into service.

At the lower end of the power range, below, say, 10 000 shp, the next few years will see a determined effort by the manufacturers of medium-speed diesels to extend their share of the marine market. The degree of success they achieve will depend in no small measure upon the ability of this type of engine to burn heavy fuels and upon reduction of gearing and transmission costs.

The future is likely to see increasing use made in ocean-going vessels of the automatic control of machinery from noise-insulated and air-conditioned control rooms situated in, or adjacent to, the machinery spaces; ultimately we may see the machinery control room combined with the navigating bridge but this pre-supposes considerable advances in the reliability of machinery and instrumentation.

REFERENCES

1. COOK, R. 'A Survey of Marine Diesel Propulsion Machinery.' 1964, *Proc. Instn mech. Engrs, Lond.*, Vol. 178, Part 3.
2. SORENSON, E. and SCHMIDT, F. 'Recent Development of the MAN Marine Diesel Engine.' 1964, *Trans. Inst. Mar. Engrs*, 76, p. 197.
3. JACKSON, P. 'Developments in the British Large Marine Diesel Engine During the Past Decade.' 1963, *Proc. Instn mech. Engrs, Lond.*, Vol. 177, p. 897.
4. HORLOCK, J. H. and BENSON, R. S. 'The Matching of Two-Stroke Engines and Turbo Superchargers.' *Congres International des Machines a Combustion (CIMAC)*, Paper A.10.
5. CRAIG, H. R. M. and JANOTA, M. S. 'The Potential of Turbo-chargers as Applied to Highly Rated Two-Stroke and Four-Stroke Engines.' 1965, *CIMAC*, Paper B.14.
6. SCOBEL, H. 'Investigations Concerning Turbocharging and Injection Methods with a View to Boosting the Power Output of Large Two-Stroke Engines.' 1965, *CIMAC*, Paper B.15.
7. GYSSLER, G. 'Problems Associated with the Turbocharging of Large Two-Stroke Diesel Engines.' 1965, *CIMAC*, Paper B.16.
8. PINOTTE, P. L., JONES, D. R., and SVENSSON, S. A. 'Radio-Active Piston Ring Wear Studies in a 6000 Horsepower Marine Diesel Engine.' 1962, *CIMAC*, Paper A.15.
9. Symposium on 'Thermal Loading of Diesel Engines.' Birmingham, 1964, *Instn mech. Engrs*.
10. FRENCH, C. C. J., HARTLES, R. E., and MONAGHAN, M. L. 'Thermal Loading of Highly Rated, Two-Cycle Marine Diesel Engines.' 1965, *CIMAC*, Paper A.8.
11. GRAY, K. H. 'Hydraulic Fuel Injector for Diesel Engines.' 1962, *Shipping World*, 146, p. 353.
12. BROWN, K. C., COOK, R., JAMES, G. J., and PALMER, K. N. 'Crank-case Explosions in Marine Oil Engines: The Efficacy of Gauze and Crimped-ribbon Flame Traps.' 1962, *Trans. Inst. Mar. Engrs*, 74, p. 261.
13. COLLINS, H. H. 'Pitting of Diesel Cylinder Liners.' July 1961,' *Diesel Engine Users Assoc.*
14. THORNEYCROFT, C. H. 'The BICERA Variable Ratio Piston.' 1960, *Mechanical World*, 140, p. 507.
15. MANSFIELD, W. P., TRYHORN, D. W., and THORNEYCROFT, C. H. 'Development of the Turbocharged Diesel Engine to High Mean Effective Pressures Without High Mechanical or Thermal Loading.' 1965, *CIMAC*, Paper A.6.
16. COATS, R. 'Pametrada Standard Turbines, Present Position and Future Outlook.' Paper read at Jt. Meeting of Inst. Mar. Engrs., Schiff. Gesellschaft e V., and Inst. Engrs and Shipbldrs, Scotland, May 1965.
17. PLATT, E. H. W., and STRACHAN, G. 'Post War Developments and Future Trends of Steam Turbine Tanker Machinery.' 1962, *Trans. Inst. Mar. Engrs.*, 74, p. 411.
18. BROWN, T. W. F. 'Application of Research to the Design of Marine Steam Turbines.' 1957, *Trans. Inst. Mar. Engrs*, 69, p. 65.
19. BROWN, T. W. F. 'Shore Trials of Marine Steam Turbine Machinery.' 1963, *Trans. Inst. Mar. Engrs*, 75, p. 261.
20. HUTCHINGS, E. G. 'The Design and Development of Two-Drum Marine Boilers.' 1963, *Trans. Inst. Mar. Engrs*, 75, p. 37.
21. TIPLER, W. 'Some Results Obtained from a Fuel–Oil Water-Washing Plant in Studies of the Fouling of Marine Superheaters.' 1964, *Trans. Inst. Mar. Engrs*, 76, p. 37.
22. McCLIMONT, W., and RICHARDSON, H. M. 'The Properties and Performance of Quarl-Block Materials. 1962, *Trans. Inst. Mar. Engrs*, 74, p. 1. ,
23. JUNG, I., and LUNDSTRÖM, T. 'Recent Developments in Propulsion Gears for Steam Turbines.' 1963, *Soc. NAME*, Sept. (see also *Mar. Engr*, 86, (1963), p. 585).
24. NEWMAN, A. D. 'Load Carrying Tests of Admiralty Gearing.' 1958, *Int. Conf. on Gearing I. Mech. E.*
25. PAGE, H. H. 'Advances in Loading of Main Propulsion Gears.' 1958, *Int. Conf. on Gearing I. Mech. E.*
26. TREWBY, G. F. A. 'Marine Gas Turbines in the Royal Navy.' 1962, *Trans. Inst. Mar. Engrs*, 74, p. 347.
27. DUGGAN, R. M., and HOWELL, A. T. O. 'The Trials and Operation of the Gas Turbine Ship Auris.' 1962, *Trans. Inst. Mar. Engrs*, 74, p. 89.
28. McMULLEN, J. J. 'The Trials and Maiden Voyage of the Gas Turbine Ship *John Sergeant*.' 1957, *Soc. NAME*, 65, p. 25.
29. HERBERT, C. W., and MILNE, G. F. 'The Application of Free-piston Gas Turbine Machinery to Marine Propulsion.' 1963, *Trans. Inst. Mar. Engrs.*, 75, p. 401.
30. TANGERINE, C. C. and SPECHT, D. H. 'Operating Experience with Gas Turbine Ships of the US Maritime Administration.' 1959, *Motor Ship*, 40, p. 18.
31. MACMILLAN, H. J., MACMILLAN, D. C., and OTHERS. *NS Savannah* Operating Experience.' 1963, *Soc. NAME*, Nov.
32. *Proc. of the Symposium on Nuclear Ship Propulsion*. Taormina Italy, 1960. Published by Int. Atomic Energy Agency, Vienna, Austria.
33. HILDREW, B. 'Problems of Merchant Ship Nuclear Propulsion.' 1962, *Trans. Inst. Mar. Engrs*, 74, p. 501.
34. KLAASSEN, H., and ARNOLDUS, W. 'Actuating Forces in Controllable Pitch Propellers.' 1964, *Trans. Inst. Mar. Engrs*, 76, p. 173.
35. SAVAGE, A. N. 'Details and Operating Data of Recent A.C. Installa-tions.' 1961, *Trans. Inst. Mar. Engrs*, 73, p. 141.
36. BROWN, A. S., and BENISTON, P. T. 'Cargo Handling Equipment for Dry Cargo Vessels.' 1964, *Trans. Inst. Mar. Engrs*, 76, p. 1.
37. MUNTON, R., McNAUGHT, J. and MACKENZIE, J. N. 'Progress in Automation.' 1963, *Trans. Inst. Mar. Engrs*, 75, p. 297.
38. JOHNSON, A. J., and McCLIMONT, W. 'Machinery Induced Vibra-tions.' 1963, *Trans. Inst. Mar. Engrs*, 75, p. 121.
39. COUCHMAN, A. A. J. 'Axial Shaft Vibration in Large Turbine-powered Merchant Ships.' 1965, *Trans. Inst. Mar. Engrs*, 77, p. 53.
40. THIRD, A. D., and OWER, E. 'Funnel Design and the Smoke Plume.' 1961, *Trans. Inst. Mar. Engrs*, 73, p. 245.
41. MORRISON, J. 'The Measurement of Shaft Torque and Thrust.' Paper to be read before Inst. Mar. Engrs, November 1965.
42. LOGAN, A. 'Corrosion Control in Tankers.' 1958, *Trans. Inst. Mar. Engrs*. 70, p. 153.
43. HOLDSWORTH, M. P., VAN DER MINNE, J. L., and VELLENGA, S. J. 'Electrostatic Charging During the White Oil Loading of Tankers.' 1962, *Trans. Inst. Mar. Engrs*, 74, p. 29.
44. LOGAN, A. 'Experience with Gas Control in Crude Oil Carriers.' 1964, *Trans. Inst. Mar. Engrs*, 76, p. 25.
45. NORRIS, A. 'Developments in Waste Heat Systems for Motor Tankers.' 1964, *Trans. Inst. Mar. Engrs*, 76, p. 397.

Mechanical Aircraft Design

by D. Howe, MS, MIMechE

The borderline between aeronautics and astronautics is rapidly disappearing with a consequent multiplication of research and development problems. A separate article in this series will deal with power units so that, in his assessment of recent progress, the author concentrates on structural and ancillary features of modern air- and space-craft design. He selects a number of items which are of particular interest to mechanical engineers outside the aircraft and missile industries.

Fig. 1. The rocket-powered North American X-15 research aircraft has flown at about five times the speed of sound for short periods; such conditions are verging on space flights

Aeronautics and rocketry have developed so rapidly during the past decade that even the title for this review is outdated. The emphasis on missiles and space vehicles has necessitated a wider interpretation of the term 'aircraft' which, here, embraces missiles operating at least partly in outer space.

The author attempts to highlight some of the more important developments in this young but active branch of engineering which are necessarily closely connected with current research work. The results of aircraft research are often applicable to engineering in general and such aspects as are of particular interest to the mechanical engineer will be emphasized here.

Vertical take-off

Aircraft and guided missiles have tended to be developed in parallel; the advent of the large ballistic missile has paved the way for manned space flight. In the sphere of manned aircraft, the main design aim is now, as always, the extension of flight speed range. Consequently, there is considerable emphasis on vertical, or short, take-off and landing at one extreme and supersonic flight at the other. Vertical take-off can be achieved by one of two basic techniques. Either a direct vertical lift or thrust can be provided; or, alternatively, a horizontal thrust or slipstream may be deflected downwards.[1]

A number of prototype aircraft using conventional, as distinct from rotating, wings have recently been flown but as yet little support has been forthcoming for practical applications of these aircraft. The major difficulties appear to be the weight penalties associated with either the special lifting engines or deflection devices, and the difficulty of poor stability and control. Although certain small vertical take-off aircraft have flown in the U.S.A. without automatic devices it would appear that a closed loop gyro stabilization system will be essential for normal operation.

Supersonic flight has become commonplace for military aircraft and there is no difficulty with regard to short flights

164

at up to about 1500 mile/h. or just over twice the speed of sound. However, the temperature effects which result from compression of the air and skin friction introduce severe problems at greater speeds or for longer periods. There is the difficulty of maintaining a satisfactory environment for the crew, passengers and equipment; material properties may deteriorate and creep occur; to which must be added the ever-present problem of fatigue. The *Concorde* supersonic airliner has received much publicity of late although, in fact, such an aircraft may be somewhat premature. Nevertheless, a considerable research programme has shown that it is a feasible, if not perhaps initially an economic, proposition.[2] The *Concorde*[3] is intended to cruise at 1450 mile/h.

Manned aircraft flights at speeds about five times that of sound* have been achieved for short periods, the altitude being of the order of 30 miles.[4] The North American X-15 aircraft[5] can be seen in Fig. 1, both in flight and on the ground. Its operating conditions can be considered as being on the verge of space flight.

Guided missiles

Guided missile design has received much attention during the past ten years, and the situation has now been reached when the first generation of operational units has been perfected. Current research is mainly devoted to refinements, particularly the improvement of operational characteristics. Thus, the earlier anti-aircraft weapons are giving way to more mobile, reliable and potent developments.

The earlier liquid fuel ballistic missiles have severe operational limitations and are being replaced by smaller, more versatile solid fuel weapons designed for launching from mobile platforms. The relative roles of manned aircraft and guided missiles in military operations are becoming more clearly defined and it is apparent that to a great extent the two are complementary rather than mutually exclusive.

The great majority of the achievements in space research were attained by the adaptation of ballistic missiles as booster rockets. The problems associated with space flight and, in particular, re-entry into the atmosphere by capsules, such as shown in Fig. 2, have resulted in some interesting structural and mechanical developments.

The environment in which aeronautical and astronautical vehicles operate has such a profound effect upon their design that it is necessary to summarize its more important features.

Temperature and pressure

The maximum surface temperature on a body in high-speed flight is proportional to the square of the velocity and occurs at those points where the air is brought to rest. The actual temperature experienced by the structure in supersonic flight is a complicated function of the aircraft's shape, surface finish and internal arrangement. When the duration of flight is sufficient to enable steady conditions to be reached, the greater parts of the external skin are subjected to temperatures of the order of those shown in Fig. 3. This figure also indicates the regions of current interest.

Re-entry of a space vehicle into the earth's atmosphere

Fig. 2. A Mercury manned space capsule, largely constructed of titanium, is launched by an Atlas rocket vehicle at Cape Canaveral

◄

Fig. 4. Thermal conditions during the re-entry of space vehicles

——— Re-entry from 18 000 M.P.H. (Orbit).
— — Re-entry from 25 000 M.P.H. (Escape).

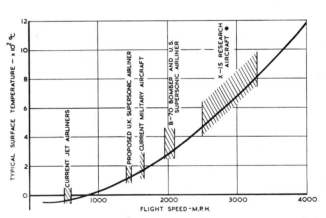

Fig. 3 How typical surface temperatures vary with airspeed: the shaded sections are of current interest

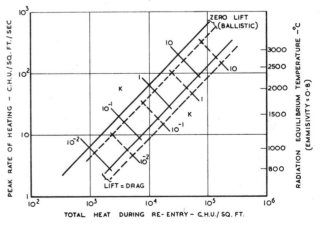

results in much higher temperatures than those so far experienced in normal flight. The form of re-entry path is such that these are transient rather than steady, and are thus a function of somewhat different parameters, although velocity is still dominant. Fig. 4 shows the temperature, maximum heating rate and total heat absorbed as functions of the main variables.[6] In the figure,

$$K = \frac{\text{Mass of Vehicle} \times \text{Dynamic Pressure}}{\text{Drag} \times \text{minimum radius of curvature of surface}}$$

The nature of the re-entry path is decided very largely by the mass-to-drag ratio of the vehicle and the ratio of the aerodynamic lift to drag. Thus a non-lifting or ballistic re-entry path is short and although the peak heating is high, the total heat absorbed is low. The use of lift increases the length of path and the re-entry time so that peak heating is reduced but the total heat is increased. For a given flight path, the temperature is inversely proportional to the local radius of surface curvature. Whilst the ballistic type of re-entry is suitable for operations from orbit velocity of 18 000 m.p.h., the very high decelerations experienced preclude its use at higher speed, such as a return velocity of 25 000 m.p.h., when some degree of lift is essential.

The U.S. manned *Mercury* capsule,[7] shown in Fig. 2, is of the ballistic type, whilst the *Dyna Soar*[8] project is intended to investigate lifting re-entry problems. The local temperature on the surface of a vehicle re-entering from escape velocity[9] may be as high as 10 500°C.

Current developments envisage civil aircraft flying at altitudes of the order of 12 to 14 miles and the required pressure differential in the cabin is in the region of 11 to 12 lb/in². The major problem associated with this is the need to guarantee the integrity of the cabin and its pressurization system in all circumstances.

In outer space a 'hard', or effectively absolute, vacuum exists. However, since only specially trained personnel are involved in space operations the pressure differential required in the cabin is less than that quoted above and 10 lb/in² is a likely maximum. The low pressures, of the order of 10^{-10} mm of mercury, are liable to affect the properties of various materials.

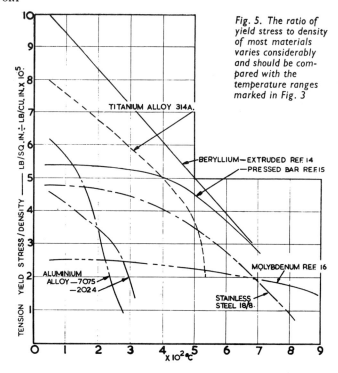

Fig. 5. The ratio of yield stress to density of most materials varies considerably and should be compared with the temperature ranges marked in Fig. 3

Radiation and meteoroids

The upper layers of the atmosphere effectively shield the earth's surface and the lower atmosphere from the various forms of radiation which exist in space. Any flight above about 12 miles in altitude will encounter these radiation effects to a greater or lesser extent, with a consequent effect upon human beings and properties of materials. Unfortunately this radiation environment is variable, being especially dependent upon unpredictable disturbances on the surface of the sun. There are, however, some clearly defined regions of lower intensity which are approximately opposite the polar regions of the earth.

A further beneficial effect of the presence of the atmosphere is that almost all of the particles of matter which collide with

Fig. 6. The prime structural material for this Bristol type 188 supersonic research aircraft is 64-ton 18/8 chromium-nickel steel. It is designed for research into surface heating problems

the planet are burned up before reaching the lower altitudes. At high altitudes, the presence of these particles constitutes a definite hazard. Although large meteors are rare, smaller fragments, or meteoroids, exist in relatively large numbers. Travelling at 30 000 to 120 000 mile/h., these meteoroids are liable to penetrate a cabin or exposed equipment with possibly disasterous results. The occurrence of meteoroids is also variable and exact data on the frequency of existence of particles of various sizes are difficult to accumulate.

Materials

Increased flight speeds have introduced higher loads, temperature effects and more severe fatigue conditions for the designer to contend with and he has been forced to consider new materials and new methods of fabrication.

Light alloy construction has been standard for the great majority of aircraft structures for many years and it is considered to be acceptable for aircraft cruising at speeds[2] up to about 1400 mile/h. The temperatures resulting from flight at higher speeds cause a noticeable deterioration in the properties of light alloys as can be seen by comparison of Figs. 3 and 5. If long periods of operation are desired, there is the additional problem of creep and, in connection with supersonic airliner design, the properties of light alloys in 30 000 to 50 000 hour creep conditions are being investigated.

Fig. 7. Light alloy structural parts are often machined from solid to avoid joints, as in the Blackburn naval strike aircraft

Both creep and fatigue can only be tolerated by designing so that actual working stresses are well below the ultimate. When speeds greater then about 1500 mile/h. are to be maintained for long periods, an alternative structural material is required. Titanium-alloy is a possibility and has found some uses both in the U.S.A. and on this side of the Atlantic. For example the *Mercury* capsule[7] is fabricated largely from titanium and experimental undercarriage leg forgings have been produced in this country.[10] The high cost of titanium has tended to restrict its use as a prime structural material and, generally speaking, high-grade steel has been preferred. It has been estimated, for example, that titanium-alloy sheet costs between 40 and 45 times as much as the equivalent 120-ton/in² high-tensile steel sheet,[11] although the trend is to reduce the difference.

Consequently it is generally anticipated that the structures of high-speed aircraft are more likely to be based on the use of steel than titanium. The Bristol Type 188, research aircraft,[12] shown in Fig. 6, is an example of this, the prime structural material being 64-ton/in², 18/8 chrome nickel stainless steel. The majority of large liquid propellant ballistic missiles are also constructed of welded steel, the skin thickness often being as low as 0·02 in. on ten feet diameter.[13]

Beryllium has also been suggested as a possible structural material.[14, 15] As shown in Fig. 5, it has outstanding properties in what may be termed the medium temperature range of 400°C to 700°C. The exact advantages of using beryllium are difficult to assess since its capabilities as a material are not yet fully established but it certainly promises well. Since it has a density which is only 65% of that of light alloy and it also possesses a high thermal capacity, one obvious application is to heat sinks. The heat shield on the *Mercury* capsule[7] consists of a 6·2 ft-diameter beryllium forging weighing 350 lb. There are, of course, severe difficulties associated with the use of this material, particularly its high cost, poor availability and toxic nature.

The very high temperatures encountered by space vehicles during the re-entry phase demand the use of different structural techniques and materials such as molybdenum. Some consideration has been given to the use of this metal in a sandwich construction[16] at temperatures up to 1700°C. One of the major problems to be overcome is its very low resistance to oxidation at high temperatures, but the development of a suitable coating seems to be possible. Other problems include the difficulty of fabrication and jointing.

Even molybdenum is not capable of withstanding the extreme temperatures which occur locally during re-entry. In these conditions a double-walled structure is necessary, the outer shell being of an insulating material which can absorb heating by an ablating, or burning, process. *Teflon*[9] is a possibility for this outer shell but phenolic nylon is more promising since it has the advantage that the char formed by burning is a good insulator and radiator. The use of a special form of graphite[17] has also been suggested and a good deal of research has been carried out.

In order to assess the properties of materials and to evaluate heat transfer rates at the high velocities and temperatures involved it has been necessary to develop special environmental facilities. For instance, velocities up to about 30 000 m.p.h. can be obtained by firing small models from a helium gas gun into an expansion tunnel.[9]

A vehicle in space can experience considerable temperature variation according to whether its surface is in sunlight or shadow. For example the *Grumman* astronomical satellite[18] is expected to have skin temperature variations over the range of −100 C to +42°C. Many materials have unfavourable mechanical properties at very low temperatures, especially when subjected to notched loading conditions, and in these circumstances glass fibre laminate appears to be very suitable.[9] The low pressure of outer space introduces the possibility of material loss due to vapourization, protective coatings and lubricants being especially prone to this.[19] Evaporation may also change the composition of alloys and adversely alter their properties.

Radiation effects upon materials will probably be most severe for transparent ones except in the regions of high intensity nuclear radiation where metals may also suffer.

Fabrication techniques

In recent years there has been a tendency to accept relatively complex fabrication techniques, with the implied extra expense, in order to produce light and efficient structures. Thus it has become the practice to reduce the thickness of sheet materials, where the loading permits, either by an overall taper or by local thinning. Very frequently this local removal of material has been achieved by a process of etching with suitable chemicals.

Etching has been used successfully on a number of light alloy constructions[20] and a similar technique may be suitable for stainless steels. Alternately, large portions may be machined from forged billets or slabs, thus eliminating to a great extent the needs for joints.[21] An example of this integral machining technique is shown in Fig. 7. It has been used on a wide variety of materials, particularly high-grade light alloys and high-tensile steels in the 100 to 120 ton/in² category.

Some examples of the latter are to be found in the structure of the Bristol Type 188 aircraft.[12] A feature of this design is the use of 'puddle' welding, which is a process whereby high tensile steel sheets down to 0·012 inches in thickness can be joined locally. The weld is made from one side only, thus maintaining a good external finish, and as little as 0·5 inch internal clearance can be accepted. One of the problems encountered in the production of the type 188 was the difficulty of obtaining large, close tolerance, sheets of the 64 ton/in² stainless steel for the skins and built-up components.

Another application of welded high-tensile steel sheet is its use for the motor cases of solid propellant rocket engines.[11] These cases are pressure vessels and often form part of a missile body structure as well. A recent development in this sphere is the use of welded, spirally-wound high-tensile steel strip,[22] thereby eliminating the need for large sheets and longitudinal welds.

Structural problems

One of the special problems in the design of large rocket vehicles has been the necessity to reduce the zero fuel weight to a minimum. This has been achieved very largely by reduction of structure weight which, for current liquid rocket vehicles, is around 2% of the total as against 14% for the German V2 rockets.[23] To achieve such outstanding results, the internal pressure in the fuel tanks is required for fuel system operation, and is used to stabilize the walls under compression and the bending loads imposed on the whole vehicle.

The structural design of space vehicles has to allow for the possibility of meteoroid penetration as well as provide insulation to minimize internal temperature variation. The high velocity of impact of meteoroids makes the problem a difficult one to assess. It has been discovered that impact behaviour at high velocities is different to that more commonly experienced,[9] and research is under way to extend the knowledge on the subject.[24] Experiments in which particles are fired from a light gas gun at models fired from a modified high velocity anti-aircraft gun are expected to provide relative velocities of the order of 25 000 mile/h. This approximates to the lower end of the meteoroid impact velocity range.

One possible method of protecting space vehicle cabins from penetration by meteoroids is the use of 'bumpers', which are simply a series of thin protective sheets, external to the main structure.[25] In practice, the section of a space cabin structure will probably be as shown in Fig. 8. The meteoroid bumpers are located in an airspace between an outer insulating shield and the main load-carrying structure. A further air-space and layer of insulation is provided internally.

A serious difficulty encountered in space flight is that, whilst large, lightly loaded structures are often required, these are most unsuitable for launching. Because of this there is considerable interest in inflatable structures. One notable achievement was the U.S. *Echo* satellite which was a large balloon of 100 ft diameter.[9] The total weight of 120 lb was packed into a small volume for launching and inflation occurred in space.

A development of this is a 135 ft diameter balloon which uses a special sandwich type skin, of 0·0002 in. thick aluminium on a 0·00035 in. thick *Mylar* core which is 20 times as stiff as the skin of *Echo*.

There have also been design studies for inflatable re-entry vehicles which have the great advantage of giving a very low mass to drag ratio.[26] Low heating rates and temperatures are natural consequences of this, and estimates show that the structure weight is about 55% of the total, comparable with that anticipated for vehicles made of more conventional materials.

The great need to produce highly efficient structures for all aero-space applications has recently lead to a reconsideration of the method of using safety factors in design. Many design conditions, such as fatigue, temperature, and creep do not lend themselves to a simple factoring process and an attempt is now being made to design components on an individual, statistical basis.[27] The idea is that the worst possible set of combinations of load, temperature, etc., should be statistically analysed and considered in conjunction with a similar analysis of material properties. Taking the worst combination and an accepted chance of failure, a definite design stress can be established for which no arbitrary factor is required.

Such a process is extremely difficult to apply due to lack of statistical information, but steps in the right direction have already been taken, particularly with regard to fatigue.[28]

Safety features

Development of aircraft has brought about a change of emphasis in the design of the auxiliary systems. Not only have considerably increased powers been called for but the correct operation of the system has become vital to the safety of the aircraft.

Thus, one major aspect of development has been aimed at ensuring system integrity and the provision of satisfactory means of emergency operation.[29]

One example of this is the hydraulic power control and operation of the tail planes of the Vickers V.C.10 airliner.[30] The elevator, which controls the fore and aft attitude of the aircraft, is made in four parts, each operated by a separate electro-hydraulic booster. Should one of these fail to produce

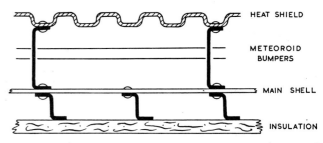

Fig. 8. A compound skin construction is necessary to protect spacecraft against the various hazards met in flight and during re-entry

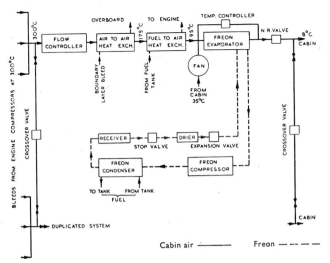

Fig. 9. Schematic of an air-conditioning system suitable for a 1450 mile/h air liner

The higher operational speeds of aircraft have extended the temperature range over which the system must operate and the choice of fluids and seals is a problem. Most of those developed can operate at higher temperatures at the expense of low temperature properties and, hence, cooling of components, such as jacks, may be necessary. Petroleum-based fluids are still generally used, since silicon fluids have the disadvantage of poor lubrication characteristics. Suggestions that the aircraft fuel be used as a hydraulic fluid meet with the same objection.

Air conditioning

Of vital importance to the operation of high-speed aircraft and space craft, the air conditioning system fulfils the purpose of providing a cabin environment which has suitable pressure, temperature, moisture and oxygen content. When the flight occurs at altitudes below about 15 miles it is possible to use an open-circuit system: the oxygen required can be obtained from outside the vehicle. This is usually achieved by means of a special compressor or offtake from the main powerplants.

The air so obtained must be cooled and the water vapour content adjusted as necessary. At greater altitudes it is necessary to carry all the oxygen required in the vehicle itself. The cooling of air taken from outside the vehicle is a major problem and closed vapour cycle refrigerating units appear to give the most promising solution.[34] The size of this part of the system can be reduced by using heat exchangers as preliminary coolers. At speeds below about 1600 mile/h. the aircraft fuel can be used as a heat sink, but at greater speeds an additional sink, such as a water reservoir, is necessary.

A diagram of an air-conditioning system[35] designed for a supersonic air-liner is shown in Fig. 9. Air is tapped from the compressors of the main power units and the initial cooling is obtained by passage through a heat exchanger located in external bleed air. Further cooling is obtained by using the fuel input to the engines as a coolant and finally a freon vapour system completes the process.

The stored fuel temperature is kept to a tolerable maximum with the aid of re-circulation fans. In practice, there would be duplication of the freon units in addition to the complete duplication shown in the diagram, and a stored oxygen supply would be available as a short-period emergency standby.

The *Mercury* capsule[7] utilizes a closed circuit system and in this case pure oxygen is circulated both in the capsule and through the astronaut's pressure suit, thus giving a measure of duplication. The oxygen pressure is maintained at 5 lb/in² with the aid of a supply carried in bottles and primary cooling is by evaporation of water. Carbon dioxide, water vapour and other impurities are extracted, the water being recoverable for drinking purposes.

full deflection, the other three can control the aircraft until the affected part of the elevator can be returned to neutral. Only one unit is required to control the aircraft, providing the other three are at the neutral position. The possibility of a failure in the signalling control valves has necessitated the introduction of such multiplex systems.[31] Electrical signalling devices are most convenient for this. A spurious signal from one wire in, say, a three wire system is then automatically rejected.

The demand for reliability and continued safe operation after a fault has been isolated may necessitate the use of four or even more parallel wires.

In the majority of aircraft the source of power for the auxiliary systems is some form of transmission from the propulsion units. Since the internal systems must continue to operate even in the event of their failure, an emergency supply is necessary. Stored power in batteries or accumulators is only sufficient for a very short time and often a power convertor is provided. One possibility is the use of turbogenerators which are lowered into the airstream and are, in effect, high-speed windmills.

Alternatively, a completely independent auxiliary power unit may prove to be the best solution.[32] The relatively low weight of constant frequency generating equipment has encouraged the development of constant speed units for taking power from the main aircraft engines. One new and promising type[33] uses a combined shaft and compressed-air transmission from a turbine. The speed is maintained constant by regulation of the air supply through a servo loop. This unit has the additional advantages of being capable of running the accessories on the ground, independent of the engine, or starting the engine by supplying compressed air from an external source.

The pressures of hydraulic systems have increased to enable greater work outputs to be obtained and components which operate at a delivery pressure of 4000 lb/in² are available.[29]

These higher pressures have increased fatigue problems so that the tendency is for pipes and components to be made of steel or titanium. Pressure vessels are frequently made of welded high-tensile steel, like solid fuel rocket motor cases.

Control and guidance

The heart of a guided missile is its control and guidance system, and this is usually electronic to a large extent. The need for accuracy and safety in manned aircraft has resulted in the introduction of control and navigation systems which are so extensive that the aircraft are almost capable of automatic flight. For example, automatic blind landing and throttle systems have been developed and aircraft such as the De Havilland *Trident* were equipped with them.[36] Automatic stabilization of vertical take-off aircraft has been tested both on simulators and in the air,[37] and the effect of a human operator in the control loop has been examined.

One interesting development in the field of guidance and navigation is the introduction of the inertial system.[38] Inertial guidance is used on most ballistic missiles and promises to provide a navigation system independent of weather conditions and terrain. In principle it is very simple, consisting of a set of three accelerometers, mounted on a platform which is stabilized by gyros. The acceleration of the vehicle in the three reference planes is recorded and, hence, the velocity and position relative to a known point can be computed. The main problem is essentially a mechanical one of producing reliable, highly accurate gyros so that drift can be minimized: production tolerances here are measured in millionths of an inch.

Fuel systems

Aircraft fuel systems have been developed mainly with a view to pumps and valves capable of dealing with large flow rates. The centre of gravity of an aircraft must be maintained within close tolerances and proportioning devices are available to regulate the fuel flow from a number of tanks simultaneously. These are usually driven either electrically or by compressed air. One type,[39] for example, passes the fuel from each tank through a separate paddle wheel arrangement, the relative speeds of the various rotors being adjusted to give the desired proportions. Large liquid fuel ballistic missiles and space boosters consume extremely large quantities of propellant in a very short time, 6000 gallons per minute, or more, being typical. The problem is complicated by the fact that the propellant is often at low temperature, for example liquid hydrogen or oxygen. Pumping liquid hydrogen is of current interest[40] and a three-stage pump has been developed for the purpose.[9] The first stage is a screw and the latter two stages are axial compressors. Since the vapour-to-liquid density ratio is low, the pump can operate with the liquid very close to its boiling point.

Undercarriages

As one of the more obvious mechanical components of an aircraft, the undercarriage warrants special consideration in this review. Much of the development work in this field has been concerned with the use of high-strength materials such as titanium[10] and ultra high-tensile steels.[41] Design refinements have enabled the weight and volume of undercarriage units to be reduced. One example is the work carried out on forced air cooling of brakes[42] which has enabled brake weight and tyre temperatures to be reduced. The time taken for tyres to cool to an acceptable level is a decisive factor in the turn-round time of an aircraft, particularly after a baulked take-off.

The unusual slender layout of many supersonic aircraft has a direct effect upon the undercarriage shock absorber design. One result of the compact arrangement is that bending moments due to airloads are small, compared with those which arise during landing impact. Thus it is essential to reduce the maximum shock absorber force as much as possible and also to arrange for it to be nearly constant for maximum efficiency. This can be done by using multi-stage shock absorbers but a possible alternative is the use of silicone fluid.[43] The silicones have highly non-linear compression curves and a careful selection may enable a single-stage liquid spring unit to be designed to have the desired characteristics.[44] An additive may be required to overcome the lubrication problem.

An appreciable part of the fatigue damage on many aircraft is connected with ground loads and taxiing of the aircraft over uneven surfaces. The response of an aircraft in these conditions has been investigated and attempts were made to correlate, experimental results obtained from tests over runways of known contours with theory.[45] Non-linearities in the shock-absorber and tyre are difficult to deal with, but some progress has been made.

When a vehicle is subjected to very high temperatures, as is the case with very high-speed aircraft and re-entry vehicles, oil-based shock absorbers and rubber tyres are inadequate. Considerable interest is being displayed in mechanical shock absorbers, two types of which are illustrated in Fig. 10. The first[46] uses a mandrel, or spike, to deform a series of washers as the means of absorbing energy whilst the second, known as a 'compliable' structure, utilizes the energy of deformation of a single member.[47] Both units are, of course, only capable of a single operation, unless the deformed parts are replaced. The former has the advantage of absorbing the energy in a series of uniform steps rather than in a single peak, but it has the disadvantage of considerably greater complexity.

The theoretical analysis of these devices involves the investigation of behaviour after initial failure has occurred as considerable energy is absorbed subsequent to this. One technique is to use the concept of a plastic hinge located at the point of failure and evaluate the energy of rotation[48, 49]. Applications of the compliable leg type to a re-entry capsule[50, 51] have been investigated. A further possibility for absorbing the landing impact of a capsule is the use of an air-bag.[52] Experiments have shown, however, that peak accelerations are only slightly less than those obtained from a compliable structure.

Skids or metal wheels could be used instead of rubber tyres and various designs have been proposed and tested.[51] These include shoe or wire-brush skids and solid or wire-brush tyres. Measurements of coefficient of friction have been made at various velocities on concrete and asphalt surfaces and these show that tungsten carbide and hard *Cermet* give the lowest friction for shoes on concrete, the coefficient being about 0·2. The rolling friction of wire

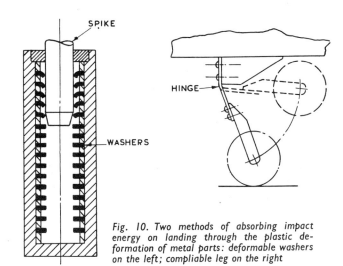

Fig. 10. *Two methods of absorbing impact energy on landing through the plastic deformation of metal parts: deformable washers on the left; compliable leg on the right*

Mr D. Howe *was educated at Watford Grammar School and became apprenticed to Fairey Aviation Co. in 1944 while attending part-time National Certificate courses at Southall Technical College. He was awarded the 1948 Hele-Shaw medal and prize. After experience as a draughtsman he won a scholarship to the College of Aeronautics and obtained a Diploma with distinction in 1951. Remaining at Cranfield for a year as a Clayton Research Fellow his studies of swept wing structures gained a Viscount Weir prize. After returning to the Fairey Aviation Co for two years as technical adviser to the project office, he returned in 1954 to Cranfield as a Lecturer in the Department of Aircraft Design. During 1959 he was granted leave of absence to read for a degree of MSc at Massachusetts Institute of Technology. His present position is Senior Lecturer in charge of aircraft design teaching and Chief Designer of ARB approved organisation at the Cranfield Institute of Technology.*

brush tyres was found to be about 0·15 with a slight decrease at high velocities, whilst that for solid metal wheels was about 0·04. In addition, high temperature pneumatic tyres using steel wire for the carcass have been suggested as another possibility.[53]

Conclusions

The aero-space field is so vast that it has only been possible to mention certain individual developments considered to be of greatest interest to the mechanical engineer. In the sphere of materials, titanium has already been used successfully and developments indicate an extension of the use of both beryllium and molybdenum for structural purposes. The design of heat-resistant structures will involve the use of new insulation materials and techniques. Development of a new hydraulic fluid with a wide temperature range is desirable and silicone fluids may have more application in the future. Mechanical shock absorbers are likely to be used extensively as a means of overcoming the temperature restrictions which limit the use of the more conventional oil types.

Future developments are hard to predict in such a rapidly changing subject but a possibility is the use of aircraft with air breathing engines as space booster vehicles.[54] The principle of such a device is that the vehicle would take off and scoop up air whilst flying at high speed in the upper atmosphere. This air would be liquified and the oxygen and nitrogen stored separately for subsequent use in either a rocket or ramjet. The advantage is that the booster could take off at low weight and return to its base after completing its mission. One obvious problem would be that of designing lightweight liquification equipment and this might be considered the ultimate in the development of air conditioning systems.

REFERENCES

1. SCHAIRER, G. S., 'Looking Ahead in VSTOL'. 1961 *Anglo-American Aeronautical Conference.*
2. RUSSELL, A. E., 'General Aspects of Supersonic Transport Aircraft'. 1961 *Anglo-American Aeronautical Conference.*
3. DES 204, 'Supersonic Airliner Project'. 1960 College of Aeronautics, Department of Aircraft Design.
4. 'New X-15 record'. 1961 *Aeroplane and Astronautics*, September 21st, p. 391.
5. MELLINGER, G., 'The design and operation of the X-15'. 1961 *AGARD* Flight Mechanics Paper, Istanbul.
6. SWANN, R. T., 'An engineering analysis of the weights of ablating systems for manned re-entry vehicles'. 1960 *Ballistic Missiles and Space Technology.* Academic Press, vol. 4, p. 65.
7. ANDERTON, D. A., 'How Mercury Capsule design evolved'. 1961 *Aviation Week*, May 22nd, p. 50.
8. 'Dyna-Soar Flight Test Schedule'. 1960 *Aviation Week*, May 9th, p. 26.
9. SILVERSTEIN, A., 'Researches in Space Flight Technology'. 1961 *Royal Aeronautical Society*, Wilbur Wright Lecture.
10. 'The production of an aircraft undercarriage component in Titanium Alloy (4Al-4Mn) forging'. 1960 *Ministry of Aviation S and T* Memo 22/60.
11. LANGSTONE, P. F., 'High Strength Rocket Motor Cases'. 1961 *Rocket Propulsion Symposium*, College of Aeronautics.
12. VICKERY, D., 'The Bristol-type 188 Research Aircraft'. 1960 *Cranfield Society Symposium*, College of Aeronautics.
13. NEWELL, A. F., 'An Appreciation of Missile Structures'. 1961 *Royal Aeronautical Society Journal*, vol. 65, no. 607, July.
14. FOSTER, R. E. and RIEDINGER, L. A., 'Future Weight Reduction with Beryllium'. 1958 *Society of Aeronautical Weight Engineers* Paper No. 159.
15. HODGE, W., 'Beryllium for Structural Applications'. 1958 Defense Metals Information Centre DMIC Rep. 106.
16. MATHAUSER, E. E., BLAND, A. S. and RUMMLER, D. R., 'Problems associated with the use of Alloyed Molybdenum in Structures at elevated temperatures'. 1960 National Aeronautics and Space Administration TN D-447.
17. SCALA, S. M. and NOLAN, E. J., 'Aerothermodynamic feasibility of Graphite for hypersonic glide vehicles'. 1960 *Ballistic Missiles and Space Technology.* Academic Press, vol. 4, p. 31.
18. ANDERTON, D. A., 'Grumman's orbiting observatory design progresses'. 1961 *Aviation Week*, 13th Feb. p. 54.
19. LAD, R. A., 'Survey of materials problems resulting from low pressure and radiation environment'. 1960 National Aeronautics and Space Administration, TN D-477.
20. 'The use of chemical etching in the production of light alloy components'. 1958 *Aircraft Production*, vol. 20, no. 7, July, p. 264.
21. NEWELL, A. F., 'Recent British Technical Progress in Aeronautics-Structural Design'. 1961 *Aircraft Engineering*, vol. 33, no. 391, p. 248.
22. 'Three-foot diameter experimental motor'. 1961 Society of British Aircraft Constructors Display, Ministry of Aviation Exhibits, p. 13.
23. 1961 *Royal Aeronautical Society Journal*, vol. 65, no. 607, July. Discussion on Ref. 13.
24. KINARD, W. H. and COLLINS, R. D., 'A technique for obtaining data by using the relative velocity of two projectiles'. 1961 National Aeronautics and Space Administration TN D-724.
25. FUNKHOSER, J. O., 'A preliminary investigation of the effect of bumpers as a means of reducing projectile penetration'. 1961 National Aeronautics and Space Administration TN D-802.
26. LEONARD, R. W., BROOKS, G. W. and McCOMB, H. G., 'Structural Considerations of Inflatable Re-entry Vehicles'. 1960 National Aeronautics and Space Administration TN D-547.
27. 1961 *Loading and Requirements Course Lecture Notes Pt. 3.* College of Aeronautics, Department of Aircraft Design.
28. 1959 *British Civil Airworthiness Requirements*, Air Registration Board. Section D, Chapter D3-1 and Appendix.
29. TROTMAN, C. K., 'Auxiliary Systems'. 1960 *Aircraft Engineering*, vol. 32, no. 379, p. 246.
30. 'First Vickers VC 10 tailplane gets functional tests'. 1961 *Aviation Week*, 21st August, p. 41.
31. ORMONROYD, F., 'Supersonic Transport Aircraft—Cockpit Design and Management'. 1961 *Royal Aeronautical Society Journal*, vol. 65, no. 602, February.
32. GREEN, J. T., 'Design and assessment of the Auxiliary Power Supplies of the Supersonic Airliner Design Project'. 1961 College of Aeronautics, Department of Aircraft Design Thesis.
33. 'Constant speed drive starter system'. 1960 Plessey Co. Ltd.
34. LEECH, R. D. T., 'Aeronautical Cooling'. 1961 *World Refrigeration and Air-Conditioning Convention*, April.
35. GOPAL, C. S., 'Air Conditioning and Pressurisation System for 1960 Supersonic Airliner Design Project'. 1961 College of Aeronautics, Department of Aircraft Design Thesis.
36. MORGAN, R. C., 'Supersonic Transport Aircraft—Engineering Systems'. 1961 *Royal Aeronautical Society Journal*, vol. 65, no. 602, February.
37. FOODY, J. J., 'Control of VTOL Aircraft'. 1961 *Anglo-American Aeronautical Conference.*
38. CAWOOD, W., 'Some design problems in Inertia navigation'. 1958 *Royal Aeronautical Society Journal*, vol. 62, no. 574, October.

39. *Fuel Flow Proportioners*. 1961 H. M. Hobson Ltd.
40. GROBMAN, J., 'A technique for cryopumping hydrogen'. 1961 National Aeronautics and Space Administration TN D-863.
41. WILLITT, A. A. J., 'Today's Undercarriage Problems'. 1961 *Interavia* No. 9 61 September.
42. 'Air Cooled Brake for Aircraft'. 1960 Dunlop Rubber Co. Ltd.
43. 'M.S. Silicones'. 1961 Technical Data Sheet, G11. Midland Silicones Ltd.
44. GARSIDE, J. F., 'Main Landing Gear, Supersonic Airliner Design Project'. 1961 College of Aeronautics, Department of Aircraft Design Thesis.
45. ARMSTRONG, K. W., 'Response Characteristics of a Landing Gear to Sinusoidal Inputs'. 1960 College of Aeronautics, Department of Aircraft Design Thesis.
46. WALLACE, G. F., 'A mechanical shock absorber'. 1960 Ministry of Aviation, Royal Aircraft Establishment Tech. Note ARM665.
47. BLANCHARD, U. J., 'Landing Impact Characteristics of Load Alleviating Struts on a Model of a Winged Space Vehicle'. 1960 National Aeronautics and Space Administration TN D-541.

48. D'AMATO, R., 'Static Postfailure Structural Characteristics of Multiweb Beams'. 1959 United States Air Force, Wright Air Development Centre TR 59-112.
49. HOWE, D., 'The Static Postfailure Behaviour of Structures'. 1960 Massachusetts Institute of Technology, Department of Aeronautics and Astronautics Thesis.
50. HOFFMAN, E, L., STUBBS, M. S. and McGEHEE, J. R., 'Effect of a Load Alleviating Structure on the Landing Behaviour of a Re-entry Capsulte Model'. 1961 National Aeronautics and Space Administration TN D-811.
51. HOUBOLT, J. C. and BATTERSON, S. A., 'Some landing studies pertinent to glider re-entry vehicles'. 1960 National Aeronautics and Space Administration TN D-448.
52. McGEHEE, J. R. and HATHAWAY, M. E., 'Landing Characteristics of a Re-entry Capsule with a torus-shaped air bag for Load Alleviation'. 1960 National Aeronautics and Space Administration TN D-628.
53. RANDALL, L. S., 'Modern Aircraft Tyre Development'. 1961 *Interavia* No. 9/61, September.
54. BOODA, L., 'Space Plane grows into a Family of Concepts'. 1961 *Aviation Week*, 19th June, p. 54.

Brakes in Transport and Industry

by T. P. Newcomb, DSc, MSc, MIMechE, and R. T. Spurr, PhD

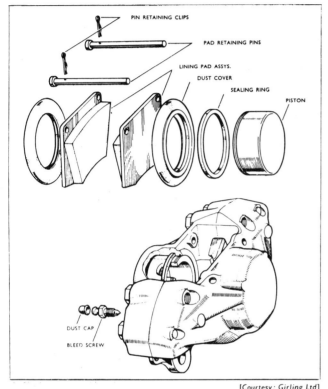

[Courtesy: Girling Ltd]

Fig. 1. The fixed caliper type needs two or more pistons

Disc brakes have spread from transport to industrial applications and the use of feedback anti-lock systems is increasing. The search continues for friction materials with better temperature and wear resistance and brakes which weigh less and fail-safe.

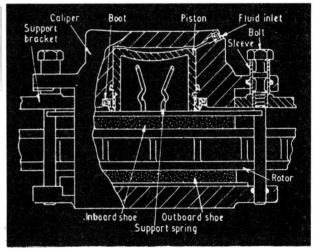

[Courtesy: General Motors]

Fig. 2. Delco Moraine single piston floating caliper disc brake

The *Apollo* lunar modules use retro-rockets to slow down when making a moon landing and the returning command module is decelerated by air resistance and parachutes before it hits the sea, but in the sub-lunary world things are slowed down by more commonplace brakes. Most are friction brakes, not at all glamorous—they get hot, they have been known to squeal and to smell, and the principle on which they work, just jamming one surface against another, is pretty crude. However, no other brake can dissipate as much energy in so little space, with so little extraneous gear and so cheaply.

A closer look shows that a lot of thought has gone into their design; the very ordinary brakes and braking system of a family car are the result of a great deal of development work. This has enabled them to keep pace with the continuing improvement in vehicle performance and with increasing safety legislation.

Friction brakes are not the only kind in use. Hydro-kinetic and electro-magnetic brakes, for example, also have their special applications but, like friction brakes, they have the obvious disadvantage that they waste energy as heat. Attempts

have therefore been made to develop regenerative braking systems, adapted most easily to electrical machinery. It has been very effective, for example, with colliery winding gear.

Car brakes

Because they are stable over a large range of operating conditions, disc brakes are widely used[1] for cars. The caliper may be fixed or movable. In the fixed caliper brake two or more pistons load friction pads against either side of the disc as shown in Fig. 1. The action of fluid seals on the pistons automatically makes up for pad wear. This type of brake is still much used but many cars are now fitted with disc brakes in which the caliper can move or float, either by sliding along pins (Fig. 2) or by rotating about a pin. Only one actuating cylinder is required, as the piston presses the movable inboard pad directly on to the disc, and the reaction on the caliper brings the other pad against the disc. A switch to these designs is taking place in the USA, where increasing use is being made of disc brakes, at least on the front wheels. Most American cars use ventilated discs which cool rather more quickly than conventional solid discs of the same weight.

Although the ideal engineering solution would be to use brakes having the same characteristics on the rear wheels as on the front wheels, some difficulty has been experienced in obtaining an adequate handbrake performance from discs. When a floating piston disc brake is fitted, an auxiliary caliper is attached to the main brake and used as the handbrake. However, a very high leverage is required from the mechanical system to obtain the necessary performance from the small pads used.

In a floating caliper brake it is not difficult to arrange for the larger pads of the main brake to be directly operated mechanically, but again a high leverage is necessary for an adequate handbrake performance. Floating calipers on the rear wheels are more common on the Continent than in England (at least in the medium power range) or the USA where drum brakes are retained on the rear wheels because, due to their inherent servo action, their performance as a handbrake is better. A number of manufacturers prefer to incorporate a drum brake in the bell section of the rear disc for parking purposes and thus utilise the good features of both types.

Where drum brakes are fitted to the rear wheels only, more use is being made of automatic adjusters to allow for lining wear; all disc brakes automatically adjust themselves for wear. Some adjusters operate only when the vehicle is moving backwards because then the drum is likely to be cool. This reduces the possibility of functioning when a hot drum expands, which would leave the brake applied when the drum cools. A gap is normally left however for thermal expansion.

To compensate for the lack of intrinsic servo action in a disc brake, the heavier cars incorporate a vacuum servo to augment the driver's pedal effort.

A few vehicles use pressure systems.[2] Hydraulic fluid is stored under high pressure in an accumulator charged by an engine-driven pump and admitted by a valve which is operated by the driver's pedal. The pressures are high so that large braking forces can be developed.

Braking force distribution

The braking force should be divided[3] between the wheels in the same proportion as the load. A major difficulty in designing a system is that the wheel-loading is affected not only by the load the vehicle carries—which can itself be variable—but also by weight transfer from rear to front wheels during braking. The weight transfer depends upon the deceleration so there is a risk that the rear wheels may be overbraked in a sudden stop, causing them to lock and send the vehicle into a skid. Ideally, the braking ratio should therefore change with the deceleration and the state of loading of the vehicle and this is now achieved on a number of vehicles.

Valves are inserted in the hydraulic line to the rear brakes which either close completely when the pipeline pressure or the vehicle deceleration reaches a pre-set figure, or which transmit only a fraction of any subsequent increase in pressure. Valves have also recently been introduced which sense the vehicle loading by determining the deflection of the suspension and change the braking ratio accordingly.[4]

Even if the braking ratio is correct, the maximum deceleration of the vehicle is limited in any case by the adhesion between road and tyre and the development of devices which enable the brakes to prevent skidding is receiving much attention.

If a wheel is about to lock, its rotational deceleration is

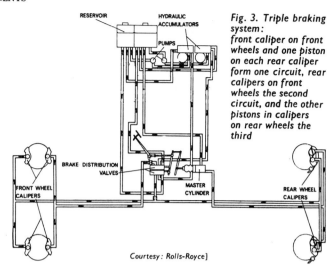

Fig. 3. Triple braking system:
front caliper on front wheels and one piston on each rear caliper form one circuit, rear calipers on front wheels the second circuit, and the other pistons in calipers on rear wheels the third

Courtesy: Rolls-Royce]

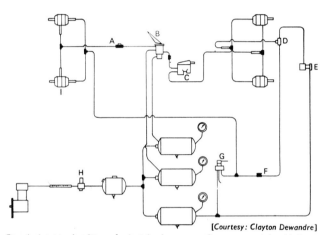

[Courtesy: Clayton Dewandre]

Fig. 4. A typical split or dual airbrake system for commercial vehicles.

A: Stoplight switch	D: Double check valve	G: Hand-control valve
B: Dual brake valve	E: Power assisted	H: Unloader valve
C: Load sensing equipment	handbrake valve	I: Triple diaphragm
	F: Stoplight switch	brake chamber

higher than the deceleration of the vehicle. In most systems[5] a sensor measures the wheel deceleration and, if it is above a predetermined value, the brake is released and re-applied when the wheel is rotating normally.

A major difficulty is to ensure that the intervals in which the brake is released are short enough not to increase the total stopping distance.

For greater safety, split or dual braking systems are mandatory in the USA. Tandem master cylinders are fitted; one cylinder may be connected with, for example, the front brakes, and the other with the rear brakes. The two circuits are quite independent so that failure of one will not affect the other. Fig. 3 shows one system in use.

Commercial vehicle brakes

Vehicles[6] at the smaller end of the commercial range are generally fitted with hydraulically operated drum brakes similar to, but larger than, those used on cars. A few models use disc brakes. The heavier vehicles, including articulated ones, use air brakes. Air is compressed by an engine-driven pump, stored in a reservoir and passed by a pedal-operated valve to the brake chambers. There it acts on diaphragms

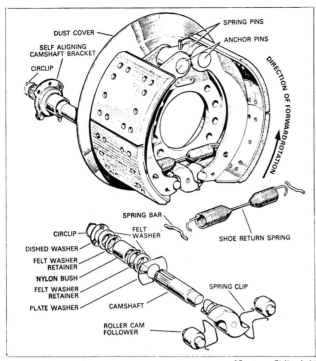

[Courtesy: Girling Ltd.]

Fig. 5. S-shaped cams, used on large brakes, keep the radius arm perpendicular to the pushrod and compensate for lining wear

and causes pushrods to move which, in turn, apply the brakes through cams. Compressed air is a cheap working fluid but air brakes are bulky and comparatively slow. This last disadvantage can be overcome to a large extent by using relay valves.

Articulated vehicles are braked by the three-line system—one line supplies air to a local reservoir, another line, in circuit with the pedal valve, operates a relay valve which controls the flow of air from the reservoir to the brakes. The third line operates the secondary braking system. By the time protective valves, load-sensitive valves, etc, are fitted, the system split, and the secondary system installed, the

[Courtesy: Centrax Gears Ltd.]

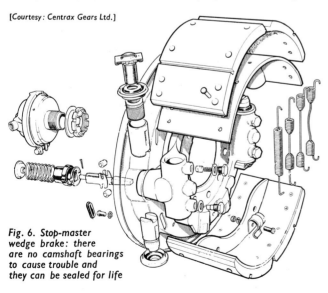

Fig. 6. Stop-master wedge brake: there are no camshaft bearings to cause trouble and they can be sealed for life

complete brakes of modern commercial vehicles are quite complicated. Fig. 4 shows a typical system.

Heavy vehicles in which adequate operating forces are available require brakes which are stable rather than have intrinsic servo action. The brake shoes are forced apart by fixed cams which may be shaped to make the mechanical advantage low at first so that lining clearance is rapidly taken up. A slack-adjuster is fitted, so that clearance caused by lining wear can be taken up. S-shaped cams (Fig. 5) are used on very large brakes fitted with thick linings. These keep the radius arm of the cam perpendicular to the pushrod of the brake chamber, and a given air pressure produces a constant force on the shoes, irrespective of lining wear. As the latter occurs more in the middle than at the ends, very thick linings are generally made to taper towards each end in order to get the maximum life out of them.

Wedge brakes, shown in Fig. 6, force the shoe tips apart by a moving wedge. They have no camshaft bearings to cause possible trouble and some are sealed for life and so require no maintenance.

Attempts have been made to develop automatic slack adjusters; the difficulty here again is thermal expansion of the drum.

Provision of an adequate hand-brake can be a problem with very heavy vehicles. Servo-assistance has the disadvantage that the air pressure might fail in an emergency. Spring brakes which fail-safe have been installed on a number of vehicles; air pressure acting on a piston keeps a very powerful spring compressed and the driver reduces the air pressure to apply the brakes.

If the brakes lock the rear-wheels of the tractor of an articulated vehicle, the outfit loses lateral rigidity and a jack-knife develops. This can be prevented by fitting a disc brake to the fifth wheel, as in the Hope device, and applying it whenever the road wheels are braked. If the disc brake rotates with the steering wheel, steering and braking will not interact. Alternatively, deceleration sensors[7,8] similar to those used on cars can be fitted. They are used on a large number of commercial vehicles and are mandatory on certain tractors to prevent jack-knifing.

Reducing heat load

A great deal of energy is dissipated by the brakes of a heavy vehicle during a sudden stop or when braking down a long gradient, and thermal troubles can arise. In a single stop the transient temperature depends on, among other things, the area of the rubbing path. Therefore this should be as large as possible. The bulk temperature during a long descent will depend upon the thermal capacity of the drum and therefore this too should also be as large as possible. This last requirement, however, conflicts with a third, that is, that the drum should be thin to minimise the thermal gradients which cause stresses. Drum diameters are limited to about $15\frac{1}{2}$ in by the size of the wheel, and drum widths by the possibility of bell-mouthing, but drums of up to 7 in wide in front and 8 to $8\frac{5}{8}$ in width on rear wheels are in use.

Conventional disc brakes are not successful on heavy vehicles because the disc has to be very thick in order to have the necessary thermal capacity and so thermal stresses can be very high and lead to disc damage. They are also exposed to the elements.

One way to reduce the load on the brakes is to fit the vehicle with a retarder.[9] There are several types. In the exhaust brake a throttle is mounted ahead of the silencer to make the engine function as a compressor on the exhaust stroke;

176 COMPONENTS

the fuel supply is cut off when the brake is in operation. The engine can also be modified so that air compressed during the compression stroke is released to the exhaust.

Eddy-current retarders can be mounted on the transmission; discs rotate at each end of a set of coils, the driver switches current through the coils and eddy currents induced in the rotor discs set up magnetic forces which produce a braking torque.

The hydraulic retarder is similar in form to a torque converter and is particularly suited to a large engine, provided it is designed as part of the engine and transmission. Retarders are generally less effective at low speeds and here the friction brake takes over. Retarders acting on the engine or transmission can of course brake the driving axle only. Fitting retarders to the axles of the trailer of an articulated vehicle adds greatly to its weight, cost and complexity.

Railway brakes

The traditional way to brake railway stock[10, 11] is to fit clasp brakes, the blocks of which are pressed against the tread of the tyre. The actuating force is transmitted from an air (or vacuum) cylinder through fairly complicated piping and losses in the latter can be considerable. Cast iron blocks are very widely used but they have the disadvantage that the μ of cast iron at high speeds and high temperatures is very low, though rising with decreasing speed. This increase is so marked that the pressure in the cylinders may have to be reduced near the end of the stop to prevent a rapid increase in deceleration and a possible wheel skid. Another disadvantage of cast iron blocks is that their wear dust constitutes a fire danger and can interfere with the track circuit.

Recently high-phosphorus blocks have been introduced as they appear to represent less of a fire hazard. Cast iron blocks are themselves cheap but maintenance is costly. Composition blocks,[12] with a longer life and a much more stable μ than cast iron, have been used for years on underground railways. Their wear dust is not dangerous, and these blocks are slowly being accepted on a wider range of railway stock. Some composition blocks may reduce rail-tyre adhesion or cause thermal damage (cracking or crazing) of the wheel tread, but there is such scope in the formulation of compositions that any disadvantages can be overcome, and such a capacity for development that they will eventually be generally used.

Disc brakes[13] were first tried on railway stock in the hope that their use would reduce maintenance costs. Much development work was carried out and two main types of railway disc brake are now in use. Where it is practicable, cast iron cheek discs are attached to either side of the wheel web, and on trailer axles ventilated cast iron discs are mounted on the axle.

In both the types the air-operated cylinder is rigidly attached to the bogie frame and supports one end of each caliper lever. The brake pads are pivoted on the other end of the lever and suspended from the bogie frame by torsionally rigid links which transmit the brake torque and restrain the pads parallel to the disc. Resin-asbestos friction materials are used for the pads.

In clasp brakes the blocks, particularly cast iron blocks, tend to clean and roughen the wheel tread and maintain the rail-tyre adhesion. There is no such action with disc brakes so that adhesion may be reduced. Slide protectors have been developed which sense incipient skids and reduce the brake-actuating force accordingly so preventing wheel damage.

Another problem is that in stops from high speed such

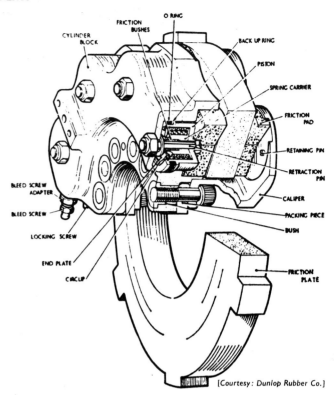

[Courtesy: Dunlop Rubber Co.]

Fig. 7. Caliper disc brake used on light aircraft: a number of pistons are housed in a cylinder-block

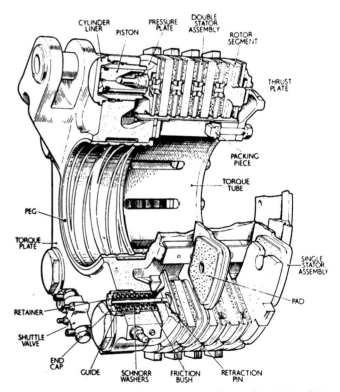

[Courtesy: Dunlop Rubber Co.]

Fig. 8. A typical multi-disc brake: a ring of pistons operates against a pressure plate

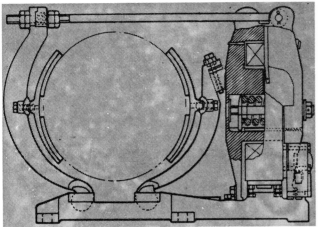

Fig. 9. Electromagnetic drum brakes are used in the industrial field, particularly on mills

large amounts of energy are dissipated that the metal mating surfaces may suffer from thermal damage.

Other forms of braking are feasible, particularly with electric stock. Regenerative braking is used down gradients on one British Rail line; the motor acts as a generator feeding the supply which must be capable of accepting such surges without unduly changing in voltage. A very simple form of regenerative braking is used on the Victoria line of the London Underground; the trains enter each station on an up-gradient and leave it on a down-gradient.

In rheostatic braking the motor again acts as a generator but the excess current is dissipated as heat: no wear dust is formed but space has to be found for the resistances and control gear.

Brakes operating on the wheel-tread cannot be used on high-speed stock because changes in tyre profile caused by tyre wear may affect the stability of the vehicle. Hydrokinetic brakes are to be fitted to the British Rail Advanced Passenger Train. They are less effective at lower speeds where friction brakes will take over.

Aircraft brakes

Both air braking and friction braking are used by aircraft in normal stops. Air brakes may be in the form of wing flaps or parachutes. Braking is also obtained by reversing the pitch of the propellers or the thrust of the engines.

The energy that has to be dissipated in one way or another is very much greater than in most other brake applications. In the extreme case of a rejected take-off, approximately 70×10^6 ft.lb must be absorbed by the brakes of a large aircraft in less than one minute; however the brakes would be expected to make only one such stop and would then be replaced. The normal kinetic energy rating is between 40–50 per cent of this. But the brakes must maintain a consistent performance over a wide range of speeds and temperatures, otherwise undesirable effects like judder or brake chatter may appear.

The types of brakes used are fairly well established[14] but steady progress over the last decade has been made in the detailed design, in improved cooling and in developing better materials so that the brakes have kept pace with the demands on them due to increased landing speeds and higher decelerations.

Light aircraft often use a disc brake fitted with a fixed caliper which houses a number of pistons in a cylinder block as shown in Fig. 7. Each piston loads a friction pad on one side of the disc which can float axially and so react against a second set of pads in the caliper. The discs are made of steel or copper and are often chromium-plated to improve the life of the friction material.

Multi-disc brakes such as the one in Fig. 8 are used on large aircraft. Thin segmental alloy-steel rings are assembled to form a stack of alternate rotor and stator elements and pads of cermet (metal-ceramic compound) friction material are attached round and on both sides of the stator rings. Allowance is made for thermal expansion so that high temperatures do not cause undue distortion or stresses. A ring of pistons protected by heat-insulating material operate against a pressure plate to press the discs together.

Some reduction in the weight of these brakes has been made by using rotors of plated beryllium (which has a high specific heat and low density). Beryllium however has low strength, a high coefficient of expansion, is toxic, and the cost per landing is higher than when steel rotors are used.

Other materials which look promising are being developed.[15] Some can withstand extremely high temperatures and so make the brake a very effective heat-sink. The Anglo-French *Concorde* uses one friction material sliding directly against another and this gives a more uniform friction and superior life than alloy-steel.

Much activity is being devoted to increasing deceleration so that congestion can be alleviated. Thus the *Maxaret* anti-skid device senses an incipient skid and reduces the actuating force as described above. Forced air cooling of the brakes and increasing the gap between brake and inner rim of the wheel have greatly reduced the turn-round time of aircraft. Liquid-cooled brakes and turbo-brakes have also been designed.[15]

Industrial brakes

Friction brakes are widely used in industry to control the motion of all types of equipment varying from small tape recorders to large colliery winders, as well as off-the-road vehicles such as excavators and tractors. Because of the diverse nature of the applications it is only possible here to mention some of the more important.

Well-established designs of drum brakes have been steadily improved together with the friction materials (often specially developed for the particular application). Where there is a limited actuating force available and very accurate control is not required, band brakes are often preferred to drums because of the high servo effect that can be introduced. They can also be easily manufactured and are less expensive than drum brakes.

Various methods are used to actuate industrial brakes. For example, in all types of mills it is usual to install electromagnetic drum brakes, such as shown in Fig. 9, fitted with hinged shoes internally lined with friction material. On other machinery which requires to be frequently started and stopped under precise control, for example, machine tools, combined electromagnetic clutch and brake units, either in the form of single or multi-plate units, are generally used.[16]

As well as conventional air-operated drum brakes, tyre-pneumatic drum brakes are used successfully on a wide range of equipment, provided braking temperatures are not too high. The 360° braking surface is built up from a series of segmental pads of friction material attached to a tyre of rubber and cord construction. In the contracting type, the tyre is enclosed in a fixed steel ring and when the tube is inflated the friction material is thrust uniformly against the drum, the tyre withstanding the stresses developed. These units avoid the use of linkages and little maintenance is required as compensation for lining wear is automatic.

Disc brakes are increasingly used in industry, following their success in the transport field, particularly on high-speed machines which are braked at short intervals and reach high temperatures.

When a machine has been in existence for some time and is then up-rated, it is easier to up-rate its brake if a disc and not a drum brake has been fitted.

Problems can arise with disc brakes if high torques are required which necessitate very high loads on the caliper with a possible reduction in pad life.

A straightforward design of industrial disc brake uses a floating caliper. The latter consists of two arms and each can rotate about a central pivot or abutment on a bridge piece straddling the disc. Friction pads are attached to the ends of the levers and a cross-rod and actuating cylinder connects the other ends, together with springs to facilitate release of the brake.

These units either use air pressure to operate the brake directly, or the brake is applied by springs and air pressure is used to hold it off so as to make it fail-safe.

Other industrial disc brakes are similar to automotive brakes except that the hydraulic pressure is used to control springs which apply the brake. Allowance for pad wear is often made manually but automatic adjustment can be obtained from positive retraction devices similar to those used on car brakes.

Electromechanical and magnetically-actuated[17] disc brakes are also available; the latter use sintered-metal friction materials to advantage.

One interesting use of disc brakes is in the marine field where their effectiveness has reduced the time needed to complete the reversal of the propeller shaft from full-speed-ahead. Ships can thus more quickly carry out their manoeuvres. On bigger vessels, aircraft-type disc brakes are sometimes fitted to cope with the large amounts of energy dissipated.

On large machinery such as colliery winding gear there is a general trend towards the use of regenerative braking on electrically-driven winders, but a friction brake is retained as an emergency brake and for 'parking' purposes. The friction brake must fail to safety and must be capable of smooth application under full emergency conditions without thermal damage or over-run of the cage at the bottom and top of the shaft.

Where a friction brake is used as the main brake, the diameter of the friction surface has remained constant (about 10 ft) but the width of path has steadily increased from about 18 in to 34 in to maintain the rate of heat generation per unit area at the nominal friction surface constant, since greater energies are dissipated now that mining is being carried out at deeper levels.

Because of their flexibility and advantages in cost, woven friction materials are exclusively used on these large brakes.

Despite the increasing use of hydraulics on both excavators and tractors there is still a large number using friction brakes. On excavators, such as the one in Fig. 10, there are travel brakes to stop the vehicle itself, swing brakes to control slewing and hoist brakes to control the raising and lowering of the boom. These brakes are often of the external contracting type.

Friction materials

The rotating part of a brake is generally made of cast iron or steel, and the stationary member is surfaced with a thin layer of material[18] which has a high coefficient of friction and a low wear rate. This friction material must not unduly damage the opposing surface, for replacement of the latter could in many instances be expensive. Wear must be largely confined to the friction material which is made in the form of a detachable lining, pad or block so that it can be easily replaced.

The most widely used type of friction material is made from thermosetting polymer resins which are reinforced by fibres (generally asbestos) and small amounts of other ingredients or fillers added to give the material the required characteristics. The resin, its polymerising conditions, the proportion of asbestos and the nature and proportion of the fillers, can all be varied so that it is possible to tailor friction materials for specific applications. For example, the disc pads of a Grand Prix racing car, the brake of a colliery winder, and the brakes of a bicycle have to contend with very different operating conditions; therefore quite different ingredients and manufacturing processes are used for each.

Inorganic friction materials are also in use. Sintered metal and ceramic[19] brake pads have been used in oil-immersed tractor brakes and aircraft brakes.

All friction materials, however, have many requirements in common. Their μ should preferably be high, for the actuating force can then be reduced, it should be very stable, not change with temperature, or speed, or use. The life of the material should be in keeping with its application, for example, the pad of a racing car brake need only last the length of a race, but the linings in the brake of a washing machine should last the life of the machine.

Dr T. P. Newcomb is 44. He studied at Nottingham University and graduated in Mathematics in 1951. In 1953 he was awarded the MSc for work on aerodynamics and in 1964 a DSc for his work on braking. Since 1954 he has worked at Ferodo Ltd on the thermal aspects of braking, brake usage and braking problems generally. He has published a number of papers on these subjects and, in collaboration with Dr R. T. Spurr, has written two books — 'Braking of Road Vehicles' and 'Automobile Brakes and Braking Systems'.

Dr R. T. Spurr was born in South Australia in 1922. He studied Physics at Adelaide University, graduating in 1947; he obtained a PhD at Imperial College, London, in 1951. Since then he has worked at Ferodo Ltd on friction and wear, squeal and the psychology of braking. He has written a number of papers on these subjects and has collaborated with Dr Newcomb on a number of publications.

[*Courtesy: Ruston Bucyrus Ltd.*]

Fig. 10. Excavator fitted with external contracting brakes on its LH side

The resin used as the bonding medium plays a large part in determining the friction and wear properties of the material. For example, if an unsuitable resin is used, the lining or pad may fade, that is, its μ may fall to a very low value, making the brake largely ineffective.

A continual search goes on for bonding media which will withstand higher temperatures. Sophisticated polymers have been tried but they are generally expensive. The more conventional phenol-formaldehyde resins are widely used, and much research and development work[20] has been done in order to give them the required high-temperature properties. Pads incorporating present-day phenol-formaldehyde resins can withstand temperatures above 700°C for stop after stop without the μ of the pad falling, or the pad showing excessive wear.

Sintered metal, cermet and ceramic friction materials in which the bond is inorganic, can withstand higher temperatures than resin-based materials and last considerably longer but they are much more expensive.

Because of its high μ, cotton is used to some extent as the reinforcing fibre in linings for industrial brakes, but asbestos is otherwise almost universally used. The advantages of asbestos are that it withstands high temperatures, that it can be easily broken down into very fine fibres, and individual fibres also have remarkably high strengths. It therefore makes an excellent reinforcing agent. Despite sensational articles in the press, the asbestos in wear dust from pads and linings is so broken down and so changed chemically that it represents no hazard to health.

Fillers can be divided into three main types; abrasives, metals and lubricants. The abrasive fillers are mineral powders added to increase or stabilise the μ. Metal powders are also used to stabilise the μ. Solid lubricants like graphite or molybdenum disulphide are also added to reduce wear.

Future trends

It is not difficult to predict the progress that will take place in automotive brakes and to some extent these act as pace-makers for brakes in other fields. Other countries will follow the lead of this country and more and more cars will be fitted with disc brakes. Split or dual systems should eventually become general and methods of varying the braking ratio with loading and deceleration will be more widely used.

Antiskid devices will only be used on the more expensive and sophisticated cars. More cars will be fitted with servos to augment the drivers' pedal effort, and, we hope, that components and pipelines immune to corrosion will eventually be developed.

Drum brakes will continue to be used on the heavier commercial vehicles. Antiskid devices may eventually become a legal requirement on articulated lorries, as may the fitting of valves to alter the braking ratio according to the loading on heavy vehicles generally.

It is likely that hydraulic systems will be used on heavy vehicles (which seem to be increasing in size) instead of air brakes. Retarders will not be used extensively unless their price is right.

High-speed railway stock will be braked with disc brakes, fitted with slide protectors.

Disc brakes may also displace drum brakes to some extent in the industrial field.

The development of friction materials to withstand higher temperatures and to last longer and longer will continue but it is likely that the drum or disc material will soon be the limiting factor, as far as temperatures are concerned, if the brake is to be of reasonable weight. New heat-conducting alloys may have to be developed for this purpose.

Acknowledgement

The authors wish to thank Girling Ltd for permission to use Figs 1 and 5.

REFERENCES

1. NEWCOMB, T. P., AND SPURR, R. T. 1969, *Automobile Brakes and Braking Systems*. Chapman and Hall Ltd.
2. KING, F. R. B., 'Brakes'. 1969, *Auto. Eng.*, vol. 59, p 231.
3. ROUSE, J. A., 'Distribution of Braking in Road Vehicles'. 1963, Symposium on *Control of Vehicles*, p. 80. IMechE.
4. HASTINGS, H., 'Safer Braking: Girling's New Antilock Braking System'. December 1969, *Motor*, p. 26.
5. 'Development of Use of Antilocking Devices on Road Vehicles'. 1966, *MIRA Bulletin*, no. 1, p. 16.
6. WILSON, A. J., 'Brakes and Brake Systems'. 1967, *Commercial Vehicles Engineering and Operation*. Chptr 7. IMechE.
7. WILKINS, H. A., 'An Assessment of the Dunlop "Maxaret" Antilocking Braking System Fitted to an Articulated Vehicle'. 1968, *Road Research Laboratory*, LR 161. Ministry of Transport.
8. WILKINS, H. A., AND CHINN, B. P., 'An Assessment of the Automotive Products "Antilock" Antilocking Braking System Fitted to an Articulated Vehicle'. 1968, *Road Research Laboratory*, LR 191. Ministry of Transport.
9. NEWCOMB, T. P., AND SPURR, R. T. 1967, *Braking of Road Vehicles*. Chapman and Hall Ltd.
10. BROADBENT, H. R. 1969, *An Introduction to Railway Braking*. Chapman and Hall Ltd.
11. BROADBENT, H. R., 'Trends in Railway Braking'. *Modern Railways*.
12. WISE, S., AND LEWIS, G. R., 'Composition Brake Blocks and Tyres'. To be published by the Railway Division, IMechE.
13. TOMPKIN, J. B., 'Development of the Disc Brake, with particular reference to British Railways application'. 1969. *Journal Inst. Locomotive Engrs*, p. 84-129.
14. 'Landing Gear Ancillary Equipment'. 1968, *Aircraft Engineering*, p. 9-13.
15. McBEE, L. S., 'Effective Braking—A key to Air Transportation Progress'. April, 1969, S.A.E. paper No. 690376.
16. ROBSON, G. C. 'Single Disc Electromagnetic Clutches and Brakes'. 17th August 1966, *Machinery*, pp. 357-362.
17. HAMMAR, L., 'Magnetic Disc Brakes for 0·2—90 kgm'. 1965, *ASEA Journal*, Volume 38, No. 9, p. 140.
18. NEWCOMB, T. P., AND SPURR, R. T., 'The Application of Tribological Principles to Brakes and Clutches'. 1969, Symposium on *Savings from Tribology*. PERA, Melton Mowbray.
19. JENKINS, A., 'Powder-Metal-Based Friction Materials'. 1969, *Powder Metallurgy*, Vol. 12, No. 24, p. 503-518.
20. 'Developing F2430'. 4th September, 1969, *Autocar*.

Gearing and Transmissions

by H. J. Watson, BSc, MIMechE and W. H. Harrison

The principal advances in the engineering of gears have been in measurement, inspection and manufacturing techniques. These, and a new lubrication theory, have contributed to the improved performance required from modern high-speed transmissions. Unfortunately, gear case design has not kept up with the design of gears themselves.

Toothed gears play a prominent role in nearly every aspect of power and motion transmission in our present-day economy and it is difficult to imagine the modern world without them. They are by far the most numerous mechanical means of connecting together two shafts running at different speeds. In terms of versatility, positive connection, efficiency, and speed range, no other form of mechanical transmission can compare with toothed gearing.

The users of gears have benefited immensely from improved design and manufacturing techniques by way of reduced weight with increase in load-carrying capacity, quiet operation and low costs. In spite of growing interest in hydraulic and hydrostatic transmissions for special applications, toothed gears are likely to continue in use in road vehicles for some time to come.

On the railways, the steam engine has been replaced by internal combustion engines and electric motors, resulting in the much more extensive use of toothed gearing to actuate locomotive driving wheels. The reciprocating steam engine was sufficiently flexible to couple directly to the driving wheels and still provides adequate starting torque; but in the interest of economy in size the later prime movers are coupled with toothed gearing. Some of the most noticeable progress in gear transmissions has been accomplished in this field.

Main propulsion gears for ships have also received considerable attention and their accuracy closely approaches that of the best mechanical measuring equipment. In some types of marine turbine gear quite revolutionary changes in materials have been made, resulting in smaller and lighter gear units.

In the field of gearing generally there has been a steady and generally slow progress towards the ever unattainable ideal, which has contributed to the improvement of many products and processes, including gear cutting itself.

Design

Involute gear design based on load capacity, has not changed greatly since before the 1939–45 war. The criterion for surface load capacity is the limiting fatigue stress, calculated on the basis of the Hertz equations for contact stresses.

There is, at present, no real cause to doubt the suitability of the Hertz analysis for this purpose, but perhaps the most difficult problem facing the gear designer, is that of determining the actual forces transmitted by the gear teeth in any given transmission. From test results, it appears, however, that, contrary to the previously held notion, the surface load capacity of gears is not reduced with increased speed. Timms[1] reported that with gears tested in a power circulator the permissible cycles before fatigue pitting set in increased with speed at a given load.

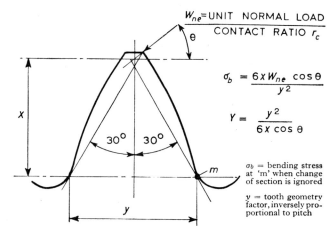

W_{ne} = UNIT NORMAL LOAD

CONTACT RATIO r_c

$$\sigma_b = \frac{6 \times W_{ne} \cos \theta}{y^2}$$

$$Y = \frac{y^2}{6x \cos \theta}$$

σ_b = bending stress at 'm' when change of section is ignored

y = tooth geometry factor, inversely proportional to pitch

Fig. 1. The critically stressed section is determined by the tangents to the fillets, intersecting at 30 degrees

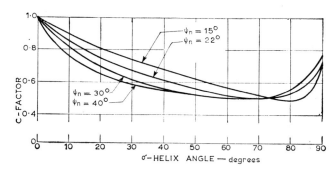

Fig. 2. Tooth bending strength, showing how C-factor varies with bending angle

Niemann and Richter[2] tested a range of accurately ground gears with different tooth proportions in a power circulator at peripheral velocities of 2, 8, 18 and 32 metres per second. In each series the tendency was the same; for N load cycles, the pitting fatigue stress increased with speed. The approximate relationship between tooth load per unit length F_i and peripheral speed v was $F_i Cv^{0.4}$, where C is constant. The authors suggest that an explanation might be found in the reduction of stress intensity as a result of the support provided by the lubricant film, which spreads the load over an area increasing with speed.

Similar tests[3] were carried out with hobbed gears as it was considered likely that the dynamic load increment resulting

Fig. 3. Surface fatigue figures for rolling discs made of different steels are obtained in this type of machine

Fig. 5. The complete equipment for subjecting test gears to full load at speed. Both wheels and pinion can be run at up to 8 in. centre distance

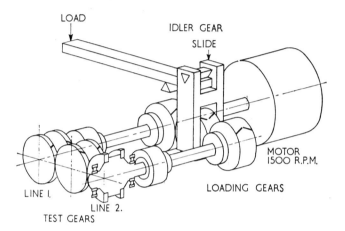

Fig. 4. A power circulating mechanism for simulating the effect of loading on gears

from the tooth errors would increase with speed. It was found that there was little change in surface load capacity at speeds ranging from 2·8 to 27 metres per second, but at the highest speed, tooth breakage occurred. This does not conflict with the proposition that the hydrodynamic lubricant film between the meshing gear teeth provides support increasing with speed, since it may be argued that this film effect offsets the speed-dependent dynamic load increment. The load capacity as limited by surface fatigue resistance might, therefore, be expected to remain constant, but the bending stress would increase with the dynamic load increment; for a more effective lubricant film, would significantly influence contact stresses but not bending stresses.

Bending strength of spur gear teeth

The bending stresses in straight spur involute gear teeth have been extensively investigated.[4-6] Dolan and Broghamer[4] gave formulae for obtaining stress concentration factors for 20 degrees normal pressure angle and $14\frac{1}{2}$ degrees normal pressure angle gear teeth. The nominal bending stress, as calculated by the Lewis formula, is multiplied by the appropriate stress concentration factor to obtain the maximum gear tooth bending stress. Jacobson[5] gave bending stress factor curves for 20 degrees pressure angle gear teeth generated from the British Standard rack tooth profile, which enables the maximum bending stress to be obtained directly for the teeth of gears generated with specified amounts of profile shift.

Winter[6] tested steel gears with different tooth proportions in a power circulating rig; also, by applying a pulsating load to single teeth in addition to determining maximum bending stresses with the aid of photo-elastic models, he correlated his results and those of others, with different variations of the Lewis formula. He found that, when the minimum root fillet radius exceeded 0·2 module, variations in this had little material effect on the intensity of maximum stress. The best correlation of all test results was obtained when the nominal bending stress was calculated in accordance with the Lewis formula, simplified by assuming load application at the tooth tip, the load normal to the tooth surface being divided by the contact ratio.

It was found sufficiently accurate to assume that the critically stressed section was determined by those tangents to the root fillets which intersected the axis of symmetry at an angle of 30 degrees, as shown in Fig. 1.

The work done by Dolan and Broghamer and also by Jacobson attempts to relate the bending strength of gear teeth to the physical properties of the material as determined by conventional tensile tests while Winter suggests that suitable material constants should be obtained by testing gears in power circulating rigs.

The latter procedure, while time-consuming and expensive, is likely to provide the most reliable basis for the calculation of gear tooth bending strength.

Helical gears

The bending strength of helical involute gear teeth was first systematically investigated by Wellauer and Seireg.[7] These authors provide a rational procedure for calculating the maximum bending moment for loaded helical gear teeth, based on the graphical superimposition of bending moment distribution curves, for point-loaded cantilever plates. The calculated values of bending moment were confirmed by strain measurements in cantilever plates with point loading, and with line loading inclined at several angles to the fixed edge. The process is suitable for calculating bending moments on the teeth of Wildhaber–Novikov,[8] the V.B.B. and other types of gears.

For involute gears the maximum bending stress may be calculated from:

$$\sigma_b = C\,\frac{W\sec\sigma}{r_c F}\cos\Psi_n\,\frac{Pn}{Y}$$

Where W = total tangential load at the pitch cylinder; P_n = normal diametral pitch; r_c = contact ratio; F = face width; Y = the tooth geometry factor of the virtual spur gear corresponding to the helical gear; σ = the helix angle;

Table 1. Materials that have proved satisfactory

Pinion Material	Surface Stress (Const. Sco.)	Wheel Material	Surface Stress (Const. Sco.)
		En. 8 normalized	1800
En 9 normalized	2300	En 146 normalized	1800
En 25T	3000	En 9 normalized	2300
En 40C. N–H	8000	En 19C N–H	6000
En 11C. I–H	9000	En 19C I–H	8000
		En 11C I–H	9000
		En 40 N–H	8000
En 34 C–H	10500	En 34 C–H	10500
En 36 C–H	11000	En 36 C–H	11000

Symbols: N–H = Nitride Hardened (Gas)
I–H = Induction hardened tooth by tooth
C–H = Carburized and hardened
En 25T — Hardened and tempered, 55 ton/in U.T.S.

Note: The pinion material on a given line is suitable for engaging with wheels in all materials listed on and above that line.

and Ψ_n = normal pressure angle. C is the factor taken from the chart in Fig. 2.[7]

The procedures for calculating the bending strength of spur gears by Winter and that of helical gears by Wellauer and Seireg can, therefore, be combined to form a completely rational approach to the calculation of the bending strength of involute gear teeth and it is to be hoped that a programme of tests on helical gears, similar to that carried out on spur gears by Winter, will be put in hand to provide further test information.

Resistance to surface stress

Chesters[9] produced surface fatigue curves for a wide range of gear steels, based on rolling tests in a disc machine which is shown in Fig. 3. His tests confirmed that surface fatigue resistance increases approximately as the square of the U.T.S. of the material, when the mating roller is in case-hardened alloy steel. It was found, however, that at over-stress the performances of normalized medium carbon steels compared favourably with those of hardened and tempered alloy steels and the unpublished results of later tests showed that, for steels in the latter category, there was a marked variation in performance with different mating materials.

Gear tests carried out in the power circulating rigs shown in Figs. 4 and 5, also those reported by Page,[10] Newman[11] and Niemann,[12] confirmed this tendency and emphasized its importance in the choice of gear materials.

Unequal distribution of load in the contact zone of gear teeth has the result that, for all practical purposes, the performance of gears in hardened and tempered alloy steels up to 75 ton/in² U.T.S. is little better than that of gears in normalized medium carbon steels, up to 55 ton/in², when mated with pinions of equal or greater hardness. However, when pinions in the higher tensile range are mated with wheels in low tensile, normalized medium carbon steel, their performance appears to be influenced by the low yield point of the mating material. This probably reduces the degree of over-stress in the contact zone, with beneficial results.

In general, test results and experience suggest that there is little, if anything, to be gained by using hardened and tempered alloy steels of up to 75 ton/in U.T.S. except for pinions to mate with normalized medium carbon steel wheels, or for wheels when their load capacity is limited by bending strength.

A convenient range of materials selected on the basis of Chesters' test results, which have performed satisfactorily in service, is given in Table 1, which is arranged to show suitable combinations of materials for pinions and wheels.

Novikov gears

Publication by Fedyakin and Chesnokov[22, 23] of work pioneered by Novikov in Russia, has revived interest in gears with concave/convex flanks on the respective teeth of the wheel and pinion of the gear pair. The tooth profile bears a vague resemblance to that of V.B.B. gearing[24] in current usage (see Fig. 6). Davies[25] and Harrison[8] discussed in detail the theoretical load capacity of Wildhaber–Novikov gearing and the latter reported results of tests of such gears.

That theoretical tooth contact stresses are lower than those in involute gears of similar proportions was amply confirmed by the test results reported by Harrison and others.[8] The frequent tooth breakages in W–N and V.B.B. gearing, however, suggest that tooth bending strength may be the limiting criterion for such gears. Relative positions of the mating teeth must be maintained within close limits, so as to avoid edge contact with unfavourable load distribution. This creates certain manufacturing problems and limits bearing clearances and the permissible differential thermal expansion between gear and case.

It must be pointed out that, for a given centre distance, the bending strength and surface load capacity of W–N gears do not decrease to the same extent as in involute gears, with a corresponding increase in gear ratio. It follows, therefore, that involute gears will compare most favourably with W–N gears when the gear ratio is 1, as in the tests reported by Harrison.

The application in service of a limited number of W–N gears is considered as an essential before the introduction into general use of a system of gearing which has certain advantages to offer, along with the inevitable disadvantages, some of which may remain hidden for a long time, like the sting in a wasp's tail.

Heat treatment

Medium-frequency, tooth-by-tooth progressive induction hardening, largely sponsored by the Admiralty, has been extensively developed during the past eight years. In full-scale load tests, induction-hardened second reduction marine gears in nickel-chrome-molybdenum and medium carbon steels have completed endurance runs of 50×10^6 wheel cycles, transmitting 16 000 h.p., corresponding to a K-factor of 900 and 2×10^6 wheel cycles, transmitting 24 000 h.p., and

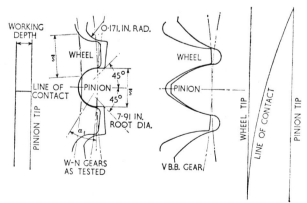

Fig. 6. Enveloping profiles of W-N and VBB gears and their contact loci

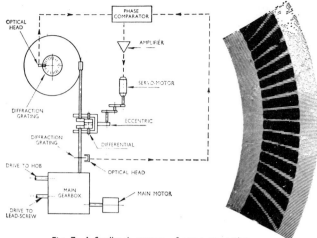

Fig. 7. A feedback system of error correction, based on the moiré fringe position measuring device shown on the right

to K corresponding 1360 as reported by Page[10] and Newman.[11]

The load-carrying capacity of gears hardened by this process to a minimum surface hardness of 550 D.P.N. is about 80 per cent of that of carburized and hardened gears; and, because of reduced distortion in heat treatment, a large induction-hardened gear can be profile-ground in about 60 per cent of the time required for a similar carburized gear.

The distortion of induction hardened gears depends on their design and particularly on their face width. Production batches of railway traction gears, about 28 in. diameter × 5 in. face width, have been induction hardened, and the gear element dimensional tolerances after hardening have remained within the limits prescribed for ground traction gears by BS 235. However, the amount of distortion increases with face width so that it is at present necessary, after induction hardening, to profile-grind gears with face widths greater than 8 in. to secure suitable distribution of load over the face.

The application of salt-bath nitriding and *Sulfinuz* treatment for gears is described by Müller[13] and Finnern[14] and treated gears have been tested extensively in France[15] and Germany.[16] Rettig[17] found that the surface fatigue resistance of low carbon steel was increased approximately nine times by a 90 minutes' salt bath treatment and that of a carbon chrome steel was increased $3\frac{1}{2}$ times. Similarly, *Sulfinuz*-treated En.9 steel gears, tested in a power circulating rig by the authors, have shown no sign of distress after completing endurance runs of 50×10^6 cycles. During these the calculated contact stresses were more than twice the value which resulted in the fatigue pitting of un-treated gears during endurance runs of about $2\frac{1}{2} \times 10^6$ stress cycles.

Müller[18] suggests that an explanation for the increase in load-carrying capacity of nitrided plain carbon steels may be found in the hypothesis that plastic working of the nitrided layers will result in the precipitation of iron nitride needles. It is considered that these will form in, and effectively block, the slip planes of the crystals, thereby increasing their fatigue strength. Maximum fatigue resistance is obtained when the components are quenched in water on removal from the salt-bath, so as to retain the greatest possible percentage of nitrogen in solution, as a source of nitride needle precipitation.

These processes, particularly the *Sulfinuz* treatment, increase resistance to abrasive wear and scuffing; in addition they offer dramatic increases in the surface fatigue resistance of gears in low carbon steels which are easily machineable. The salt-bath temperature needs to be only 550 degrees C and a stabilizing treatment before finish machining permits thermal distortion to be sufficiently limited so that the need for post-heat-treatment machining is obviated.

Chrome-molybdenum alloy steels such as En40, En18, En19 and En24 may be gas-nitrided by disassociation of ammonia vapour at 510 degrees C.

Newman[11] and Chamberlain[19] reported the testing of a nitride-hardened En40C, first-reduction marine test gear whose performance closely approached that of a similar case-hardened one. Further testing has since been carried out on this gear. It was damaged as a result of tooth breakage on a mating En40C pinion, also nitrided, at a load 25 per cent higher than the maximum at which the comparable case-hardened gear had been tested.

Published and unpublished results of disc tests by Chesters[9] indicate that the fatigue resistances of En19C and En40C nitride-hardened steels are, respectively, about 70 and 85 per cent of that of case-hardened En36 steel.

Nitrided En24 steel gears tested in the 8 in. centre-distance power circulator shown in Fig. 5, have not given such encouraging results as the nitrided En40C steel test gears.

Improved machining accuracy

Following the establishment of BS.1498 in 1948, the precision of the best quality hobbing machines has markedly improved. New master wheel generators have been developed from which it is possible to cut wheels up to 140 in. diameter in which the spacing errors over any arc do not exceed 0·001 in. or 3 seconds of arc.

A moiré fringe continuous circular measuring system, developed at the N.E.L., feeds error signals into a servo loop controlling the indexing of a 30 in. capacity hobbing machine, as shown in Fig. 7. This has reduced the maximum cumulative tooth spacing error from 0·0015 in. to below 0·0004 in. on a 30 in. diameter test gear. A multi-polar induction transducer has been similarly used to reduce spacing errors in a 140 in. precision hobbing machine for high-class gears. The arrangement is shown in Fig. 8. A 140 in. diameter test gear was cut on this machine and the maximum cumulative tooth spacing error did not exceed 0·0008 in.

Stepanek[20] developed a magnetic scale measuring system: a recording head, similar to that of a tape-recorder, traces a sine wave on a band of magnetic oxide deposited on the periphery of a disc or of a hobbing machine table. The system permits circular measurement accurate to within one second of arc; and a mechanical system employing a cam, designed in accordance with measured errors, can be installed in the production machine's indexing transmission to correct its errors. This measuring system is particularly elegant and has great flexibility. However, both the moiré fringe system and the inductive transducer have the distinct advantage that position errors of individual lines on the gratings (or of different radial elements in the inductive device), are averaged out in the respective measuring systems.

Höfler[21] describes a range of inductive pick-ups designed for the automatic measurement of tooth thickness and space width round a gear of any size, whilst the gear is being rotated at a speed, corresponding to the passage of one tooth per second past the measuring head. Additional heads measure the spans of pitches or single pitches, whilst the gear is rotated slowly on a hobbing machine. Devices mounted on the hob saddle are designed for measuring tooth flank undulations and the accuracy of tooth profiles.

The measurements are recorded automatically on a paper strip chart and, compared with current inspection methods, save up to 80 per cent of the time. Other attachments are provided for measuring the accuracy of hobs and worms; and index errors in hobbing machines. These, while useful to any organization at present lacking the means to carry out such measurements, are not as useful as some instruments already available.

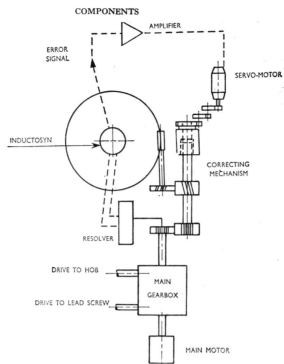

Fig. 8. A similar feedback system for index control based on the 'inductosyn' transducer shown beside it: the stator is at the top

Finishing

The improvement in gear accuracy and surface quality that resulted from the policy, instituted immediately after the end of the 1939–45 war, of building more accurate gear producing plant, has become more obvious during the past few years. No sudden advance can be reported but, whereas a few years ago the production of gears to comply with BS.1807 class A1 standard was a severe tax on resources, this has now become easier. The use of large hobs integral with their shafts and carried on closely spaced bearings in hobbing machines, built to the higher standards imposed by BS.1498, has gone far towards the production of hobbed gear profiles that deviate very little from the theoretically ideal modified involute. Surface finish obtained with such hobs is now nearly as good as that resulting from the most carefully shaved finish.

Feed rates have little influence on hob life and can be increased to 0·100 in. without adverse effect but they do modify the finish. The rather rough hobbed flank surface conforms closely in shape to the required profile but requires a shaved finish. Hobbing, followed by shaving prior to heat treatment or case-hardening, has therefore almost entirely replaced previous methods such as profile grinding or lapping after hardening.

A process that seems to have evolved from a combination of shaving and grinding is gear honing, in which an abrasive gear-like tool is meshed with the work at crossed axes.[26] The tool is the driving member while the work is reciprocated past it.[27, 28] Selective honing can be carried out to improve contact or to impart small deviations to the tooth spirals by repeated passes. Pressure between tool and work is applied either radially or tangentially to remove burrs and bruise projections, to improve finish or to modify profile or helix. It has also been successfully used to correct small amounts of distortion[29] due to heat treatment. The process can also be used on soft gears, for example, marine turbine gears *in situ*. In this application the pinion is usually treated and the hone is carried on a hob, capable of engaging with both helices. This is rapidly oscillated as it is fed slowly in an axial direction while the gear is turned by the prime mover.[30]

Grinding has been used for a long time as a gear finishing

operation but its application to large gears is a more recent introduction and its use has been fostered by the employment of case-hardened gears for marine main propulsion units. The carburizing and hardening processes applied to such gears result in distortion far exceeding what can be tolerated in a precision product and grinding is the only known practical means of removing this. The finished gear is highly accurate in tooth spacing, profile and helix, although both the latter usually deviate from the geometrically correct form to compensate for deflection under load.

A more recent application of widespread interest concerns the gears coupling together the two diesel engine banks in British Railways Type 2 12 LDA 28 locomotives, where grinding of the case-hardened teeth is the only practical method of attaining the BS. 1807 Class A1 standard.

Machines have been introduced for deburring gears by continuous grinding. Spur, helical and bevel gears are dealt with by a grinding wheel like a worm which produces a slight chamfer on the edges of the gear teeth.[31, 32]

Lapping has been applied as a finishing process to bevel and hypoid gears as part of the normal manufacturing procedure; also to spur and helical gears of all sizes, but it can so easily destroy a tooth profile unless it is carefully controlled. Further, with soft materials there is a tendency for the abrasive to become embedded in the tooth surfaces with unfortunate consequences. For these reasons, lapping is not generally used, but when it can be applied for a few minutes only, as a final finishing operation to gears normally finished by some other process, some benefit from a reduction of noise in running has been found.[33]

Surface treatment

The great majority of gears are not given any surface treatment after manufacture because they function satisfactorily without one. When treatment is applied for a specific purpose, it has to be economically justified; for example, phosphating. This is a chemical process which

Table 2. The effect of mineral oil viscosity on scuffing

Viscosity, secs. Redwood at 140°F.	98	215	450	750	5000
Time to produce scuffing, secs.	33	120/130	200/300	210	970/1420

imparts an iron or manganese phosphate coating to ferrous surfaces and assists running in. Originally intended to prevent scuffing in hypoid axle gears, it has more recently been applied to other types, notably high-speed turbine gears. A somewhat similar result is obtained from a salt-bath-produced magnetic iron oxide surface. Such surface treatments permit the use, if necessary, of a lubricating oil one grade lower in viscosity than that normally employed for untreated gears.

Plating gear teeth with a thin coat of copper or cadmium assists running in, particularly with heavy loads. It appears to reduce the effect of asperities and permits the maintenance of an oil film. As surface asperities are removed by abrasion, plastic flow, or deformation during bedding in, the plated surface slowly wears off. This process, originally introduced for vehicle axle driving gears, has recently been applied to a wider range of gears. Heavily plated tooth surfaces are usually unsatisfactory since the plating flakes off during normal tooth action.

A surface treatment that is really a lubrication adjunct is a coating of molybdenum disulphide. In many cases this is carried in a plastic film which is baked on to the gear tooth flanks and, while it remains in position, provides protection against scuffing. This type of surface finish enables bedding in of the tooth profiles to be carried out but it does not appear to be so effective as the chemical treatments. It is useful, however, with dry running or with inadequate fluid lubrication. Test results show that such surface treatments do not enhance scuffing resistance of oil-lubricated steel discs, but scuffing loads are much less scattered and approach the maximum attained without treatment.

Lubricants

Plain mineral oils have remained by far the most widely used lubricants for gears. Some measure of their relative performance was given by a recent test. The effect of viscosity was shown by running different oils in a power circulator where two 8 in.-centres helical gear units were connected in parallel by torsion shafts. The gear ratio was 14/34, the pitch circle speed was 180 ft/min. and a torque of 500 lb in. on the pinion was circulated. After smearing the tooth flanks with each of the lubricants to be tested in turn, the gear units were run until a change in noise level or the appearance of oil vapour indicated that the oil film had been disrupted and scuffing was beginning. The running time for each run with the different oils, and the results, are shown in Table 2.

A well-defined increase in the scuffing resistance obtained with mild types of e.p. oils has been noted in recent years. Previously, their advantage over plain mineral oils was most pronounced at low speeds. Lately, however, an improvement in load capacity of about 30 per cent over that of a plain mineral oil has become usual. They also retain their e.p. properties for a long period of operation. Active e.p. lubricants, too, have in some instances given greater scuffing resistance and more stability. Some vehicle manufacturers have adopted the American practice of fitting into their products, back axle hypoid units sealed for life.[34] This has involved the use of more stable additives to the e.p. lubricants. Some, however, have been reported to break down in axle units at temperatures above 200 degrees F and to cause gear scuffing with bearing corrosion and wear.[35]

Examination of lubricants removed from the axle units of U.S. Ordnance Department vehicles after 5000 to 10 000 miles showed that all gears had suffered loss of load carrying capacity.[36] The conclusions to be drawn indicate that e.p. additives are still expendable. But, with some of the latest types, their life approaches that of the vehicle.

For a number of years, attempts have been made by operators of road vehicles to use the same oil in all axle units, whether fitted with hypoid, bevel or worm gears. In practice this has meant that an oil suitable for hypoid axle gears has been used and, although it will also be satisfactory for bevel gears, it may not be equally satisfactory for worm gears. Recently, oils of multi-purpose type have been produced.

The period under review has seen the more extensive use of synthetic lubricants for gears. In worm gears the characteristically low friction with polyalkalene glycols and oxides, equal to that with castor oil, has resulted in a considerable increase in operating efficiency. It has been found possible to run identical worm gear units with temperature rises of 75 to 80 per cent of those obtained with plain or compounded mineral oils.

In rear axle worm gear differential units, the use of a synthetic oil of S.A.E.90 or preferably S.A.E.140, viscosity has resulted in a fuel saving of about 2 per cent.[37] When vehicles are subjected to continual starting and stopping, as in urban passenger transport service, synthetic oils do not seem to improve pitting resistance of the wormwheel teeth; the fuel saving, however, is still apparent and the reason for this is being investigated; in industrial units no such difficulty has been encounted.

Another feature of synthetic lubricants is that they have a low pour point which means that they can be used in ambient temperatures down to − 20 degrees F and still remain fluid. This is very useful in cold climates.

Within the past few years the elastohydrodynamic theory of machine lubrication, with particular reference to gears, has been considerably expanded, and experimental evidence that the theory is basically tenable has been obtained by some carefully controlled work.[38–40] Briefly, the theory postulates that the action of gear teeth is such that a thin film of oil is drawn between them. Because viscosity and density of a mineral oil are increased very considerably when it is under pressure, this film is maintained when under very high pressures. The load on gear teeth is transmitted through the oil film which, because of the elastic deformation of the tooth surfaces, remains of substantially constant thickness throughout the loaded zone.[41–43]

Measurement of oil film thickness by experimental means has shown that, for very high surface loading, it is very thin and largely independent of the load.[40, 44] This result emphasizes the desirability of high geometrical accuracy.[45] Further work still remains to be done but the progress made in elastohydrodynamic theory has brought a gear rating formula based on scuffing resistance significantly closer.[46]

Cutting oils

Fig. 9 illustrates typical results of cutting En.9 steel with a single hob tooth on which the amount of wear was recorded for a measured length of material cut. It shows the variation that can be obtained with two different neat cutting oils of the same viscosity.

No spectacular change has been made in cutting oils used for gear production during the past few years but a feature of their performance has been consistency in results. The intermittent type of cutting in the hobbing, shaping and milling processes is apparently not appreciably affected by the extensive use of additives in the straight cutting oil. Cutters do not appear to have a longer life between grinds but there is usually an improvement in surface finish. It has been shown, however, that viscosity is a more important feature of a cutting oil than was previously realized.

Generally, plain oils, usually not highly refined, have been found to be at least as effective gear cutting coolants as some of the more highly refined types containing additives. On the other hand, a shaving coolant should possess a low coefficient of friction and also carry an anti-welding additive for it to produce a good surface finish and be most efficient.[47]

The growth of high-speed hobbing for automotive gears has introduced the problems of excessive vapour formation which can be mitigated to some extent by choosing the correct type of cutting oil, but the most effective method is to ensure a copious supply of coolant to the hob and the workpiece.[48]

High-pressure jets possess certain attractions for gear cutting but little or no influence on tool life; and they create problems of spray. The same general comments apply to cutting oils used in the form of mist.

Future trends

In this review a few items out of many possible ones have been chosen for more detailed examination. Over the period considered, these have influenced the thoughts and actions of gear technologists and in that respect they must be credited with importance.

For a better understanding of gear tooth loading both tooth strength and surface fatigue are of importance. The work reviewed forms a serious attempt to arrive at a rational procedure for determining the bending strength of spur and helical gears. Although not yet widely used, it will undoubtedly influence future thought and practice on this subject. As regards resistance to surface stress, results from both laboratory tests and service experience have demonstrated the importance of mating gear materials. The reasons for this are by no means fully understood and the search for the true explanation may well form the nucleus of future work on permissible ratings for gears.

In the field of heat treatment, induction hardening has demonstrated its advantages and limitations. Induction hardening tooth by tooth has many potential applications and when a marked increase in fatigue resistance over that obtained with through-hardened materials is required, the method is likely to prove attractive for both vehicular and industrial use, as a substitute for case-hardening. Other heat treatment processes that have demonstrated their usefulness are salt bath nitriding and the *Sulfinuz* treatment. If these processes are more extensively applied in the future, as they deserve, there is a good reason to expect an increase in permissible wear ratings for low and medium carbon steels that will have useful economic results. Gas nitriding too is a useful substitute for carbon case-hardening, particularly for turbine and epicyclic gears. Since, after nitriding, correction of distortion by grinding is seldom necessary, considerable future use may be confidently anticipated.

The outstanding advance in accuracy has been achieved by precision hobbing machines and by the continuous automatic correction systems. Perhaps the next steps will lie in the improvement of hobbed gear tooth profiles and surface finish.

It seems probable also that reductions in gear noise can be expected as the result of further progress in accuracy; in any case, the reduction of noise in machines should be an obligation of mechanical engineers.

The elastohydrodynamic concept of gear lubrication has explained rationally some of the observed phenomena but there are still many more aspects of gear lubrication to be explored. Plain mineral oils have held their own by increased consistency and stability. It seems probable that they will continue in use for separating and removing heat from contact surfaces.

Gear units will probably be equipped with oil of a life expectancy not shorter than that of the gears themselves. Some advance towards this end has already been made with the introduction of more durable lubricants in sealed back-axle units. Finally, synthetic substances seem to have an assured place as gear lubricants. They are still in the early stages of development but, because desirable characteristics can be built into them, they will be used increasingly.

If the advances made in the design, the accuracy of manufacture, the heat treatment and the lubrication of gears are to be exploited to best advantage, gear cases and bearings must be capable of transmitting the increased loads. Unfortunately, the principles of gear case design have not

Mr H. J. Watson *was educated in Grimsby and at University College Nottingham; he received practical training with Ruston and Hornsby, Ltd, Lincoln and J. S. Doig, Ltd, Grimsby. In 1929 he joined the Research Department of David Brown and Sons, Ltd, Huddersfield and held a number of senior technical appointments with David Brown Gear Industries. In his career he has been concerned with many aspects in design, manufacturing methods, materials and tribology in the continuing evolution of toothed gearing. He has served on NAVGRA and Institution Committees concerned with gearing and tribology.*

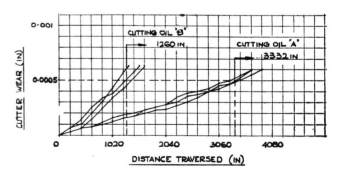

Fig. 9. *Rate of wear, measured on a single hob tooth cutting En. 9 steel with two neat cutting oils of the same viscosity, shows the difference in performance. Three tests were made with each oil*

Mr W. H. Harrison *is the chief Research and Development Engineer of David Brown Gear Industries Ltd, Huddersfield. He was educated at Coventry Technical College and joined the Coventry Gauge and Tool Co Ltd, in 1938. In 1940 he served in France with the British Expeditionary Force and became a prisoner-of-war until May 1945. He rejoined the Drawing Office Staff of his former Company and subsequently represented them on B.S.I. Gearing Committees. In 1956 he joined the Research Department of David Brown Industries Ltd, and has been closely associated with the B.S.I. Gearing Committees.*

been developed as greatly as those of the gears. Deflection noise, and vibration during operation need reduction.

Acknowledgments

The authors wish to acknowledge their indebtedness to David Brown Industries Ltd, and to Mr P. Walker, Head of Research, for permission to publish information based on results of work carried out in the Research Department at Huddersfield.

They also acknowledge the help given to them by their colleagues in the preparation of this review.

REFERENCES

1. TIMMS, C., discussion on paper by A. SYKES. Nov. 1959. 'Allowable Loads on Helical Gears'. 1959 British Gear Manufacturers Association.
2. NEIMANN, G. and RICHTER W., 'Einfluss der Umfangsgeswindigkeit auf die. Zahnflanken Tragfahigkeit.' 1960 *Konstruktion* 12 Heft 5.
3. NIEMANN, G., and RICHTER, W., 'Minderung der Zahnflanken Tragfahigkeit durch Verzahnungsfehler'. 1960 *Konstruktion* 12 Heft 7.
4. DOLAN, T. J., and BROGHAMER, E. L., 'A photo-elastic Study of Stresses in Gear Teeth Fillets'. 1942 Univ. Illinois *Bull.* 335/1942.
5. JACOBSON, M. A. I., 'A Photo-elastic Investigation of Bending Stresses in Spur Gear Teeth'. *Proc. Int. Conf. on Gearing. Instn mech. Engrs, Lond.,* p. 99.
6. WINTER, H., 'Gear Tooth Strength of Spur Gears'. 1961 British Gear Manufacturers Association.
7. WELLAUER, E. J., and SEIREG, A., 'Bending Strength of Gear Teeth by Cantilever-Plate Theory'. 1959 *A.S.M.E.* paper 59–A50.
8. HARRISON, W. H., 'Belastungsproben an Stirnrädern mit Zähnen nach dem Wildhaber-Novikov-System'. 1961 *V.D.I. Berichte* Nr 47.
9. CHESTERS, W. T., 'Study of the Surface Fatigue Behaviour of Gear Materials with Specimens of Simple Form'. 1958 *Instn mech. Engrs, Lond., Proc. Int. Conf. on Gearing.*
10. PAGE, H. H., 'Advances in Loading of Main Propulsion Gears'. 1958 *Instn mech. Engrs, Lond., Proc. Int. Conf. on Gearing.*
11. NEWMAN, A. D., 'Load Carrying Tests of Admiralty Gearing'. 1958 *Instn mech. Engrs, Lond., Proc. Int. Conf. on Gearing.*
12. NIEMANN, G., 1960 *Maschinenelemente*, Band 2., Springer-Verlag pp. 90, 120.
13. MÜLLER, J., 'Das Weichnitrieren und das Sulfinuzieren, zwei neuere Verfahren zum Behandeln Verzahnter Bauteile'. 1958 *V.D.I.-Z* 100 Nr 6.
14. FINNERN, B., 'Badnitrieren, ein Verfahren zur Erhöhung der Verschleiss und Wechselfestigkeit von Bauteilen'. 1959 *Maschinenschaden* 32 Nr 7/8.
15. CHEVALLIER, J., and NOEL, R., 'Essais D'Engrenages Reducteurs Marins'. 1958 *Assoc. Tech. Maritime et Aero.* Paris.
16. NIEMANN, G., and RETTIG, H., 'Weichnitrierte Zahnrader'. 1960. *V.D.I.-Z*, 102 Nr 6.
17. RETTIG, H., 'Tragfahigkertsunterserchungen an Weichnitrierten Zahnradern'. *V.D.I. Berichte* Nr 47.
18. MÜLLER, J., 'Beobachtungen zur Frage der Wälzfestigkeitserhöhung bei Zahnradern aus unlegiertem Stahl'. 1961 *V.D.I. Berichte* Nr. 47.
19. CHAMBERLAIN, A., 'Developments in the Heat Treatment of large (Marine) Gears'. 1958 *Instn mech. Engrs, Lond., Proc. Int. Conf. on Gearing.*
20. STEPANEK, K., 1960 *Machinery*, Lond. vol. 96 pp. 646–652.
21. HOFLER, W., 'Die Ursachen der Verzahnungsfehler beim Walzfräsen und Wälzstossen sowie ihre Ermittlung mit neuen electronischen Verzahnungs-Messgeräten'. 1959 *V.D.I. Berichte* Nr 32.
22. FEDYAKIN, R. V., and CHESNOKOV, V. A., 'Principles of the Novikov Gear System'. 1958 *Vestnik Machin* No. 4.
23. FEDYAKIN, R. V., and CHESNOKOV, V. A., 'Design of Novikov Gears'. 1958 May *Vestnik Machin.*
24. *Marine Eng.* Dec. 1933 'V.B.B. Enveloping Tooth Gearing'.
25. DAVIES, J. W., 'Novikov Gearing'. 1960, 13th January. *Machinery.*
26. PELPHREY, H., 'Abrasive Finishing of Hard Gears'. 1958 *A.S.M.E.* Paper No. 58-A-226.
27. BREGI, B. F., 'Gear Tooth Honing—A new approach to improving Gear Surface Finish'. 1958 *Amer. Soc. Tool Engrs,* Paper No. 77.
28. JONES, J. H., 'Honing set-up Processes Gears in Fast Adaptable cycle'. 1959 16th April, *Iron Age*, p. 124.
29. HOOPER, E. A., 'Honing Specified to Salvage Hardened Gears'. 1960 *Metalworking*, Aug.
30. de Laval Information Bulletin No. 23 1961 Oct.
31. BOEHRINGER, GEBR., G.m.b.H., 'Boehringer Type Z.400 Gear De-burring Machine'. 1961 13th Sept. *Machinery*, Vol. 99.
32. Alfred Herbert, Ltd, 'D-Burr Gear Deburring Machine'. 1961 *Metalworking Production*, vol. 105, No. 6, p. 70.
33. OPITZ, H., 1961 *V.D.I. Berichte* No. 47.
34. HUNSTAD, N. A., 'General Motors Looks at Rear Axle Lubricants—Sept. 1958'. 1958 National Lubricating Grease Institute, U.S.A., Spokesman, 1958, Feb. p. 12.
35. ELLIS, W. E., HILL, R. L., and DOWNS, E. M., 'Breakdown of E.P. Lubricants in Pinion Bearings'. 1961 *J. Amer. Soc. Lub. Engrs*, 1961, Nov. p. 529.
36. NOLL, C. R., 'Status of Automotive Rear Axle Lubricants'. 1958 *Soc. Auto. Engrs Inc.*, paper S 108.
37. WHITTLE, J., 'Lubrication of Automotive Worm Gears'. 1960–61 *Proc. Instn mech. Engrs, Lond.*, (A.D.) No. 3. p. 119.
38. ARCHARD, G. D., GAIR, F. C., and HIRST, W., 'The Elasto-hydrodynamic Lubrication of Rollers'. 1961 *Proc. Roy. Soc. A.* vol. 262, p. 51.
39. CROOK, A. W., 'Elasto-hydrodynamic Lubrication of Rollers'. 1961 24th June, p. 1182, *Nature.*
40. CROOK, A. W., 'The Lubrication of Rollers II and III'. 1961 *Phil. Trans. Rov. Soc.* vol. 254 p. 223.
41. SIBLEY, L. B., and ORCUTT, F. K., 'Elasto-hydrodynamic Lubrication of Rolling-contact Surfaces'. 1961 *Amer. Soc. Lub. Engrs.*
42. DOWSON, D., and HIGGINSON, G. R., 'The Effect of Material Properties on the Lubrication of Elastic Rollers'. 1960 *J. mech. Eng. Sci.* No. 3. p. 188.
43. DOWSON, D., and HIGGINSON, G. R., 'New Roller-Bearing Lubrication Formula'. 1961 4th Aug., p. 158, *Engineering.*
44. MacCONOCHIE, I. O., and CAMERON, A., 'The Measurement of Oil-film Thickness in Gear Teeth'. 1958 *Trans. Amer. Soc. Engrs.*
45. SMITH, F. W., 'Lubrication Behaviour in Concentrated Contact—Some Rehological Problems'. 1960 *Trans. Amer. Soc. Lub. Engrs.* Vol. 3. No. 1. p. 18.
46. MERRITT, H. E., 'Gear-tooth Contact Phenomena'. 1961 *Proc. Instn mech. Engrs, Lond.*
47. TAYLOR, C. J., and HARRIS, B. V., 'Cutting Fluid Selection and Special Cases'. 1959 Nov. p. 43, *Scientific Lubrication.*
48. MORTON, I. S., 'Methods of Cutting Fluid Application'. 1959 Nov. *Scientific Lubrication*, p. 39.

Heat exchangers are used in most industries and the ways in which thermal energy is transferred continues to be the subject of much research. Although advance has been slow in the design of some heat exchangers, nuclear, cryogenic and aerospace applications have stimulated the development of new ideas which may find wider uses. Progress has been made in manufacturing methods and in the use of newer materials.

The Ubiquitous Heat Exchanger

by M. K. Forbes, MA, CEng, MIMechE

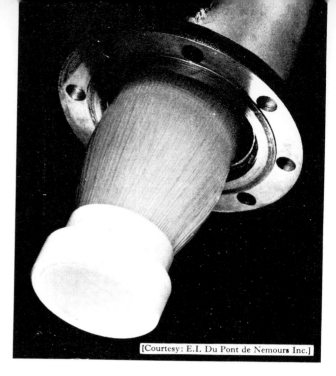

[Courtesy: E.I. Du Pont de Nemours Inc.]

Fig. 1. A Teflon tubular heat exchanger of honeycomb construction with a ring replacing the tube plate

Heat exchangers have—in a multitude of forms—as wide a range of applications as almost any other engineered product. From central heating systems and internal combustion engines through to chemical processing plants and electricity generating stations, heat has to be transferred economically.

In some instances a heat exchanger may be made an integral part of the equipment it serves, such as on the cylinders of an air-cooled IC engine or the fuel cans of a nuclear reactor but in the majority of cases heat exchange between two fluids takes place in a separate unit designed specifically for that purpose.

Sometimes complete mixing of the two fluids is permissible, such as air and water in a cooling tower. Here there is direct transfer of heat (often accompanied by some mass transfer), the heat exchanger's prime function being to promote contact between the fluids. In yet other cases where complete mixing may be unacceptable (perhaps because of a substantial difference in pressure) a small degree of mixing may be tolerated. Then the regenerative type of heat exchanger may be employed, in which a porous, heat-retentive 'matrix' is alternately heated by one fluid and cooled by the other. Although I shall touch upon 'direct contact types', I shall deal mainly with exchangers in which the fluids are kept separate.

The wide range of operating conditions for which heat exchangers are required has produced a corresponding array of designs, ranging from robust pieces of ironmongery to light-weight constructions made almost entirely from foil. The development of new materials and manufacturing techniques have had their impact and the extension of the knowledge of heat transfer also plays an important role.

The mechanisms of heat transmission, under the headings of condensation, boiling, forced and free convection, radiation and conduction, are still far from being completely understood so that heat transfer remains very much an empirical science. Its more theoretical aspects are continuously dealt with in an international periodical[1] and of recent years several international technical conferences[2] have

been devoted to the subject, providing a wealth of theoretical and practical papers on many of its facets.

References to many published papers and articles in the field are published by the NEL.[3] Amongst the available reference books, the work of McAdams[4] probably still remains the most widely used in the general field and that by Kern[5] in the specific area of process plants. A survey of heat transfer equipment is published annually.[6]

From the point of view of their application, heat exchangers may be divided into those which are essential for the operation of a process (eg, by maintaining the right temperature) and those whose use increases thermodynamic efficiency. An example of the former is the jacket water cooler of an IC engine and of the latter a combustion air preheater in a boiler.

While cost is of course important in the selection of an essential heat exchanger, it is the decisive factor in the 'desirable' category. Thus, most gas-turbine-driven generating sets do not employ an air preheater, even though overall efficiency could be increased substantially by its use.

Reliability ranks high, if not at the head of the design requirements. A heat exchanger is usually an ancillary but nevertheless vital part of a larger and far more costly plant. Other design factors are weight and compactness (particularly for aerospace uses), accessibility for cleaning (especially for food and drink processing) and life span.

Tubular types

Amongst the most widely used forms of heat exchanger is the shell-and-tube type. For processing plant its design standards are largely prescribed in the American TEMA Standards[7] of which 1968 saw a new edition and to which the British Standard[8] closely corresponds.

Partly because of the widespread acceptance of these standards, recent development of shell-and-tube heat exchangers has been confined mainly to improved manufacturing techniques and the use of new materials. For joining tubes to tube-plates the tungsten inert gas (TIG) technique has become more widely used and special-purpose automatic equipment has been developed for controlling electrode movement and the weld current cycle.

To combat corrosion from chemicals, titanium, and more recently tantalum, have found increasing use, for the tubes or linings. Whilst clad materials have been employed for some time, particularly for the tube-plates, cladding by weld deposition and explosion welding are making such designs more economical. For high-temperature applications, such as in steam superheaters for electrical generating plant, further development of metallic cladding appears to be the most likely way forward[9] rather than the employment of exotic and expensive metal alloys or, conceivably, non-metals such as ceramics.

While the design of the product has altered little in recent years, considerable knowledge has been gained of the performance of shell-and-tube units. Work at the University of Delaware[10] and by Tinker[11, 12] on heat transfer and pressure loss for single-phase operation on the shell side has produced correlations which increase the accuracy of performance predictions. The corresponding data for the inside of round tubes have recently been evaluated and summarised.[13]

Non-metal constructions

From the USA comes the use of fluorocarbon materials.[14] The ends of small-diameter, thin-walled tubes are joined together in honeycomb fashion, inserted within, and secured to, an annular ring, also of fluorocarbon, which effectively forms a tube plate at each end of the tube bundle as shown in Fig. 1. The unit so formed may be inserted into a shell or, fitted with headers to conduct the fluid flowing through the tubes, immersed in a tank to heat or cool its contents.

The flexibility of the tubes can, of course, be a great advantage in this application. The inertness of fluorocarbons, particularly PTFE, is well established and makes them very suitable for corrosive media. There are, however, strict limitations of temperature and of stress, and therefore operating pressure. The comparatively low thermal conductivity of the material is to a great extent offset by the large heat exchange surface packed into a given volume through the use of small diameter tubes.

Another interesting feature of this design is that, instead of using plate baffles with the tubes laced through them, which is normal in shell-and-tube types, the tubes are spaced apart and the fluid guided over them by transverse strips. This may

be done with the tube array flat, after which it is rolled up, swiss-roll fashion.

In the high-pressure field, some recent designs of feed water heater[15] for fossil fuel power stations have adopted a toroidal form of header, welded to the flat tube-plate of a shell-and-tube unit employing hair-pin tubes, as shown in Fig. 2. This is claimed to economise on material. In those heaters a bore welding procedure is being used to join the tube to the tube-plates. High integrity of these joints is essential, bearing in mind the cost of outage of the associated steam-raising plant which could result, should a leak develop.

Also in the power station field, we hear of another use of a non-metal: glass tubes are proving successful in combustion air pre-heaters,[16] where the products of combustion frequently cause corrosion of metal tubes. Borosilicate glass has been selected because of its high corrosion resistance and its low coefficient of expansion. Some care has naturally to be exercised when cleaning these tubes but, provided hot water is used to clean hot tubes, no disaster results. In the event of a mishap, individual tubes can be replaced, as each is secured into the tube-plates by impregnated asbestos or bonded glass fibre grommets. The glass tubes are somewhat thinner-walled than the metal ones they replace and in practice the lower thermal conductivity of the glass has little effect on performance. While this application of glass tubes is novel, glass is no newcomer to the heat exchanger field and is used for handling very corrosive and very pure fluids in the chemical, food and soft drinks industries.[17]

In considering non-metals we should not omit graphite.[18] The thermal conductivity of this material is in the same bracket as that of common metals, a property not shared by most other non-metals; it has generally very high corrosion resistance and can withstand high temperature. These properties offer advantages in some applications but its comparatively low tensile strength and high brittleness impose limitations. While shell-and-tube units are made of graphite, it is more usual to have a composite rectangular graphite block into which are built layers of passages through which the two fluids pass in cross-flow. Operating pressure can be increased by keeping the block in compression.

Turbulence and finning

In almost all instances, increased turbulence results in better heat transfer between a fluid and the surface over which it flows. In a shell-and-tube heat exchanger this may be achieved by indenting the tubes with a pattern of dimples, as in the aircraft engine-fuel-cooled oil cooler illustrated in Fig. 3. This unit employs 3 mm diameter, thin-walled

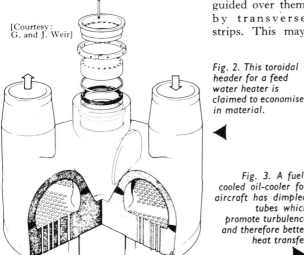

[Courtesy: G. and J. Weir]

Fig. 2. This toroidal header for a feed water heater is claimed to economise in material.

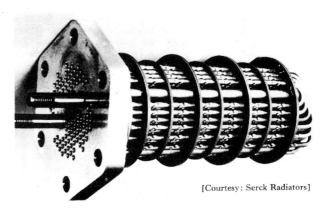

Fig. 3. A fuel-cooled oil-cooler for aircraft has dimpled tubes which promote turbulence and therefore better heat transfer

[Courtesy: Serck Radiators]

aluminium tubes, secured into aluminium-alloy tube-plates by means of expanded ferrules.

The aircraft industry needs ever lower weights and greater compactness and yet the utmost reliability. Since the days of piston engines, the aircraft engine oil cooler has shrunk in volume by a factor of about 15, while the engine power output has risen perhaps $2\frac{1}{2}$ times. By far the biggest step in this development came with the gas turbine, the fuel flow to which is sufficient to act as the coolant for the engine lubricating oil, whilst the heat absorbed by the oil is less than in piston engines. Indeed the fuel is now often used as the coolant for the hydraulic control systems and other heat-generating equipment on board.

Water is an excellent medium for heat transfer (hence its widespread use) and oils generally are much inferior. For compactness, extended surface finning may be used on the oil side of the separating material. For the cooling of engine lubricating oil, particularly when using the jacket water as the coolant, fins bonded to the tubes of a shell-and-tube heat exchanger provide a neat solution. In the design illustrated in Fig. 4, segmental (D-shaped) baffles are used to guide the oil over the tubes in several passes and the fins are circular but with two equal segments removed on parallel chords to provide the necessary channels for flow of oil between the passes.

Another type of extended surface now used in similar applications takes the form of a matrix of spherical metal particles occupying the space between the tubes. The particles are metallically bonded to each other and to the outer surfaces of the tube walls, whilst the intercommunicating interstices between the spheres are available for the flow of fluid. Very high rates of heat transfer are achievable but, of course, resistance to flow is correspondingly high.

Occasionally an engine layout is such that it is more convenient for the lubricating oil to pass through the tubes of its cooler and the cooling water outside the tubes. Some designs of internal finning have been developed for this purpose from wire[10] or foil, these being inserted into, and bonded to, the tubes. Another approach is to increase the heat transfer between the oil and the tube walls by fitting into the tubes turbulence-promoting devices such as a twisted strip.[20] In other designs, suitable for very small oil-coolers, an extended-surface member lies in the annulus between two concentric tubes, the oil flowing through the annulus whilst the water passes both over the outer and through the inner tube.

Plate types

Originally employed in the processing of milk, the demountable plate type of heat exchanger has found increasing use in other plants for the processing of food[21] and chemicals[22] because it can be taken apart with comparative ease, making every surface accessible for inspection and cleaning. Most plate designs, such as the one in Fig. 5, have a corrugated heat exchange surface. To withstand the difference in pressure between the fluids handled, judiciously placed indentations in the plates are necessary for strengthening, eg, round the ports. These features, together with the grooves round the periphery and round each port, which house the inter-plate seals, combine to give a pressing of rare complexity. The design of suitable tools demands a deal of expertise and not a little development.

Whilst stainless steel is the most common material for such plates, further manufacturing problems have been introduced by the demand for titanium plates, because of the greater 'spring-back' experienced with this material.

[Courtesy: Serck Radiators]

Fig. 4. Fins bonded to the tubes increase the compactness of an oil cooler

[Courtesy: Serck Radiators]

Fig. 5. Most plate-type heat exchangers use corrugated plates for improved performance and strength. This demountable layer construction makes for easy cleaning

Recent trends have been towards the greater use of this type of heat exchanger for shipboard cooling duties, with plates of aluminium-brass and the development of designs with increased temperature and pressure capabilities. Generally, higher heat-transfer rates can be obtained from plate-type than from tube heat exchangers, mainly because of the greater turbulence in the sinuous flow passages.

In the past, performance data for commercial models have been closely guarded by their manufacturers. Recent reports[23] from the NEL give the thermal and hydrodynamic performance characteristics of three such units. These results are commented upon by Lohrisch[24] who compares their efficiencies to those obtainable from shell-and-tube designs.

Where pressures are too high for the demountable type and separation of the surface for cleaning is unnecessary, plates with a regular pattern of abutting indentations may be welded together round their edges. This provides the support between plates necessary to withstand the pressure difference.

An interesting design, illustrated in Fig. 6, is somewhere between the fixed-plate and the shell-and-tube type. The heat transfer surface is constructed from sheet material, formed to provide fluid flow passages where the sheets are joined. At their ends, the sheets are secured by welding into a skirt which provides a member to be sealed to the shell. One feature of this design is that the two fluids handled in the heat exchanger pass throughout in strict counterflow which gives the highest obtainable thermodynamic efficiency.

The hot roll-bonding technique of joining metal (particularly aluminium) sheets together and forming interconnected flow channels between them has been used for some years in the manufacture of evaporators for domestic refrigerators.

A pattern printed in a special ink on one sheet corresponds to the desired flow passages; a second sheet is placed on top and both are passed through reducing rolls at high temperature. Those areas not inked bond together with a true weld, but not the others, through which flow passages are then formed by hydraulic inflation.

More recently this method has been applied to the manufacture of larger units. Where the material required is not amenable to roll-bonding, the sheets may be preformed and welded together. Such units are used either for jacketing or as heat exchangers within liquid processing vessels.[25] This type of construction is also used to transfer heat by contact between three separate units in a three-sheet sandwich, with the outer two handling the one fluid and the centre one the other as shown in Fig. 7. By clamping the sandwich together, good contact between the plates is provided and each unit can be made of the material best suited to the fluid which it handles. Perforation of any one plate will not cause the fluids to intermix.

Where the nature of the fluids permits it, advantages in compactness, weight or both, can be gained by the addition of extended surface in contact with both fluids. Here the 'brazed' surface exchanger, shown in Fig. 8, may provide the best solution. Manufactured in a variety of materials but most commonly in aluminium, this type of construction has found favour particularly in air-to-air heat exchangers for aircraft cabin air-conditioning and in the cryogenic field.

The search for a better ratio of heat-transfer/pressure-loss has led to a multitude of extended-surface shapes. The performance characteristics of several have been determined by Kays and London[26] and, more recently, those of a widely-used design by London and Shah.[27] The metallurgy of brazing and the manufacturing problems due to the thin materials used are now well enough understood to permit the fabrication of consistently reliable units for operation at moderate pressures.

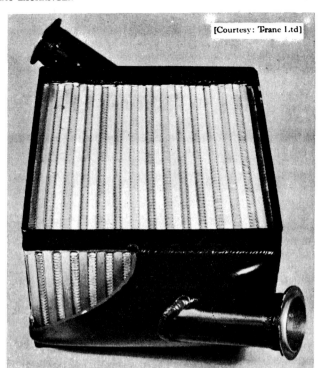

[Courtesy: Trane Ltd]

Fig. 8. This brazed aluminium type offers extended surface to both fluids

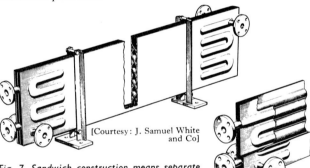

[Courtesy: J. Samuel White and Co]

Fig. 7. Sandwich construction means separate elements with metallic contact for heat transfer but mixing due to a leak is almost impossible

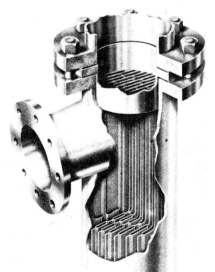

[Courtesy: APV Co]

Fig. 6. This welded sheet-metal design combines aspects of both shell-and-tube and plate types

◀

Air-cooled types

There are probably more air-cooled heat exchangers in existence than any other type; one is to be found under the bonnet of every motor vehicle. Where heat must be dissipated, rather than conserved, atmospheric air is often the most convenient and, for all transport except ships, the only available ultimate heat sink.

Air at atmospheric pressure is, in common with most gases, a comparatively poor heat transfer medium and, used in conjunction with a liquid, demands extended surface on the air side. This has led to a range of finned heat exchangers primarily for this purpose.

The familiar vehicle radiator (which principally uses forced convection) has for many years now been constructed from copper and brass foils.[28] In view of the large quantities, we might have expected some substantial advances in design. Increasingly economical use has been made of the comparatively expensive materials by reduction of thickness and improvement in fins. Manufacturing methods too have become more sophisticated but otherwise designs today look much like those of 20 years ago. Some attention has, however, been paid to the aerodynamics of the air side.[29]

Despite sporadic contrary claims, it appears that no-one has yet solved the problems of producing a satisfactory vehicle radiator in a considerably cheaper and more price-stable material, namely aluminium, at a price which can compete with the traditional materials. In the USA only one popular model—and that a high-performance and comparatively low production type—is fitted with an aluminium radiator.

At the heavier end of the scale, in process plants, the trend towards the use of air-cooled, rather than water-cooled, heat exchangers continues, stimulated by the increasing shortage of cooling water supplies in several parts of the world and the increasing cost of maintenance of water systems. Here, the ribbon tube, in which an aluminium ribbon is wound' in a helix edgewise on to a round tube, has established itself as the industry-wide standard heat transfer element.

Progress has been concerned mainly with improved machines for its manufacture, giving increased reliability and reduced production cost. A few designs of ribbon tube employ slits in the ribbon to interrupt the boundary layer of the air flowing over its turns and thereby increase performance, with some penalty in pressure loss.

Some of the largest air-cooled heat exchangers are used to cool water for a jet condenser. One of the CEGB's generating sets at Rugeley operates on such an installation, which was built largely to gain experience.[30] The heat exchangers themselves, constructed from round aluminium tubes expanded into holes in aluminium fins are arrayed round the base of a reinforced concrete tower for natural convection.

Some other designs

The regenerative heat exchanger is perhaps the oldest type of any, the process of heating stones or bricks for the subsequent heating of air being employed for many centuries in man's culinary activities. The essential features of the matrix of such a heat exchanger are high thermal capacity and reasonably free passage of fluids.

In its rotary form, the matrix is usually a flat circular, rotating disc, connected with sliding seals between two semi-circular ducts, each of which carries one of the fluid streams. Each element passes alternately through the hot and cold streams. Although this type has been employed in steam raising plant for the preheating of combustion air by exhaust gas, using a metal wire or foil matrix, the design of the sliding seals has been a problem: cyclical expansion of the matrix requires that the seals can tolerate distortion.

The recent development of a ceramic matrix has made possible the use of the rotary regenerative type for small gas-turbine applications,[31,32] preheating the compressed combustion air by means of the turbine exhaust. The increase in efficiency of the turbine set due to this heat exchanger is, of course, considerable.

The matrix is made in the form of nesting small triangular passages and the low coefficient of expansion has eased development of suitable seals.

Where two fluids are immiscible or their mixing is permissible, the two may exchange heat by being brought together in intimate contact. Water may be heated by mixing with steam[33] or a liquid heated by bubbling a hot gas through it,[34] for both of which purposes equipment has been developed. A familiar form of direct-contact heat exchanger is the cooling tower to be seen cooling large flows of water by atmospheric air on many power station sites. Whilst in these large towers the draught of air is normally induced naturally, forced draught units are often used in industrial applications.

The design of the packing used in cooling towers,[35] in order to provide an enlarged area of contact between air and water, is a matter for constant development. In recent years thermoplastics have come into the picture, complex patterns being vacuum-formed from sheet. The life of these materials in comparison with the conventional impregnated timber packings still remains somewhat in doubt.

Invariably a high proportion of the total cost of a heat exchanger is due to the actual heat-exchange surface so that the search for better forms of surface is never-ending. In the

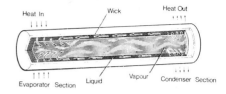

Fig. 9. The heat pipe: a volatile liquid boils at one end, condensing at the other and carrying far more heat per unit area than any metal

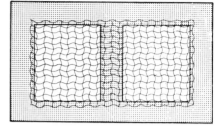

Fig. 10. Unorthodox materials have been tried: this cryogenic heat exchanger is made of resin-impregnated paper and wire-mesh

[Courtesy: Philips Gloeilampenfabrieken]

nuclear reactor field[36] more work towards this end has probably been done than elsewhere and this is resulting in advances that show promise. A very worthwhile prize can be won by achieving even relatively small increases in the transfer rate from a reactor core while minimising the mass of material in the heat-exchange surface. Complex, twisted, interrupted finning on fuel cans has been used in the past but now it seems that 'rough' surfaces can be a better solution, the roughness being provided by circumferential rings[37] acting as small hurdles over which the fluid must jump. This knowledge could well be applied with advantage to the design of other heat exchangers.

A novel American heat exchanger,[38] mainly, at present, for aerospace applications, is the 'heat pipe', shown in Fig. 9. It provides a 'rod' of thermal conductivity many times greater than that of any metal. The principle is simple enough: the pipe is filled with a volatile liquid which boils at the hotter and condenses at the cooler end: the condensed liquid is returned by the capillary action of a layer of gauze or other porous material in contact with the inner surface of the pipe. The liquid must suit the temperature of operation and for high temperatures liquid metals are often the most suitable media.

The idea of constructing a heat exchanger from paper and wire may seem somewhat far-fetched but has become a reality[39] and the principle, developed for cryogenic applications, is illustrated in Fig. 10. Its module of construction is a sheet of resin-impregnated paper in which is cut a regular pattern of rectangular holes. An embedded copper wire gauze covers the whole sheet. Layer upon layer of these sheets are stacked and bonded together. The two fluid streams flow through alternate adjacent passages formed by the pattern of holes, each fluid being in contact with the layers of wire gauze.

[Courtesy: The Hymatic Engineering Co]

Fig. 11. A miniature gas liquefier: the spiral tubing wrapped round the outside is only 0.5 mm diameter

Heat is transferred from each passage by conduction through the wires to each of the four adjacent passages.

The design of a miniature gas liquefier, illustrated in Fig. 11, is of interest in that the extended-surface tube wrapped around the outside is made from 0.5 mm OD tubing!

Heat transfer fluids

Although one of the fluids handled in any heat exchanger is frequently prescribed by the process, the other is sometimes a matter of choice. In the simplest instance the choice may lie between air or water. When an intermediate fluid circuit is used to convey heat from one heat exchanger to another, the choice of that fluid is often open. Several liquids have been developed specifically for this purpose, particularly for use when the range of operating temperatures is above or below that which can conveniently be handled by water. Some are hydrocarbon oils, some salts and some aqueous solutions—an ethylene glycol solution is, of course, one very widely used in engine cooling. A gas may also be used to

advantage in some instances, eg, hydrogen to cool an electric alternator. The best fluid to use in any given application will depend not only on its operating temperature and compatibility with the equipment through which it passes but also on its effect upon the initial and running costs of the whole system.[40]

For a given consumption of power in pumping them, liquid metals can provide a greater rate of heat transfer from a surface than any other fluids. Unfortunately, unusual and costly equipment is required to pump them, expensive materials to contain them, and rather special safety measures to handle them. It is therefore scarcely surprising that liquid metals are used only in applications where very high heat-transfer rates yield very great advantages.

Of commonly available fluids, water ranks first as a heat transfer medium, followed by aqueous solutions, salts, oils and gases, roughly in descending order: the quality of any gas as a heat transfer medium is, however, progressively improved with increasing pressure.

REFERENCES

Mr Forbes was educated at University College School, London, and Cambridge University, where he graduated in Mechanical Sciences in 1945. Following a graduate apprenticeship at Vickers Armstrong, he remained a further two years, working on control devices for naval armaments. In 1949, he joined the R and D Department of Serck Radiators and in 1963 he became engineering director where he was responsible for design and development of a wide range of heat exchangers. Since 1967 he has been a Director of Serck R and D, which serves the Group in the control, conditioning and transport of fluids.

1. *International Journal of Heat and Mass Transfer.* Pergamon Press Ltd.
2. *Proceedings* of the 1951, 1961 and 1966 International Heat Transfer Conferences. ASChE, ASME, IChE, IMechE.
3. NEL: *Heat Bibliography* 1958 et seq. HMSO.
4. McADAMS, W. H. *Heat Transmission;* 3rd edition, McGraw Hill Book Co. Inc. New York, 1954.
5. KERN, D. Q. *Process Heat Transfer.* McGraw Hill Book Co. Inc, New York, 1950.
6. *Chemical Process Engineering.* Heat Transfer Survey. Aug. 1962.
7. *Standards of Tubular Exchanger Manufacturers Association:* 5th edition. 1968, Tubular Exchanger Manufacturers Association, Inc, New York.
8. British Standard 3274, 1960, *Specification for Tubular Heat Exchangers for General Purposes.* BSI, 1960.
9. HOLMES, D. R. 'New Corrosion-Resistant High Temperature Heat Exchanger Materials' *Corrosion Science,* Vol. 8, 1968, pp 603-622.
10. BELL, K. G. 'Exchanger Design based on the Delaware Research Programme' *Petrol.Engr.* 1960, 32 (ii) C 26-36, C 40a-40c.
11. TINKER, T. 'Shell-side Characteristics of Shell-and-Tube Heat Exchangers', Parts I, II and III; Proc. gen. Discussion on Heat Transfer 1951, pp 89-116. 1952 IMechE *Proceedings.*
12. TINKER, T. 'Shell-side Characteristics of Shell-and-Tube Heat Exchangers, a Simplified Rating System for Commercial Heat Exchanges' *Trans. ASME* 1958 Vol 80, 36-60.
13. Engineering Sciences Data Unit: *Forced Convection Heat Transfer in Circular Tubes;* Part I, Item 67016, 1967; Part II, Item 68006, 1968; Part III, Item 68007, 1968.
14. GITHENS, R. E., MINOR, W. R. and TOMSIC, V. J. 'Flexible Tube Heat Exchangers'. *Chemical Engineering Progress,* Vol. 61 No.7, July 1965, pp 55-62.
15. SPENCE, J. R., RYALL, M. L. and McCONNELL, G. 'The Development and Prôduction of High Pressure Feed Heaters in Modern Central Power Stations'. *Proc. IMechE,* Vol. 182, Part I, No.36, pp 735-756.
16. BENDER, R. J. 'Glass Tube Air Heater Designs Show Promises'. *Power,* December 1968, pp 68-69.

17. DRAPER, R. F. 'Glass Heat Exchangers'. *Chem. Proc. Eng,* Heat Transfer Survey, August 1968.
18. HILLS, B. E. G. 'Delanium Graphite Heat Exchangers with Particular Reference to their Application in the Metal Finishing Industry'. *Electroplating and Metal Finishing,* September 1966, pp 326-332.
19. MAYER, E. 'Improving the Heat Transfer in Tubes by Internal Fittings'. *Sulzer Technical Review,* Research Issue 1968 pp 21-24.
20. SMITHBERG, E. and LANDIS, F. 'Friction and Forced Convection Heat Transfer Characteristics in Tubes with Twisted Swirl Generators'. ASME *Journal of Heat Transfer,* February 1964, pp 39-44.
21. SELIGMAN, R. J. S. 'The Plate Heat Exchanger in the Food Industries' *Chem. Ind.,* September, 1964.
22. JONES, W. H. and USHER, J. D. *'Plate Heat Exchangers in Chemical Engineering—A Comparative Study;.'* 4th Congress of the European Federation of Chemical Engineering; June 1966.
23. EMERSON, W. H. *Thermal and Hydrodynamic Performance of a Plate Heat Exchanger;* Part 1—Flat Plates: NEL Report 283; Part II—An APV Heat Exchanger: NEL Report 284; Part III—A de Laval Heat Exchanger: NEL Report 285; Part IV—A Rosenblads Heat Exchanger: NEL Report 286.
24. LOHRISCH, F. W. 'Efficiency of Plate Heat Exchangers' *Process Engineering* December 1968.
25. POTTER, D. S. 'Use of Embossed and Dimpled-Plate Heat Exchange Jacketing of Pressure Vessels'. *Process Engineering,* April 1968.
26. KAYS, W. M. and LONDON, A. L. Compact Heat Excahngers; 1964. McGraw Hill Book Co. Inc, New York.
27. LONDON, A. L. and SHAH, R. K. 'Offset Rectangular Plat-Fin Surfaces—Heat Transfer and Flow Friction Characteristics'. *Trans. ASME Jl. of Eng. Power;* July 1968, pp 218-228.
28. FORBES, M. K. 'The Manufacture of Vehicle and Industrial Radiators' *Tin and Its Uses,* Nos. 67, 68 and 69, 1965.
29. PAISCH, M. G. and STAPLEFORD, W. R. 'A Rational Approach to the Aerodynamics of Cooling System Design'. *Proc. IMechE,* Vol. 183, Part 2A, 1969 (to be published): IMechE paper AD P3/69.
30. CHRISTOPHER, P. G. 'The Dry Cooling Tower System at the Rugeley Power Station of the GEGB'. *English Electric Journal,* Vol. 20, No.1. January/February 1965, pp 22-33.
31. WALTER, L. 'Glass Ceramic Heat Exchanger for Gas Turbines'. *Engineering Materials and Design,* Vol. 8 No.3, March 1965 pp 176-177.
32. PENNY, N. *The Development of the Glass Ceramic Regenerator for the Rover 25/150R Engine.* Paper presented at the Society of Automotive Engineers meeting, Detroit, June 1966.
33. KULJIAN, H. A. and FADDEN, W. G. 'New Power Plant Idea—Feed Water Heating Towers'. *Power,* (Engineering and Management Section). July 1956.
34. HARRIS, B. R. *The Condensation of Organic Vapour by Direct Contact with Static Water.* Paper presented at NEL Direct Contact Heat Transfer Discussion, January 1969.
35. BUTLER, R. G. 'Cooling Towers—Materials of Construction'. *Process Engineering,* October 1968.
36. WEBB, T. B. and WESTON, W. R. 'The Design and Construction of Heat Exchangers for Gas-Cooled Nuclear Power Plant. *Proc. IMechE,* Vol. 177, No. 15. 1963, pp 383-405.
37. WILKIE, D. 'Forced Convection Heat Transfer from Surfaces Roughened by Transverse Ribs. *Proc. 3rd Int. Heat Transfer Conference,* Chicago, 1966 pp 1-19.
38. EASTMAN, G. Y. 'The Heat Pipe'. *Scientific American,* Vol. 218, May 1968 pp 38-57.
39. VONK, G. A Compact Heat Exchanger of High Thermal Efficiency'. *Philips Technical Review,* Vol. 29, No.5, 1968. pp 159-162.
40. KASPER, S. 'Selecting Heat Transfer Media by Cost Comparison.' *Chemical Engineering,* Vol. 75, No.20, December 1968 pp 117-120.

Hydraulic Servomechanisms

by R. Hadekel, BSc, AFRAeS, AMIEE, FIMechE

Hydraulic servos are assuming increasing importance in automatic systems. This has been made practicable by the introduction of reliable and reasonably priced electro-hydraulic servo valves. It is to the further improvement of valve characteristics and system stability that recent research work has been mainly directed.

Fluid pressure servos are today assuming increasing importance in automatic control installations of all kinds. The reason for their adoption may vary, and may—according to the case—lie in high performance, low cost, or greater convenience. Any advantages possessed by them would, however, have remained largely theoretical, were it not for the fact that sound electro-hydraulic servo valves have become available as commercial off-the-shelf items in recent years.

Hydraulic servos are capable of performance unmatchable by other means, on account of their very high torque (or force) to inertia ratio. This quality is inherent in hydraulic actuators of all kinds, but is further enhanced by the fact that it is very often possible to use jacks instead of high speed motors followed by reduction gearing, and thus avoid the high effective inertia associated with high speed machinery. The use of jacks also circumvents the backlash problems which plague servo drives that incorporate reduction gearing.

Jacks have their limitations, since the elasticity of the large volume of fluid which they contain in turn sets limits to the performance obtainable. This effect is of particular importance in the case of pneumatic servos, whose performance is consequently lower, though still often better than that of their electrical counterparts. On the other hand, pneumatic servos can often be made very substantially cheaper than any other device capable of fulfilling the same function.

A valuable account of the origins of fluid pressure servos has been given by H. G. Conway[1].

The earliest fluid servos were ship's steering gears, consisting of a jack and valve mounted in a follow-up arrangement. Originally, steam was used as the pressure fluid, but was soon replaced by water (and later oil) under pressure. This elementary type of servo survives today in vehicle power steering, powered flying controls for aircraft, and copying attachments on machine tools; it is too simple and well known to warrant much discussion here. Given good detail design, its performance is excellent, indeed often more than adequate for the purpose, which is quite simply the amplification of force in the mechanical transmission of motion. Modern developments have resulted not so much from a desire to improve performance as from the need to perform less simple functions.

Power dissipation

Servo-control of a jack or motor by a valve is essentially a power-dissipating process; the consequent wastage of energy and heating of the working fluid may pose serious problems where the powers involved are high, as in marine steering gears, rolling mill control, and gyro-stabilization of naval gun mountings, to quote a few outstanding examples. The

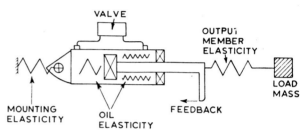

Fig. I. This diagram shows the places in a servo of this kind where elasticity is present; this, combined with the load masses, forms an oscillatory system, tending to be underdamped

advent of the variable delivery pump made possible another kind of hydraulic servo, in which the flow to the actuator was controlled by varying pump stroke instead of valve opening, and in which power dissipation was relatively low. The essentials of variable delivery pump servos have often been described in the literature[2], and need not be dwelt on here. In any case, nearly all variable delivery pump servos are preceded by a valve servo stage which provides the forces necessary to vary the pump stroke; to this extent, much of what will be said later applies to them also.

Typically, variable delivery pump servos might come into their own in the relatively rare cases where the output required is, say, 20 h.p. or more, depending on the importance of power economy and the difficulty of cooling the working fluid. Valve-controlled hydraulic servos overlap this range, and are in widespread use for powers down to perhaps 0·25 h.p. or less. At the bottom of the power range come instrument and computing servos, the work output of which is so small as not to be a design criterion; in this range, electrical servos are supreme today and likely to remain so, although pneumatic equipment has been used for computing purposes in various applications. Indeed, the pneumatic process controllers in widespread use today embody what may be termed a computer stage, of which more later. Generally speaking, pneumatic servos show up at their best where powers are fairly low (say below 1 h.p.), where low cost is important, and performance demands modest.

In recent years, however, there has been strong interest in the application of relatively high power, high pressure, gas servos to missiles, and perhaps eventually to high speed aircraft, the attraction being availablity of the gas (in the case of missiles) and the ability to work at high temperatures. An interesting development in this connection is a high speed gas motor based on the principle of Hero's turbine[3]. This device has a remarkable power to weight ratio, but its use for servo purposes appears to be limited by the fact that the torque to inertia ratio is not particularly good.

Part of a control loop

As used in the majority of modern applications, a servo-mechanism is no longer a plain force amplifier in a mechanical linkage (as are ships' servo steering gear and their immediate relations), but rather one item in a chain of devices which, together constitute a control loop, often of great complexity, such as an aircraft automatic pilot or a numerically controlled machine tool. The servomechanism is a small closed-loop system in a larger loop, and may be regarded as the 'muscle' of the system, the remainder of which comprises sensing or measuring devices, and a 'brain' in the form of a computer. Finally, there is the device or plant to be controlled, which usually closes the overall feedback loop.

In most cases, the servo is required to produce a given position; this it achieves with the help of position feedback (usually electrical) which cancels a demand signal given by the computer. In some cases the required output is a velocity and velocity feedback from a tacho-generator is then used. There are also cases where actuating devices may be used without feedback, either because their inherent construction makes it unnecessary, or because the nature of the overall control loop does not demand it. Such actuating devices are not servomechanisms according to textbook definitions but, for practical purposes, this point may very often be ignored, since much the same techniques are applicable to both kinds of apparatus, as long as what is involved is a more or less proportional response to some command signal, and not on-off action.

The typical modern fluid pressure servo consists of a valve, a jack, or a motor followed by reduction gearing, which is controlled by the valve, and some feedback device. The valve must respond to some specified signal, which may be pneumatic but is more often electrical. In the latter case the valve is usually driven from an amplifier, which may be part of the servo loop.

Stability considerations

The stability problem lies at the heart of all automatic control technology. Any servo, or more complex control loop, must have some specified response to human or automatically generated commands and/or disturbances from external causes; practically all servos tend to become unstable if an attempt is made to increase their response beyond a certain point. In practice, there are relatively few servos whose performance requirements are so modest that no stability problems arise in their development.

When a servo-actuator is used as part of a larger automatic control loop, attention must be paid to the stability of the actuator regarded as an independent component, and to the stability of the complete loop with the actuator as part of the chain. In theory, a control loop may contain unstable components and still be stable, but few engineers would be bold enough to design a system on this basis.

An excellent insight into the stability problems of fluid servos can be obtained by studying the case of the simplest hydraulic position servomechanism, consisting of a valve, jack or motor, feedback, and load. In most studies a pure inertia load is assumed, this being usually the most severe as regards stability and a realistic approximation to actual conditions in many cases[4]. The broad conclusions of this and similar analyses can be reached by quite simple reasoning, as follows:

Consider the position servo shown in Fig. 1. Elasticity is present in the oil in the jack, in the jack mounting, and in the connection between jack and load. The combination of the load mass and these elasticities forms an oscillatory system, leading to a tendency for the servo to be underdamped. In the case of oil elasticity, and sometimes also mounting elasticity, the resulting oscillatory motion is fed back to the servo input in an adverse phase relationship, and this may lead to instability. Now, assume that a signal proportional to the acceleration of the load mass can somehow be derived, and fed back to the valve. The valve and jack combination acts as an integrator, since jack velocity is at least roughly proportional to the valve opening, and thus to the acceleration; this has a damping effect. Since the load is pure inertia, its acceleration is proportional to jack force, and hence (if friction is disregarded) to the pressure difference across the valve output. Thus a stabilizing effect is obtained if valve output flow is sensitive to load pressure, decreasing as the latter is increased.

The important effect of this principle on servo valve design will be discussed later. There is some evidence however to the effect that fluid servos are rather more stable than simple analysis would indicate, and it has been surmized[5] that this is due to friction in the jack, which is seldom negligible. This has been confirmed by analogue computer studies.[6]

Simple analysis also shows that the maximum response obtainable from a servo is closely related to the natural frequency of the load mass and elasticity combination, as indeed might well be guessed intuitively. The volumetric elasticity modulus of a gas is γp, where p is the pressure and γ the adiabatic index (about 1·4 for air). For any practical working pressure, this is at least two orders of magnitude lower than the compressibility of oil (about 200,000 p.s.i., for reasonably well de-aerated oil). In practice this factor governs the performance obtainable from pneumatic servo-actuators of the jack type.

Looking at the matter more closely, the linear elasticity of a jack is $\gamma p A$, where A is the cross sectional area. Now, if p_L is the maximum pressure required to overcome the peak load F, $p_L = F/A$, and the linear elasticity is $\gamma p F/p_L$. This shows that raising the working pressure brings no benefit unless the ratio p/p_L is increased, i.e. unless the jack is over-dimensioned.

Signal processing

The response of hydraulic servos is usually so high that phase advance is rarely called for. Indeed, the characteristics of a hydraulic servo-actuator, which tend to include an underdamped quadratic lag, are usually such that phase lag is more effective in improving stability. One example of a phase lag device in this kind of application is found in the valve dampers which have been fitted to a number of powered flying controls. Their action depends on elasticity in the actuating linkage.

There are some applications which require particularly high steady-state accuracy when following a constant velocity input. Two examples are machine tool copying devices, and gyro-stabilized naval gun mountings. As is well known in servo theory, zero steady-state velocity error is obtained by using double integration, plus a phase advance term to achieve stability in the face of this. The well-known 'creeping relay' used in gun mounting servos[7] is an example of a device

pressure drop; and a transient component proportional to the rate of opening. The latter is usually unimportant but, in some cases, capable of destabilizing the valve. The transient component is easily eliminated. Lee and Blackburn of M.I.T. have devised valves which very nearly eliminate the steady-state component, enabling valves for quite high powers to be driven directly from electric torque motors. Much practical work on valves of this type has been done in this country by Eynon[12], but in the U.S.A. this line of development appears to have been dropped in favour of two-stage valves, which are considered later in this review.

Valve design

Apart from first-stage valves in two-stage arrangements, the standard device is the cylindrical slide valve, broadly similar to those long used for directional control in hydraulic systems, though considerably refined. Orifices in the form of complete annuli are seldom used, as these would usually give excessive gain, and are difficult to produce with sufficient accuracy. The usual orifice is a row of holes, either circular for lowest cost, or rectangular, for greater linearity. Rectangular holes are usually produced by making the valve sleeve in several pieces, which are furnace-brazed together. Alternative processes, such as spark erosion machining, do not appear to have lived up to expectations so far.

The main problem is the accurate control of valve lap. Nominally zero lap is usually aimed at in high grade servos, with accuracies greater than 0·0005 in. One type of slide valve, with flat mating surfaces in the preferred version, has been invented by Lee of M.I.T.[8] This allows nominally zero lap and effectively rectangular orifices to be obtained by relatively simple means. In one particular version this valve is free from friction forces. This construction, however, has not been widely adopted so far.

An alternative approach to the accuracy problem is described in a recent paper by Bahniuk and Lee[13].

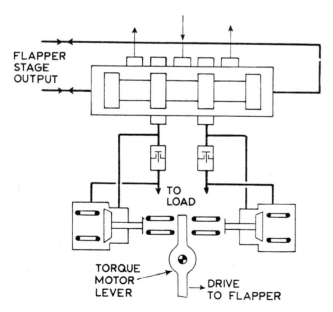

Fig. 5. In this type of servo the feedback signal is provided by a poppet-valve flowmeter and the flow is practically independent of load pressure, up to a saturation point. Advantages claimed are cheapness of construction and low standing power loss

Two-stage valves

As mentioned earlier, the hydraulic servo-actuator forms part of some larger system and, in such a case, the servo valve is almost invariably driven through some form of electric torque motor, which can only provide quite limited forces. It thus becomes necessary to keep valve operating forces at low levels, and the most widely adopted solution is to use two-stage valves. Torque motors are mostly variants of moving-iron meter movements, capable of giving work outputs which are relatively large for this kind of device. Moving coil movements have also been used on occasion, at least in the past.

The function of the first stage in a two-stage valve is to drive the second stage, so that it only handles very small powers. One of the oldest arrangements is to use slide valves for both stages, with position feedback between them, as shown in Fig. 2. Precautions must be taken to keep friction very low in the first stage valve. The latter need not necessarily be of the slide type; it could be a jet pipe or a flapper valve.

The position feedback system reduces the force, but not the stroke, required from the driving device. To achieve still lower demands on the torque motor, the travel must be reduced also, since it is the work output that counts. A typical second stage stroke may be between 0·01 in. and 0·03 in., while the travel of a first stage valve need only be about 0·001 in., if position feedback is not used. Further savings can be made if the first stage valve is of the flapper type, which requires quite low forces to drive it.

Thus, a typical motor to drive a two-stage valve with position feedback and a slide valve first stage may consume 2 to 4 watts, as opposed to 100 milliwatts or even less for a typical flapper-spool valve. Various valves of this type have been developed in the U.S.A., and have achieved an astonishing degree of compactness in relation to the power handled.

The earliest (and still probably the most popular) valve of this type is that using the open-loop arrangement of Fig. 3. The output of the flapper stage is a variable pressure, which is applied to the spring-centred spool; this therefore moves by an amount proportional to the flapper output pressure. To help make the latter independent of supply pressure, the torque motor driving the flapper is made to have zero rate, so that its output is a force rather than a displacement. Friction in the second stage is still quite important (as opposed to the case where some form of feedback is used) but, in practice, it would seem that this is taken care of by some natural dither, arising probably from low-amplitude instability.

A more recent design, becoming increasingly popular in the U.S.A., is the flapper-spool valve with force feedback (Fig. 4), in which the second stage is not spring-centred. Owing to the definite stiffness characteristics of the torque motor, force feedback is very similar in its action to position feedback (the feedback spring generates a force proportional to relative travel between the two stages), but does not equate the travels of the two stages, and thus requires about the same work output from the torque motor as the open-loop type shown in Fig. 3.

Some recent designs of two-stage valves feature a jet pipe first stage in lieu of a flapper. Very little has been published about jet pipe valves, though they have been known and used (mostly in the process control field) for many years, but there is no reason to doubt the main claim made on their behalf, which is very low sensitivity to dirt. A systematic study of jet pipe (or 'nozzle' or 'Askania') valves, or the publication of any work at present withheld, is long overdue.

which obtains this characteristic by means of hydro-mechanical networks i.e. combinations of springs, dashpots, and suitable mechanism. Similar devices have been used in steam engine governor gear, for the purpose of obtaining specific characteristics in the overall control loop.

With modern techniques, devices of this kind may still have their use occasionally, if electronics are barred for one reason or another. One difficulty with such apparatus is that of providing hydraulic resistances which are both linear and insensitive to temperature, whereas effective electrical networks are easily constructed from cheap and readily available components.

Non-electrical signal processing is, however, still in widespread use in pneumatic process controllers, usually for the purpose of adding integral and/or derivative terms to the basic error signal. The temperature problem is much less severe with air than with liquid, and effective devices are readily and cheaply constructed from combinations of capillaries or other resistance elements, bellows, and levers. Even in this field, however, there is a trend towards electric signal processing.

Valve characteristics

The control valve is the really significant component in a fluid pressure servo actuator. Hydraulic control valves are perhaps almost as old as engineering itself, but it was not until the great development of hydraulic servos in quite recent years that some of their less obvious (and hitherto less important) characteristics began to receive serious study. The most significant published work in this field is that done at the Massachusetts Institute of Technology, which has been collected in a recent book[8], and it is therefore unnecessary to quote original papers in which results were first published.

Two main topics are of interest: the relation between valve opening, flow, and load pressure, and the forces necessary to actuate the valve.

Details of flow phenomena in valves have received recent attention in this country[9]. One of the main findings of this research is that two distinct flow patterns can exist, which may involve a discontinuous change in the valve discharge coefficient, and hence in the output flow. There is however little if any evidence of such an effect occurring with valves in practical use, the explanation being perhaps that the conditions for it to happen are outside the normal range of the valve. Of more immediate practical application is the digest of available information on discharge coefficients, produced recently by Shearer[10].

For most design purposes, the assumption of a constant discharge coefficient, of the order of 0·7, would seem to be good enough. The calculation of valve output flow on the basis of such constant coefficients has been rationalized by Blackburn[8], who gives expressions for different valve configurations, and curves showing the flow as a function of load pressure, for various valve openings. This presentation is now accepted as a standard by most workers in the field.

Friction and Bernouilli forces

It has been known for a long time that slide valves were subject to friction forces even though nominally in balance, and that other forces related to flow ('Bernouilli forces') were also present. It is only during the last decade or so that these effects have been rationally explored. Friction forces have been studied in this country[11], and at the M.I.T. It has

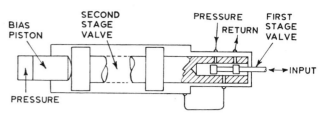

Fig. 2. In two-stage valves the first stage, using much less power, controls the second: one of the oldest arrangements uses slide valves for both, with position feedback between them

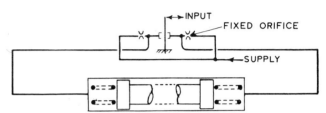

Fig. 3. To obtain accurate control with very little power, a flapper-spool valve can be used. An open loop arrangement of this is here shown schematically

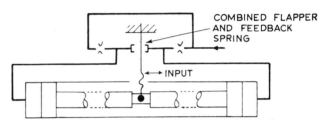

Fig. 4. A more recent design than that shown in Fig. 3 is the flapper-spool valve with force feedback. Unlike position feedback, this does not require equal travel of both stages

been established that these forces are caused by side loads generated by adverse departures from the nominally symmetric pressure pattern over the valve lands, which are caused either by departure of the valve surfaces from geometrical perfection, or by the presence of dirt, or both.

One conclusion of this work is that favourable conditions can be obtained by giving the valve lands a deliberate taper (of an order of magnitude comparable to the clearance) in the right direction, and in the present writer's experience this effect can be utilized to give friction-free slide valves, provided due care is taken with filtration.

Friction effects are of course all-important if the valve is driven by a device of limited stiffness, in particular by an electric torque motor. In the past, friction forces have been overcome by the use of dither, or by rotating the valve sleeve, and these methods are still in use to some extent. Eynon[12] reports good results in single-stage slide valves obtained by keeping the valve lands sufficiently narrow to reduce friction to quite low levels. In the present writer's experience, this is inadequate for some of the more demanding applications. Current American practice using two-stage valves in which the first stage is inherently friction-free will be discussed later.

Bernouilli forces have been studied at the M.I.T.[8], and have been shown to consist of a steady-state component proportional to flow and to the square root of the orifice

Mr R. Hadekel, *C.Eng. B.Sc.(Eng).*
A.C.G.I., D.I.C., F.I.Mech.E.,
F.R.Ae.S., M.I.E.E., was born in
Russia and educated in Germany,
France and England. After studying
at the Imperial College, London,
he joined Miles Aircraft Ltd in 1936,
and later worked with de Havilland
Aircraft Ltd. During the war he was
Chief Designer at Messier Aircraft
Equipment Ltd (now Electro-Hydrau-
lics Ltd). From 1946 to 1948 he was
Assistant to the Technical Director,
B. & P. Swift Ltd, and from then
until 1954 he worked on his own
account as a Consulting Engineer.
From 1954 until 1961 he was
with the Sperry Gyroscope Ltd, where
he was Engineering Consultant. He
is now Chief Engineer at Trico-
Folberth Ltd. He is the author of two textbooks on Hydraulics, and of
a number of papers on this and various subjects. He has also been a
prominent member of the Institution's Automatic Control and Hydraulic
Groups.

Yet another type of two-stage valve has been described in a recent paper by Bahniuk and Lee[13]. The first stage is again a flapper valve, to which a feedback signal proportional to the second stage output flow is applied. The signal is generated by an 'area' flowmeter, *i.e.* a poppet valve, biased by a low rate spring, and hence working at a practically constant pressure drop. Poppet lift is thus proportional to flow. This lift is converted to force by a spring, and then fed back to the torque motor, as shown in Fig. 5. The inner loop gain is made high, and flow is thus practically independent of load pressure, up to the point where saturation occurs. This might seem a doubtful benefit from the standpoint of stability, but very substantial benefits are claimed in other respects. Thus the second stage (spool) valve need only be made to relatively low accuracy, and can have round ports and high diametral clearance, plus some overlap, resulting in cheapness of construction and low standing flow loss.

Stabilizing devices

Theoretical aspects of fluid servo stability have been treated earlier. It should perhaps be stressed that fluid servos do not differ in principle from other kinds as regards their stability problems. Whereas, however, in most closed loop systems instability results from an accumulation of several lags, in fluid pressure servos the basic trouble is most often due to the presence of an underdamped quadratic lag term, which arises from the high output stiffness of the servo valve. The most effective, and sometimes the only, way to deal with this is to increase the damping in this term which means increasing the sensitivity of valve output flow to load pressure.

One method of obtaining the desired effect is to use a valve with underlap, but this has the disadvantage (in most cases) of excessive power wastage. Another method, which also wastes power but to a much lesser degree, is to introduce a shunt resistance across the valve output. It is also possible to use a shunt-resistance-capacity line, by inserting an accumulator in the path. This reduces power wastage, and has the additional advantage of making the pressure sensitivity transient. The required size of accumulator may, however, be prohibitive.

The most potent, and least wasteful, method of improving stability characteristics is to feed back a signal proportional

to load pressure. By this means the sensitivity to load pressure can be made as high as desired. On the other hand, low load-sensitivity (or high output stiffness) is a distinct merit in fluid pressure servos (particularly in hydraulic ones), since it means that actuator position can be accurately determined in spite of the presence of external loads, particularly friction.

It is in fact possible to get the best of both worlds, by using transient pressure feedback, *i.e.* by feeding back a signal which, on a short-term basis, varies with load pressure, but which vanishes if the load stays constant. To obtain benefit from this system, the feedback must be fully effective at the natural frequency of the servo actuator, and virtually ineffective, or at least greatly reduced, at the (inevitably lower) frequencies which are of interest in the overall control loop; such as for instance the natural frequencies of an aeroplane in the case of an automatic pilot servo. A method of providing transient (or 'dynamic') pressure feedback in a flapper-spool valve has been described by Garnjost[5], the valve in question being shown diagrammatically in Fig. 6.

Future trends

The excellent results obtainable from modern fluid servo equipment are not due to any startling innovations, but rather to detailed refinement of components based on fairly old-established principles. Such refinement has been brought about in large part by commercial development of components in industry, which has of course been greatly helped by the increased understanding of valve characteristics resulting from recent research. The process is continuing, but with diminishing returns, since the most glaring areas of ignorance have been cleared up. Indeed it is now becoming fairly difficult to suggest new fields for basic research on fluid

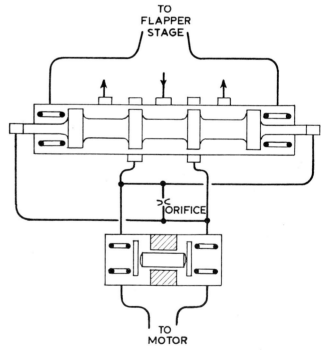

Fig. 6. To improve stability under dynamic conditions, this flapper-spool valve has been equipped with transient pressure feedback. The feedback must be fully effective at the natural frequency of the actuator but greatly reduced at the lower frequencies at which the overall control loop operates

servo components, although the scope for development (as opposed to research) remains great.

Probably the most important factor which limits the application of servos, both hydraulic and electric, and indeed the application of automation techniques in general, is price. There should therefore be good scope for some really cheap design of electro-hydraulic servo valve, selling at say £30 to £40, which the present writer believes should be quite feasible for moderate powers. One quite significant factor in the cost of such valves is that of the torque motor. Some time ago, experiments were made in America with piezo-electric driving devices, which do not appear to have given any useful results so far. Perhaps some day a new piezo-electric material will come along, which might represent an appreciable step forward in simplicity and cost.

There should be good scope also for a much wider use of low-pressure pneumatic servos, particularly in applications where performance requirements are relatively modest, and an air supply is already available; for instance, in machine shops, and in turbine-engined aircraft. Further work is needed to develop satisfactory seals requiring no lubrication. Apart from process control gear, however, there is very little off-the-shelf equipment available in this field, and little technical 'know-how', as compared with that available for hydraulic equipment.

Another limiting factor in the spread of servo devices is the effort and cost required for development of the particular application, even when the basic components are already available. Apart from trivial cases, no servo system is ever quite like the one before, and performance problems must be solved for every application. Admittedly, servo theory is less of a black art today than it was, say, ten years back, and many engineers are available who can discuss transient responses, transfer functions, Nyquist diagrams, and root loci, with the greatest fluency, and even make good use of them. The man who can take a broad view of a 'systems'

problem, who appreciates the potentialities and limitations of the different techniques, and uses them to produce a sound engineering solution is, however, still rather rare. The question of how such people are to be trained is a recurrent topic of discussion. One prominent American engineer has gone so far as to say that 'systems engineering is a management function', a statement with which the present writer is inclined to agree, but which he must refrain from defending for fear of being driven still further from the immediate object of this review.

REFERENCES

1. CONWAY, H. G., 'Some notes on the origins of mechanical servo-mechanisms'. 1954 *Trans. of the Newcomen Society*, vol. 29, p. 55.
2. HADEKEL, R. 1953 *Proc. of Conference on Hydraulic Servo-Mechanisms*, I.Mech.E., 'Hydraulic Servos'.
3. SCHER, R. S. 1960 *Proc. Symposium on Recent Mechanical Developments in Automatic Control*, I.Mech.E., 'Development of gas-generated reaction-jet servomotor'.
4. HARPUR, N. F. 1953 *Proc. Conference on Hydraulic Servo-Mechanisms*, I.Mech.E., 'Design considerations of Hydraulic Servos of the jack type'.
5. GARNJOST, K. D. 1960 *Proc. Symposium on Recent Mechanical Engineering Development in Automatic Control*, I.Mech.E., 'Design of an electro-hydraulic servo actuator with integral mechanical position feedback'.
6. GLAZE, S. G. 1960 *Proc. Symposium on Recent Mechanical Engineering Development in Automatic Control*, I.Mech.E., 'Analogue technique and the non-linear jack servo-mechanism'.
7. PORTER, A. *An Introduction to Servo-mechanisms*, Metheun. 1950.
8. BLACKBURN, J. F. *et al. Fluid Power Control*, John Wiley. 1960.
9. MACLELLAN, G. D. S., MITCHELL, A. E., TURNBULL, D. E. 1960 *Proc. Symposium on Recent Mechanical Engineering Development in Automatic Control*, I.Mech.E., 'Flow characteristics of piston type valves'.
10. SHEARER 1960 *Proc. Sumposium on Recent Mechanical Engineering Development in Automatic Control*, I.Mech.E., 'Resistance characteristics of control-valve orifices'.
11. MANNAM, J., 'Further aspects of hydraulic lock'. 1959 *Instn mech. Engrs, Lond.*, vol. 173, no. 28.
12. EYNON 1960 *Proc. Symposium on Recent Mechanical Engineering Development in Automatic Control*, I.Mech.E., 'Developments in high performance electro-mechanical servomechanisms at the Royal Aircraft Establishment, Farnborough'.
13. BAHNIUK and LEE, 'The design and analysis of a servovalve with flow feedback'. 1960 *Jrnl of Basic Engineering*, March.

INDEX